NATO Science for Peace and Security Series C: Environmental Security

NATO Science for Peace and Security Series

Sub-Series C. Environmental Security

Volumes in the Environmental Security Sub-Series deal with scientific and technological aspects of innovative and promising projects related to environmental security, including regional environmental problems and natural and man-made disasters. Floods, droughts, landslides, desertification, earthquakes and industrial accidents are examples of topics covered, as are regional environmental problems related to transboundary water resources or transfrontier air pollution. Other areas covered relate to climate change, environmental chemistry, disaster prevention and forecast, remote sensing and detection, and decision support systems/risk assessment.

Special offer: For all clients with a print standing order we offer free access to the electronic volumes of the Series published in the current year.

Matthew Huber · Marthie Grobler
Editors

Critical Infrastructure Systems and Extreme Threat Nexus: Preparing for Resilience and Security

Springer

Editors
Matthew Huber
Earth, Atmospheric, and Planetary Sciences Department
Purdue University
West Lafayette, IN, USA

Marthie Grobler
CSIRO's Data61
Melbourne, VIC, Australia

ISSN 1874-6519 ISSN 1874-6543 (electronic)
NATO Science for Peace and Security Series C: Environmental Security
ISBN 978-3-032-13169-0 ISBN 978-3-032-13170-6 (eBook)
https://doi.org/10.1007/978-3-032-13170-6

Proceedings of the NATO Advanced Research Workshop on Unravelling the Cyber-Physical-Social Infrastructure Climate Change (CPSICC) Nexus Workshop, held in Washington DC, USA, 29 July–1 August 2024.

This Springer imprint is published by the registered company Springer Nature Switzerland AG
The registered company address is: Gewerbestrasse 11, 6330 Cham, Switzerland

Foreword by David J. Alexander

The New Front Lines of Security

We are living in an era where the most urgent threats to national and homeland security do not arrive on the backs of tanks or through traditional battlespaces. They arrive as cascading power failures triggered by a cyber-intrusion into an aging control system or through misinformation campaigns that erode public trust in science during wildfire evacuations. They arrive through compromized sensors misreading or spoofing flood thresholds or through burned-out substations in the wake of a heatwave. The terrain of modern security is not just physical or digital; it is climatic, algorithmic, infrastructural, and profoundly interconnected. This is the new frontline of security, and it is already here.

This book emerges from a recognition that we need to be better prepared to defend this front. Our technical, institutional, and social systems of protection remain fragmented in the face of these new and converging threats. The Cyber-Physical-Social Infrastructure Climate Change (CPSICC) Nexus Advanced Research Workshop, from which these chapters originate, was not designed to be an abstract academic exercise. It was conceived as a strategic and deliberate attempt to understand, anticipate, and shape the future of security in a world where risk is no longer siloed, and threats are no longer linear.

At the heart of this effort lies a simple but radical premise: that security in the twenty-first century cannot be meaningfully achieved without understanding the compound and interconnected nature of emerging threats. Climate change, once thought of as a distant environmental issue, has now become a threat multiplier across every domain of national and homeland security. Cyber-vulnerabilities, often relegated to technical experts, now sit at the core of our physical resilience. Critical infrastructure, long treated as a fixed and stable foundation, is now revealed to be a dynamic, interdependent, and often fragile system exposed not only to sabotage and failure but to misperception, misinformation, and neglect.

This volume brings together the voices of scholars, technologists, security experts, and policy leaders from across NATO members and partners. It synthesizes the insights and outputs from the CPSICC workshop that was co-led by Purdue University, Sandia National Labs and Australia's Commonwealth Scientific and Industrial Research Organisation (CSIRO) with funding support from the NATO Science for Peace and Security Programme and the US DHS Science and Technology Directorate. The workshop was held in Washington, DC, during July/August 2024, a year that was marked by record-breaking floods, targeted cyberattacks on infrastructure, and political unrest fueled by deepfake disinformation campaigns. It was a fitting, if not sobering, backdrop for a conversation about the fragility of our connected systems and the urgency of thinking differently about resilience.

Over the course of four intense days, participants engaged in scenario-based exercises, strategic gaming simulations, and facilitated expert dialogues. More than 136 R&D gaps were identified across six core domains, including systems modeling, socio-technical resilience, risk assessment, governance, and misinformation. These gaps are technical and symptomatic of a deeper institutional lag: our risk models still treat climate, cyber, and infrastructure as separate domains. They also magnify critical themes that emerged during the workshop, such as our governance systems are optimized for linear threats rather than the rapidly evolving and nonlinear landscape of risks, our funding structures reward response over anticipation, and our public discourse, often shaped by partisan division, struggles to articulate what is at stake.

This book is an attempt to correct that course. It begins by mapping the landscape of converging risks. Many chapters delve into how cyber-vulnerabilities can magnify the impact of extreme weather events and how, in turn, climate change opens new vectors for cyberattack. Others explore the cascading failures that occur when human and social systems are ignored: from misinformation-induced panic to institutional inertia during crisis. The energy grid is treated not just as a technical asset, but as a complex system of cyber-physical dependencies that can be disrupted by adversaries or overwhelmed by simultaneous hazards. Several contributors examine the resilience of offshore wind infrastructure, the implications of climate variability on deterrence postures, and the potential risk that AI systems, designed to help us, might fail us in the very moments we need them most.

Running through the book's chapters is a clear message: our adversaries understand the CPSICC Nexus better than we do. Nation-states and non-state actors are developing tactics designed to exploit the convergence of natural disasters and digital vulnerability. As climate pressures intensify, producing migration surges, resource scarcity, and geopolitical flashpoints, threat actors will increasingly see opportunity in moments of compounded crisis. Injecting a cyberattack during a flood, conducting a misinformation campaign during a blackout, or prosecuting a kinetic strike masked as infrastructure failure. These are not hypotheticals. They are signals already in the noise.

For the US Department of Homeland Security and for allied agencies across the NATO alliance, the challenge is not simply to defend against these threats, but to reimagine what security means in a climate-disrupted, digitally entangled world. It means designing infrastructures that can adapt dynamically to both cyber and climate stresses and embracing transdisciplinary science to understand how social norms, economic behavior, and institutional trust shape resilience. It means embedding scenario-based foresight into everyday planning. It means recognizing that critical infrastructure is no longer just steel and concrete but also code, cognition, coordination, and community.

David J. Alexander
Senior Science Advisor
US Department of Homeland Security,
Science and Technology Directorate
Washington, DC, USA

Foreword by Eyup K. Turmuş

NATO's 2022 Strategic Concept identifies the importance of investing in cross-cutting opportunities and challenges to its collective security efforts through its three core tasks: deterrence and defence; crisis prevention and management; and cooperative security. The Alliance recognizes that vulnerabilities from climate change and cyber-defence gaps are destabilizing forces that can have compounding ripple effects with devastating implications. NATO's focus on national and collective resilience remains critical for all core tasks and is the foundation for safeguarding Allies and their shared values. Strengthening resilience at both the national and collective level requires investment in research, partnerships, and foresight. NATO's Science for Peace and Security (SPS) Programme provides a framework for advancing this work.

The SPS Programme supports cooperation through established grant mechanisms that provide funding for multi-year projects and events, such as the Cyber-Physical-Social Infrastructure Climate Change (CPSICC) Nexus Advanced Research Workshop. Through events such as this, researchers, academics and experts are able to provide their expertise and foresight and play an important role in helping the Alliance identify, understand, and respond to vulnerabilities and threats. As such, the outcomes from this workshop reinforce SPS's role as connector and ensures that knowledge exchange does not remain siloed. This event is a concrete example of how SPS helps NATO move from awareness, to knowledge building, to building resilience strategies. By bringing together academia, government, and industry, it also demonstrates how SPS fosters the innovation and cross-sector collaboration needed to anticipate and prepare for complex security risks.

NATO, through SPS, and its broader mandate, remains committed to ensuring that Allies can anticipate and withstand emerging, interconnected threats. Its strength lies in fostering innovation between Allies and with NATO partners, ensuring the Alliance can collectively deter and defend, bolstering its commitment to collective defence. This reaffirms the Alliance's role in shaping security for a future with new, complex, nonlinear challenges. As a result, the workshop and this book meaningfully contributes to NATO' strategic preparedness, reinforcing the Alliance's enduring mission to safeguard peace and security in Europe, North America, and beyond.

To be clear, the future we are confronting is not one of decline but rather critical decision. The CPSICC Nexus, while exposing vulnerabilities, also reveals leverage points. Advances in digital twins, quantum-secure infrastructure, decentralized energy systems, and trust-based data architectures all offer opportunities for transformative resilience across the NATO members and partners. But these will not emerge automatically; they require deliberate investment, policy reform, and cultural shift. And they require, above all, that we stop treating climate, cyber, and infrastructure as technical problems to be solved independently and begin treating them as inseparable components of a shared security ecosystem.

The chapters in this book represent not just an agenda for research, but a call to action for policy, strategy, and leadership. They also demonstrate that NATO has the knowledge, the talent, and the tools to chart a different course but only if we are willing to embrace uncertainty, to challenge assumptions, and to act boldly across traditional boundaries.

This book arrives at a time of profound instability. But it also arrives at a moment of possibility. If our homelands are to be secured not just against enemies, but against entropy, erosion, and epistemic collapse then it must be reimagined. Not as a fortress, but as a network of adaptive systems. Not as a fixed geography, but as a dynamic nexus of risk, trust, and resilience. The CPSICC Nexus is not just a technical frontier. It is a conceptual one. And it is time we cross it deliberately, strategically, and together.

Eyup K. Turmuş
Head, NATO Science for Peace
and Security (SPS) Programme
NATO Headquarters
Brussels, Belgium

Introduction to the Cyber-Physical-Social Infrastructure Climate Change (CPSICC) Nexus

Abstract The security landscape of the twenty-first century is defined by convergence and cascades. Risks once were treated as isolated: climate hazards, cyber-intrusions, supply-chain disruptions, information manipulation, and hybrid attacks, were treated as separate threats. Now we see how they interact to create complex, cascading threats across interconnected systems. The Cyber-Physical-Social Infrastructure and Climate Change (CPSICC) Nexus initiative, launched through a 2024 NATO Advanced Research Workshop, addresses this reality by examining how technological, environmental, and societal domains overlap to shape systemic resilience and security. The workshop and subsequent volume explore how these interdependencies manifest in energy, water, agriculture, information, social, economic, and built systems. Furthermore, these show how vulnerabilities in one domain propagate across others and the ways in which resilience built in one component can lend stability to the whole. Contributors from NATO members and partner nations worked together to identify leverage points for transformation, e.g.: improved system modeling, integrated governance, resilient infrastructure design, and investment in public trust and cooperation. Together, these insights form a framework for advancing from awareness to action—linking science, policy, and strategy to strengthen collective security. The CPSICC Nexus thus represents both an analytical framework and a community of practice aimed at ensuring that NATO and its partners can anticipate, adapt, and thrive amid accelerating environmental, technological, and social change.

Keywords Cyber-, physical-, social-systems infrastructure and climate change nexus · National security · Critical materials · Critical infrastructure · Resilience · Climate change · Interdependencies · Systems integration · Hybrid threats

Introduction

The Converging Threat Landscape and CPSICC Nexus Thinking

The twenty-first century has produced a security environment defined not by isolated hazards but by their interdependence. Climate change, rapid digitalization, and social fragmentation interact to generate a new class of systemic risks that transcend borders and institutions. As Alexander and Turmuş noted in their preface, these convergent risks are not distant or hypothetical concerns, but the new frontlines of security, where the boundaries between digital, physical, social, and climatic domains have all but disappeared. At the same time, systemic opportunities are presented by continuously improving technology, including potentially transformative technologies such as AI, and enhanced cohesion developed by strengthening institutions and alliances. Thus, convergence defines what we call the Cyber-Physical-Social Infrastructure and Climate Change (CPSICC) Nexus—an analytical and operational framework for understanding how cyber, physical, and social systems co-evolve with and under intensifying stressors and how they may operate together to create positive futures.

Earlier doctrines separated "natural" and "human-induced" threats, but these distinctions are increasingly blurred. The same technologies that knit societies together—such as automation, artificial intelligence, and global logistics systems—also introduce systemic and shared vulnerabilities. For example, a power grid failure during a heatwave may result not only from physical stressors but also from the complex interplay of algorithms, market dynamics, and public expectations across digital, physical, and social domains. Infrastructure systems are simultaneously physical, digital, and social/political, all nested within and interacting with the natural environment and contributing to the blurring of the doctrine distinction. Specifically, power grids, water networks, health systems, and markets operate through cyber-physical architectures governed by institutions of trust. Climate hazards impose external stress; cyber-threats exploit internal complexity; social dynamics determine both vulnerability and recovery, and ultimately, the state of the other systems depends on these social-political dynamics.

Our view and the perspective articulated further in this book is consistent with NATO's *Strategic Concept* which labels climate change as a "crisis multiplier," amplifying humanitarian emergencies, migration, and deterrence challenges and which states clearly that "resilience underpins our efforts to safeguard our nations, societies, and shared values" [1]. Thus, the concept of security has evolved and one can redefine security as systemic resilience—the capacity of interconnected infrastructures and societies to anticipate, absorb, and adapt to compound stress [2]. From this perspective, risk and resilience are not independent; instead, *risk-and-resilience feedback networks* are built from interlinked nodes through which perturbations propagate. For example, a cyberattack can disable power; flooding can sever communications; disinformation can corrode governance. Digital transformation multiplies both capacity and exposure, with AI and quantum computing improving efficiency yet expanding the attack surface. While risk is seldom confined to a single domain, improving resilience within individual subsystems can benefit the entire system, if the interdependencies and feedback mechanisms between those subsystems are well understood and effectively managed. Resilience, therefore, is not a static condition but an emergent property of interaction quality among recovering systems. Deterrence depends on credible resilience—the assurance that societies and systems can absorb shocks without systemic collapse.

The CPSICC Nexus reframes the challenge from *incident response* to *adaptive-system design*: which necessitates the development of learning systems that evolve as fast as the threats they confront. It demands coordination among scientists, engineers, policymakers, and the military. It becomes not only a research program but an emerging doctrine of integrated resilience and security. Building out this program requires bringing together a community of practice that thinks and acts at the convergence of these systems and forging a new way of thinking about their security within an integrated and layered framework.

The purpose of this chapter is to provide an overview of the CPSICC Nexus Workshop and this SPS volume which resulted, summarizing the key contributions developed from workshop presentations, scenario sessions, and follow-on collaborations, as well as the chapters that comprise this book. Some recommendations for actionable emerging research areas are made at the end.

The CPSICC Nexus Initiative

To meet the need for identifying key R&D needs in this convergent topic, the Advanced Research Workshop (ARW) *Unravelling the Cyber-Physical-Social Infrastructure Climate Change Nexus* emerged through a collaborative effort among Purdue University's Institute for a Sustainable Future (ISF), Sandia National Laboratories, and Australia's Commonwealth Scientific and Industrial Research Organisation (CSIRO). Sponsored by the NATO Science for Peace and Security (SPS) Programme and the US Department of Homeland Security Science and Technology Directorate (S&T), the initiative sought to integrate research communities that had not interacted enough, e.g.: climate science, cybersecurity, critical materials and infrastructure, and social-system analysis, which lie at the heart of the convergent risk facing NATO [3], the US [4], and partnered nations.

Hosted at the American Geophysical Union Headquarters in Washington DC (29 July–1 August 2024), the CPSICC Nexus workshop gathered over fifty experts from NATO nations, partner countries, US agencies, and allied laboratories. The workshop reflected the SPS Programme's central aim: to foster cooperative research and innovation that enhance the security and stability of NATO members and partner nations. By situating discussions at the intersection of climate, cyber, physical, and social infrastructure, the CPSICC ARW expanded NATO's resilience agenda to include environmental, technological, and societal interdependencies—an urgent need in the era of accelerating global change [5–9].

The four-day CPSICC Nexus Workshop was itself an experiment in systems integration—an innovative progression from strategic framing to scientific exploration and collaborative foresight. We employed a combination of plenary discussions, short talks, informal "banquet" talks, video presentations, analytical games, and scenario-based exercises to examine how climate change amplifies vulnerabilities across cyber, physical, and social infrastructures. Participants included experts from NATO and partner nations, federal laboratories, universities, and private-sector organizations, engaging in interactive dialogue and collaborative problem-solving across scientific, operational, and policy communities. Each session was designed to elicit insights into system interdependencies, threat multipliers, and governance challenges at the CPSICC Nexus.

Initial materials emerged from workshop presentations, and moderated discussions, which were recorded and transcribed to capture key insights. Using both designated note-takers and the self-documenting features embedded in the analytical games and scenario-based exercises, a set of summary documents was compiled. Presentation materials were collected and, with author permission, made publicly available via the workshop website (https://cpsiccnexusworkshop2024.org). More information on the workshop is available in other chapters in this book (e.g. Chaps. 1–4) and links to archived versions of the briefing, pre-workshop booklet, presentations and workshop summary report can be found in Zenodo repository listings in the Appendix.

From Workshop to Book: Contributions and Thematic Overview

The workshop served as both an incubator and a conceptual blueprint for the edited book that follows. While the workshop distilled converging insights into a shared interdisciplinary research framework, this book expands those ideas by translating core principles into applied chapters that demonstrate practical relevance across domains. Contributions to this book were developed through a structured, multi-stage process designed to ensure rigor and relevance: (1) initial workshop presentations and facilitated discussions; (2) synthesis sessions identified emergent cross-domain themes; (3) chapters emerged from workshop discussions, and were expanded by their authors, and (4) a structured peer-review process coordinated by the editorial team and reviewed internally and externally across partner institutions so as to ensure an intertwined narrative on the CPSICC Nexus. This collaborative process reflects the core principles of the NATO SPS initiative—transparency, inclusivity, and cross-sector integration—and the strategic partnership among DHS, NATO, and CSIRO, which together link North American, European, and Indo-Pacific perspectives on security and resilience.

Table 1 CPSICC Nexus thematic clusters and representative chapters

Thematic cluster	Representative chapters
Developing a Systems Understanding of the CPSICC Nexus	Chapter 1—The Cyber-Climate Nexus (Grobler et al.) Chapter 2—Information Gathering (Reinhardt et al.) Chapter 3—Risk elicitation (Johnson et al.) Chapter 4—Analysis and Key Findings (Reinhardt et al.)
Policy, Social, Economic, and Governance Approaches	Chapter 5—The Cascading Impact of Cyber on Climate (Chamikara et al.)
Technical Resiliency and Risk Mitigation Alternatives	Chapter 6—Mapping Critical Infrastructure Interdependencies (Dias Baptista et al.) Chapter 7—Resilient Cyber-Physical Systems for Offshore Windfarms (Lehtola et al.) Chapter 8—Research Portfolio (Haddad et al.)
Cyber-Physical Process Characterization	Chapter 9—Diversified Reservoir Management (Okioga et al.) Chapter 10—Agro-Environmental Security (Středová et al.) Chapter 11—Beyond the Trend (Karamperidou)
Social and Economic Considerations	Chapter 12—Wiring Resilience (Gunda et al.) Chapter 13—From Misinformation to Climate Crisis (Aamir et al.)
Risk Assessment and Decision Making	Chapter 14—Cyber Jackal (Hammouchi et al.)

Each chapter examines a distinct aspect of the CPSICC Nexus, e.g., from systems modeling and governance frameworks to applied studies of water, energy, agriculture, and information integrity. Collectively, these chapters demonstrate how resilience science can move from analysis to implementation across scales and sectors. Table 1 summarizes these contributions, illustrating how the entire book enhances our understanding of the interconnected vulnerabilities and adaptive capacities within the cyber-physical-social-climate system. The thematic clusters are based on the emergent themes identified as outcomes of the CPSICC Nexus workshop.

A great deal of information emerged from the workshop itself, and a preliminary synthesis was led by researchers from Purdue, Sandia, and CSIRO, who consolidated themes and identified research and development gaps. An in-depth description of the information collected and the various themes that emerged is presented in Chaps. 1–4. We touch upon some of the key research and development (R&D) themes that emerged from the workshop but refer to the other chapters in this book for a more focused discussion.

Part I: Developing a Systems Understanding of the CPSICC Nexus

Part I Overview There was a broad consensus that climate acts as a threat multiplier within the CPSICC Nexus, amplifying the complex interdependencies among cyber, social, informational, and cognitive systems. Yet, there remains a critical gap in methodologies, shared frameworks, and integrated data needed to build a coherent systems-level understanding across these domains. This book presents four chapters that contribute to this thematic cluster.

Chapter 1—The Cyber-Climate Nexus: Navigating Dual Threats to Critical Infrastructure and Society by Grobler et al. examines how climate hazards and cyber-threats intersect to endanger critical infrastructure, economic stability, and societal resilience. It argues that as systems become increasingly digital and interdependent, climate-induced disruptions can compound with cyberattacks to generate cascading failures across energy, water, and food systems. The authors trace these compound risks through multiple dimensions: direct infrastructure vulnerability, disruption to supply chains and emergency responses, and broader economic, social, and political destabilization. Through global case studies, they illustrate how climate events amplify digital vulnerabilities and vice versa, with profound consequences for public health, governance, and security. The chapter concludes by identifying key research and policy gaps in cross-sector coordination, data sharing, and governance integration and proposes ten actionable recommendations emphasizing systemic resilience, digital and environmental co-design, and multilateral cooperation. By mapping these interactions across multiple systems and scales, this chapter contributes to a deeper systems-level understanding of the CPSICC Nexus, highlighting how compound risks emerge and propagate through interconnected infrastructures and governance structures, framing the positioning of the CPSICC Nexus in the rest of the book.

Chapter 2—CPSICC Nexus: Information Gathering by Reinhardt et al. documents the information-gathering methods and preparatory activities underpinning the CPSICC Nexus workshop. Employing a multi-pronged approach, the team designed a workshop structure that integrated diverse disciplinary and national perspectives. The analysis revealed that while physical and cyber-systems are relatively well-studied, social infrastructure and its interdependencies with climate and technology remain the least understood and most pressing research frontier. Drawing lessons from prior US and NATO wargaming efforts, and informed by collaboration with Purdue and CSIRO partners, the authors developed three key scenarios—Arctic Shipping, Power Grid Failures, and Solar Radiation Modification—to elicit cross-domain insights and decision challenges. The chapter identifies critical intersectional R&D and concludes with eight actionable recommendations, including the integration of existing frameworks and models, mapping cross-country infrastructure dependencies, improving climate data security, and addressing misinformation impacts. This chapter builds on the CPSICC Nexus foundation laid out in this chapter detailing the approach and methodology used to bring together a multidisciplinary event to establish a baseline for CPSICC Nexus R&D and contributes to the subsequent framing of Chap. 4.

Chapter 3—A Participatory Risk Elicitation Exercise to Identify Knowledge and Policy Gaps at the Intersection of Climate Change and Cyber-Physical-Social Infrastructure by Johnson et al. introduces an innovative, participatory risk elicitation methodology designed to uncover research and policy gaps at the intersection of climate change and cyber-physical-social infrastructure. Building on strategic foresight and wargaming traditions, the authors developed role-based, scenario-mapping exercises conducted at the 2024 NATO Advanced Research Workshop Unravelling the CPSICC Nexus. Participants from 12 NATO nations explored plausible 2030 and 2050 threat environments—such as cyberattacks on energy grids during heatwaves, disruptions to Arctic shipping, and sabotage of solar-radiation-modification technologies—to stress-test preparedness and governance systems. The exercises generated over 140 potential R&D gaps grouped into six cross-cutting themes: systems understanding of the CPSICC Nexus; socio-economic impact assessment; cyber-physical process characterization; resilience and mitigation technologies; risk management and governance; and policy frameworks. The chapter demonstrates how structured foresight methods and participatory scenario design can integrate diverse expertise, surface assumptions, and identify early indicators of systemic vulnerability, offering a replicable model for NATO and allied partners to inform future research agendas and resilience planning. This chapter strengthens the theme by demonstrating how participatory foresight and scenario-based risk elicitation can uncover hidden vulnerabilities and inform strategic decision-making, building on the systemic insights from Chaps. 1–4 to offer a holistic approach to navigating complex, compound risks across the CPSICC Nexus and conclude the book's contribution to the CPSICC body of knowledge.

Chapter 4—CPSICC Nexus: Analysis and Key Findings by Reinhardt et al. distills the analytical findings from the CPSICC nexus workshop, using a Decision Quality (DQ) framework to systematically map R&D gaps across the cyber-physical-social-climate continuum. The chapter identifies the primary challenge as improving decision-making under deep uncertainty, limited resources, and complex interdependencies among critical infrastructures. The analysis reveals that most existing research gaps cluster around information and alternatives, underscoring the need for improved data quality, system modeling, and risk understanding. 136 discrete R&D gaps are categorized into six thematic areas. Collectively, these findings define a transdisciplinary research agenda for NATO and partner nations to enhance the resilience of interconnected systems in a rapidly changing threat environment. This chapter contributes directly to the advancement of systems-level understanding of the CPSICC Nexus by identifying actionable pathways for bridging research gaps across cyber, physical, social, infrastructure, and climate domains.

Part II: Policy, Social, Economic, and Governance Approaches

Part II Overview Another theme that emerged was the urgent need for robust policy, legal, and governance frameworks capable of addressing the long-term risks posed by the complex interdependences across cyber, physical, social, infrastructure, and climate domains. A key take-away is that safeguarding digital infrastructure is inseparable from climate adaptation, and to avoid cascades and develop more secure societies require more secure cyber-infrastructure. This book presents one chapter that contributes to this thematic cluster.

Chapter 5—The Cascading Impact of Cyber on Climate by Chamikara et al. provides a comprehensive analysis of how cyber-technologies both amplify and mitigate the risks posed by climate change, mapping the complex feedback between digital, physical, and social systems. It identifies five "nexus types" that define the cyber–climate relationship: (1) how AI, IoT, big data, and quantum technologies enhance forecasting and mitigation; (2) how climate hazards disrupt cyber-physical infrastructure; (3) how cyber-vulnerabilities exacerbate climate disasters; (4) how digital systems themselves create environmental and systemic risks; and (5) how governance frameworks can manage these interdependencies. Drawing on real-world cases, the authors show that cascading disruptions arise when climate shocks and cyber-incidents interact across critical infrastructures such as energy, water, and communications. The chapter concludes with a tiered governance framework emphasizing integrated cyber-climate risk assessments, resilient infrastructure design, international policy harmonization, and workforce training. It argues that safeguarding digital infrastructure is inseparable from climate adaptation, and that resilient cyber-physical systems are now central to sustainable, secure societies.

Part III: Technical Resiliency and Risk Mitigation Alternatives

Part III Overview A key theme that emerged centers on the dual role of emerging technology in the CPSICC Nexus: while they offer promising avenues for resilience and risk mitigation, they also introduce new and often underexplored threat surfaces that can amplify systemic vulnerabilities. This book presents three chapters that contribute to this thematic cluster, starting with a broad consideration of resilience and mitigation technologies, then progressively drilling down into specific domains and technical approaches. This allows for both a systems-level perspective and detailed exploration of how emerging tools and methods can address or exacerbate vulnerabilities across the CPSICC Nexus.

Chapter 6—Mapping Critical Infrastructure Interdependencies for Building Resilience to Climate and Other Risks by Dias Baptista et al. systematically maps the interdependencies among critical infrastructure sectors to strengthen resilience against climate, cyber, and cascading risks, with a primary focus on energy, communications, and data storage or processing. Using Australia as a case study, this chapter argues that these enabling sectors underpin all others and form the backbone of national and global resilience. Drawing on a comprehensive desktop review, the study traces the bidirectional dependencies that link physical and digital systems, showing how failures in power or data networks can cascade across healthcare, transport, finance, and emergency response during crises such as bushfires, floods, and cyberattacks. The chapter advances a whole-of-system approach to risk management that integrates cross-sector collaboration, public–private partnerships, and shared data systems. Ultimately, the chapter argues that mapping critical infrastructure interdependencies is foundational for designing adaptive, data-driven resilience strategies that can anticipate and mitigate compound threats across interconnected climate, cyber, and social domains. This chapter exemplifies the theme by presenting a whole-of-system approach that maps interdependencies among critical infrastructure, to proactively identify and mitigate cascading risks from climate, cyber, and compound threats.

Chapter 7—Autonomous Uncrewed Systems for Offshore Wind Farm and Undersea Infrastructure Maintenance: Pathways to Climate and Cyber-Physical Resilience by Lehtola et al. explores how autonomous uncrewed systems (AUS) can enhance resilience in offshore wind farms and undersea infrastructure, such as communication cables and energy pipelines, under intensifying climate and cyber-stresses. Using the North Sea as a primary case study, the chapter analyzes the challenges posed by extreme weather, high maritime traffic, and hybrid cyber-physical threats to critical infrastructure. They demonstrate how AUS, equipped with advanced perception, AI-driven autonomy, and sensor fusion, can ensure continued inspection and maintenance operations even amid communication loss, harsh environmental conditions, or cyber-interference. The chapter develops a conceptual model of Autonomy-Assisted Remote Control, combining human oversight with adaptive machine decision-making to balance safety, mission success, and resilience. It further introduces the concept of Hybrid Situational Awareness, integrating human cognition with real-time autonomous sensing to counter hybrid warfare and GNSS spoofing risks. By embedding AUS within the NATO CPSICC Nexus, the chapter positions these technologies as key enablers of climate adaptation, maritime security, and cyber-physical resilience across international waters. It demonstrates how autonomous systems offer innovative, scalable alternatives for technical resiliency and risk mitigation in the face of increasingly complex and hybrid threats to critical offshore infrastructure.

Chapter 8—A Research Portfolio to Advance the Resilience of Energy Systems: The Resilient Energy Systems Mission Campaign by Haddad et al. presents the *Resilient Energy Systems Mission Campaign (RES MC),* a $50 million, seven-year research portfolio advancing the resilience, survivability, and recovery of energy infrastructure against intentional threats such as cyberattacks, kinetic assaults, and electromagnetic pulses. The RES MC integrates three major research thrusts—Science of Vulnerabilities, Materials, Devices, and Cyber-Innovation, and System-Level Threat-Informed Computational Science—to develop tools, materials, and models that strengthen grid reliability and accelerate recovery, including four marquee product areas: cyber-physical vulnerability tools, EMP mitigation, solid-state transformers, and resilience-measurement frameworks. The chapter details the delivery of novel defense-in-depth architectures, self-healing controls, and new materials for EMP protection; these are all capabilities directly translatable to NATO's energy-security and resilience goals. The chapter concludes with lessons on interdisciplinary integration, technology transition pathways, and collaborative frameworks to sustain innovation beyond the campaign's completion in 2027. This chapter aligns with the theme by showcasing how the RES MC delivers integrated, threat-informed innovations that enhance the robustness, adaptability, and recovery capacity of energy systems facing complex and evolving risks.

Part IV: Cyber-Physical Process Characterization

Part IV Overview The growing complexity of infrastructure systems as they increasingly blend physical operations with digital controls and data flows emerged as a key theme. As climate pressures and cyber-threats intensify becoming more extreme, understanding and characterizing these cyber-physical processes and their climate interactions becomes essential for identifying vulnerabilities and designing resilient, adaptive systems within the CPSICC Nexus. This book presents three chapters that contribute to this thematic cluster, each focusing on a distinct case study of a cyber-physical environment. These chapters identify key threat surfaces that are opened up at the intersections of these systems and provide insights into how to mitigate and improve them.

Chapter 9—Diversified Reservoir Management in Enhancing Water Security and Climate Resiliency by Okioga et al. presents a comprehensive case study of the Western Virginia Water Authority's diversified reservoir management strategy as a model for climate-resilient water security. It highlights how integrating multiple reservoir types creates redundancy and operational flexibility in response to intensifying droughts, floods, and water quality challenges. Through detailed analysis of three systems, the chapter illustrates how physical design, environmental stewardship, and governance coordination together support a stable regional supply. The chapter also demonstrates how "hard-path" infrastructure solutions are complemented by "soft-path" strategies such as forest management, stakeholder cooperation, and adaptive policy planning. Emphasis is placed on technological integration, including remote sensing, IoT-based water quality monitoring, and cybersecurity protections guided by NIST standards. To exemplify best practices in infrastructure resilience, concluding with economic and policy insights, the chapter argues that diversified reservoir management offers a scalable blueprint for achieving water and climate resilience under increasing environmental and cyber-stresses. This chapter contributes to the theme by illustrating how cyber-physical technologies are embedded within water infrastructure systems, offering both enhanced operational insight and new dimensions of risk that must be actively managed in climate-resilient planning.

Chapter 10—Agro-Environmental Security Under Climate Change (Central Europe): Climate-Borne Risk and Their Evaluation by Středová et al. presents a comprehensive assessment of climate-driven risks to agro-environmental security in Central Europe, using Czechia as a detailed case study. It evaluates primary and secondary threats, ranging from drought, floods, extreme heat, and frost to cascading hazards such as wildfires and landslides, through both Preliminary Risk Analysis and Multi-Criteria Risk Analysis frameworks. By comparing expert assessments, national risk analyses, and student-led evaluations, the chapter reveals shifting perceptions of agricultural risk under accelerating climate change, with newer analyses assigning higher risk scores to heat extremes, flash floods, and drought. The study emphasizes that adaptation requires both technological and ecosystem-based solutions, while noting the rising importance of cyber-physical threats as agriculture transitions toward digital systems (Agriculture 4.0 and 5.0). The chapter concludes that agro-environmental security demands dynamic risk reassessment, integration of cybersecurity into agricultural planning, and education-driven awareness, positioning Czechia's evolving risk management framework as a transferable model for climate resilience in temperate regions worldwide. This chapter advances the theme by highlighting how the digital transformation of agriculture introduces new cyber-physical dependencies that require continuous risk characterization and integrated planning to safeguard agro-environment systems under climate stress.

Chapter 11—Beyond the Trend: Climate Variability as a Threat Multiplier for Critical Infrastructure and Global Cyber-Physical-Social Systems by Karamperidou examines how large-scale tropical Pacific climate variability, particularly the El Niño–Southern Oscillation (ENSO) and related multidecadal phenomena like the Pacific Decadal Oscillation, acts as a global threat multiplier for cyber-physical-social systems. Karamperidou demonstrates that variations in ENSO "flavors" (Eastern vs. Central Pacific El Niños and La Niñas) can dramatically shift the geography and frequency of extreme heat and humidity events, magnifying risks to infrastructure, data centers, and energy systems. Using detailed climate analyses and new model-based risk metrics, the study shows that ENSO-driven humid-heat extremes can expose multiples more data centers to critical wet-bulb temperatures (> 31 °C), while cyberattackers have already exploited heat-related vulnerabilities in cooling management systems. The chapter frames ENSO as both a source of prediction opportunity and an amplifier of systemic fragility, urging improved modeling of weather-to-climate interactions, enhanced coupling between physical and cyber-resilience, and targeted technological innovations, such as secure, climate-adaptive cooling systems, to mitigate compounding climate and cyber-threats. This chapter contributes to the theme by revealing how large-scale climate variability can dynamically reshape cyber-physical risk landscapes, underscoring the need for integrated modelling and adaptive technologies that account for climate-induced stressors on digital infrastructure systems.

Part V: Social and Economic Considerations

Part V Overview Another theme centers around the critical need to better understand and quantify the socio-economic dimensions of resilience and risk across cyber-physical-social systems. Despite growing recognition of their importance, significant R&D gaps remain in effectively capturing public, institutional, and economic impacts within the CPSICC Nexus, particularly under compounding climate and cyber-stresses. Solutions proposed involved great inclusion of AI, modeling and data-driven decision-making systems as well as developing more advanced data integrity and anti-disinformation functionalities. This book presents two chapters that contribute to this thematic cluster.

Chapter 12—Wiring Resilience: Metrics for a Weather-Adapted Power Grid by Gunda et al. presents a quantitative framework for assessing how power grids can withstand and recover from extreme weather and cyber-disruptions through the development of multidisciplinary resilience metrics. The chapter argues that while reliability measures have long guided grid design, the growing frequency and intensity of extreme and often compounded events demand a shift toward resilience-based approaches. Drawing on theories from engineering, information science, and ecology, they adapt three core metrics—entropy (robustness), functional redundancy (redundancy), and global reaching centrality (resourcefulness)—to capture systemic performance under stress. A case study using a synthetic Texas power grid illustrates how a heatwave and concurrent cyberattack degrade resilience differently and jointly, with compounded threats amplifying load shedding due to diminished redundancy. The analysis demonstrates how these quantitative indicators can complement traditional reliability metrics, supporting better modeling, risk assessment, and decision-making for climate-adapted energy systems. The chapter concludes by calling for integration of artificial intelligence, agent-based models, and cross-sector data to advance real-world validation and inform the design of weather-resilient, cyber-secure power networks. This chapter contributes to the theme by demonstrating how resilience-based metrics can inform socio-economic decision-making, helping stakeholders better anticipate and manage the cascading impacts of extreme weather and cyber-disruptions on energy systems and the communities they serve.

Chapter 13—From Misinformation to Climate Crisis: Navigating Vulnerabilities in the Cyber-Physical-Social Systems by Aamir et al. explores how misinformation and disinformation act as amplifiers of vulnerability across the interconnected cyber-physical-social-climate nexus. It argues that human factors, such as cognitive biases, decision-making silos, and declining trust in institutions, are as destabilizing as technical or physical failures, often transforming localized cyber-incidents or climate shocks into systemic crises. Drawing on case studies such as the 2019–2020 Australian bushfires and the 2019 European heatwave, the chapter demonstrates how false narratives can erode public trust, distort policy priorities, and trigger cascading infrastructure failures. They assess how misinformation spreads through digital networks, influences social behavior, and undermines emergency and policy responses, especially during climate-related disasters. The chapter evaluates emerging government and international responses and surveys AI and cybersecurity tools that can detect and counter false information. Ultimately, it calls for integrating misinformation resilience into critical infrastructure governance, emphasizing public literacy, cross-sector coordination, and proactive "pre-bunking" strategies as essential components of climate and national security resilience. This chapter contributes to the theme by demonstrating how misinformation can undermine institutional trust, distort public decision-making, and exacerbate the socio-economic impacts of climate and cyber-crises, highlighting the need for integrated governance and public resilience strategies within the CPSICC nexus.

Part VI: Risk Assessment and Decision Making

Part VI Overview The final theme focuses on the urgent need for comprehensive frameworks and governance models to manage increasingly complex risk profiles at the CPSICC Nexus. As infrastructure systems become more interconnected and exhibit emergent behaviors across diverse time horizons, stakeholders identified significant R&D gaps in developing methodologies that can effectively address multi-system dependencies and dynamic risk convergence. This book presents one chapter that contribute to this theme.

Chapter 14—Cyber Jackal: Opportunistic and Malicious Behavior in Times of Extreme Climate Conditions—A Position Paper by Hammouchi and Lenzini introduces the concept of the Cyber Jackal, a metaphor for opportunistic cyber-adversaries who exploit societal and infrastructural vulnerabilities during extreme climate events. Drawing parallels between ecological opportunism and cyber-exploitation, the authors demonstrate how crises such as floods, hurricanes, and wildfires weaken socio-technical systems and create openings for malicious actors to strike. Through empirical analysis using blackhole network data from Luxembourg's 2021 floods, they detect subtle shifts in network traffic indicative of opportunistic scanning and exploitation attempts during environmental stress. The chapter highlights how cognitive overload, institutional strain, and resource scarcity compound vulnerabilities, transforming climate disasters into cyber-physical crises. To counter these risks, the authors propose an integrated resilience blueprint combining real-time data fusion between meteorological and cybersecurity feeds, adaptive response loops, and a quantifiable resilience score linking environmental readiness with cyber-posture. The chapter concludes that resilience against compound threats demands coordinated preparedness, continuous training, and adaptive governance to anticipate "Cyber Jackal" behavior in a changing climate. This chapter contributes to the theme by introducing a novel framework for anticipating opportunistic cyber-threats during climate crises, demonstrating how real-time data fusion and adaptive decision-making can enhance risk assessment and guide proactive resilience strategies in complex, compound threat environments.

Lessons Learned and Path Forward

The CPSICC Nexus Initiative is built upon the value of integrating cyber-physical, social, financial, environmental, policy, and military expertise within a unified systems framework. Workshop participants and chapter co-authors agreed that climate resilience is inseparable from cyber, physical, and social resilience, and these interdependencies demand sustained innovation across scientific, defense, and civil communities. Workshop outcomes translate conceptual synthesis into a coordinated research portfolio, showing how NATO and partner institutions can connect science, technology, and policy to confront shared CPSICC Nexus risks. This transdisciplinary model forms both a foundation for research and a roadmap for operational collaboration, advancing a resilient transatlantic community of practice. From these discussions emerged six strategic lessons.

Six strategic lessons and corresponding priorities emerged:

1. **Integration and Collaboration Across Domains and Institutions:** Integrating existing frameworks and models across disciplines and fostering collaboration among NATO member countries to address cross-country dependencies and vulnerabilities is necessary to address current and emerging challenges. For example, embedding CPSICC principles within NATO's Science and Technology Organization would institutionalize this integration.
2. **Interdependencies and convergence:** Enhancing systemic security requires increasing resilience across systems, especially against cascading risks. To accomplish this requires people and information systems that operate comfortably in this intersecting and convergent way. Effective adaptation, risk mitigation, and deterrence require linking climate science, cybersecurity, and physical infrastructure engineering, and social, political, and economic action through shared data frameworks, analytical tools, and resilience metrics. For example, developing a *Living Foresight Platform* that combines models, cyber-risk analytics, and social data would provide actionable insight and early warning across sectors.
3. **Data integrity and cyber-physical security:** Protection from cyberattack, disinformation, or "poisoning" or loss of key data sources, whether from the climate, social, physical, economic, or defense sectors is necessary to ensure decision-making capabilities. This may require the development of so-called zero-trust methodologies and systems to deploy solutions.
4. **Engagement:** Sustained engagement among research institutions, federal agencies, and international partners is essential to align investments, maintain data integrity, and reduce redundancy. This could be extended to include public-private partnerships, in energy, finance, and digital infrastructure, including AI, which should translate research into deployable technologies and operational practices. Building coherence and trust between disciplines, institutions, and nations remains the ultimate infrastructure for security and sustainability.
5. **Human capital investment and actionable foresight:** Combining social-science and physical-science methods leads to innovation. For example, analytical gaming and foresight methodologies enable anticipation of cascading threats and strengthen decision-making under uncertainty. Expanding these methods into training and education would cultivate the next generation of resilience professionals. For example, a proposed *CPSICC Nexus Fellowship Program*, modeled on NATO's SPS framework, would pair early-career scientists with mentors across academia, government, and industry. As such has been developed by DEEP for resilience [2], an analogous curriculum could be built for Nexus content.
6. **From threat multiplier to opportunity multiplier:** Through coordinated investment in resilient food, energy, and water resources, digital security and information integrity, and adaptive infrastructure, NATO and its partners can transform vulnerability into collective strength. The CPSICC Nexus provides both a framework and a community for turning complexity into cooperation. Its legacy will be an adaptable international security architecture capable of navigating accelerating change with confidence, adaptability, and shared purpose.

Acknowledgments

The authors would like to thank Leslie Proctor, Lynne Dahmen, Karan Hustedt-Warren, Tara Greene, Kayla Gurganus for project management, project support, logistics, curation and summation of workshop material and much more. None of this would have been possible without those efforts. Guidance and some heavy lifting that are much appreciated were carried out additionally by Surya Nepal, Dongyan Xu, David Johnson, Jason Reinhardt, Courtney Falk, Carly Frank, Pragathi Jha, Guy Barnett, Diana Hackenburg, Subulola Jiboye, Krista Kelley, Joel Rasmus, Alex Blauensteiner, and Stacey Connaughton. We thank the NATO Science for Peace and Security (SPS) Programme, the US Department of Homeland Security Science and Technology Directorate (DHS S&T), and the Commonwealth Scientific and Industrial Research Organisation (CSIRO) for their co-sponsorship and continued collaboration in advancing this transdisciplinary initiative. We also acknowledge support from Purdue University, especially from the Institute for a Sustainable Future, with additional support from Purdue's Center for Education and Research in Information Assurance and Security and Policy Research Institute. Huber also thanks support while writing this chapter and editing this book from Purdue's Provost Office, the Rockefeller Foundation Bellagio Center Fellows program, and the Yale Institute for Biospheric Studies Edward P. Bass Distinguished Visiting Environmental Scholars Program. We must also give our thanks to the workshop attendees and chapter authors and the broader community whose contributions are the foundation of CPSICC Nexus thinking. We would also like to thank American Geophysical Union for providing an amazing venue for the workshop. The views presented in this chapter represent those of the authors alone, not of the people or funders or affiliated organizations mentioned. AI was utilized at some stages of the data collection and curation process, including transcription of talks and sometimes as step in the summary of presentations and chapters.

Competing Interests The authors declare no competing interests.

West Lafayette, USA — Matthew Huber
Melbourne, Australia — Marthie Grobler

References

1. NATO strategic concept (2022)
2. Deep. Resilience reference curriculum (2025) https://www.nato.int/cps/en/natohq/topics_233455.htm
3. NATO (2023) Strategic foresight analysis 2023 (Rev. 2). Brussels: Allied command transformation
4. US Department of Homeland Security, strategic framework for addressing climate change (2023)

5. Loverdos A (2022) The offence–defence balance: NATO's growing cyber-challenge. NATO parliamentary assembly, defence, and security committee
6. Kairidis G (2025) Bolstering cybersecurity in NATO and beyond. NATO parliamentary assembly, Committee on the civil dimension of security
7. Teitelbaum V (2025) Protecting Allied societies from disinformation emanating from the People's Republic of China. NATO parliamentary assembly, Committee on democracy and security
8. NATO (2023a) Climate change and security impact assessment, In: 2nd ed. Brussels: NATO HQ
9. NATO (2024) Climate change and security impact assessment, In: 3rd ed. Brussels: NATO HQ

Appendix

Supplementary material including archived versions of CPSICC Nexus presentations, Briefing Reports, and a Workshop summary report are available at the following Zenodo links.

Briefing Documents

Huber, M., Grobler, M., Homan, R., Purdue University, CSIRO's Data61, & Sandia National Laboratories. (2024, July 29). Briefing: Unravelling the Cyber-Physical-Social Infrastructure Climate Change (CPSICC) Nexus. Advanced Research Workshop Unravelling The Cyber-Physical-Social Infrastructure Climate Change (CPSICC) Nexus, Washington DC, USA. https://doi.org/10.5281/zenodo.17280188

Workshop Booklet

Huber, M., Grobler, M., Homan, R., Purdue University, CSIRO's Data61, & Sandia National Laboratories. (2024, July 29). Unravelling the Cyber-Physical-Social Infrastructure Climate Change Nexus Workshop Booklet. NATO Advanced Research Workshop: Unravelling The Cyber-Physical-Social Infrastructure Climate Change (CPSICC) Nexus, Washington DC. https://doi.org/10.5281/zenodo.17280248

Summary Presentations

Huber, M., Grobler, M., Homan, R., Purdue University, CSIRO's Data61, & Sandia National Laboratories. (2024, October 21). Unravelling the Cyber-Physical-Social Infrastructure Climate Change Nexus: Summary of Speaker Presentations. Unravelling the Cyber-Physical-Social Infrastructure Climate Change Nexus, Washington DC, USA. https://doi.org/10.5281/zenodo.17280380

Speaker Presentations

Huber, M., Grobler, M., Homan, R., Purdue University, CSIRO's Data61, & Sandia National Laboratories. (2025, October 13). CPSICC Workshop Presentations 2024. Unravelling The Cyber-Physical-Social Infrastructure Climate Change (CPSICC) Nexus (CPSICC Nexus), Washington DC, USA. Zenodo. https://doi.org/10.5281/zenodo.17343798

Contents

Part I
Developing a Systems Understanding of the CPSICC Nexus

Chapter 1
The Cyber-Climate Nexus: Navigating Dual Threats to Critical Infrastructure and Society

Marthie Grobler, Lygia Romanach, Cody Christopher, Youyang Qu, Guy Barnett, and Matthew Huber

Abstract This chapter explores the dual threats of climate hazards and cybersecurity events to critical infrastructure, society, and the economy. Climate hazards such as floods, hurricanes, and heatwaves compromise cyber-physical systems, amplifying vulnerabilities in our increasingly digital infrastructure. The convergence of climate risks and sophisticated cyber threats poses significant challenges to energy, water and food security, which are essential to global security and societal wellbeing. It also has widespread economic impacts, threatening global trade, financial stability, and corporate operations. Socially, they may jeopardize public health, safety, and community resilience, disproportionately affecting the most vulnerable populations. Politically, these threats can undermine public trust, fuel social unrest, and contribute to geopolitical tensions. To combat these challenges, the chapter underscores the importance of integrated strategies to unravel the cyber-physical-social infrastructure climate change nexus and to enhance infrastructure resilience, improve cybersecurity, and foster global cooperation.

Keywords Cybersecurity · Climate hazards · Cyber-climate nexus · Critical infrastructure · Cascading failures · Resilience

M. Grobler (✉) · Y. Qu
CSIRO's Data61, Clayton, VIC, Australia
e-mail: Marthie.Grobler@data61.csiro.au

L. Romanach
CSIRO's Environment, Dutton Park, QLD, Australia

C. Christopher
Australian National University, Acton, ACT, Australia

G. Barnett
CSIRO's Environment, Black Mountain, ACT, Australia

M. Huber
Purdue University, West Lafayette, IN, USA

M. Huber et al. (eds.), *Critical Infrastructure Systems and Extreme Threat Nexus: Preparing for Resilience and Security*, NATO Science for Peace and Security Series C: Environmental Security, https://doi.org/10.1007/978-3-032-13170-6_1

1.1 Introduction

Climate hazards, such as floods, hurricanes, and heatwaves pose significant risks to cyber, physical, and social infrastructure, disrupting such infrastructure and leading to widespread consequences to human lives and societal functions. In addition, climate hazards and cybersecurity threats are converging as interconnected risks, presenting challenges to global security, infrastructure resilience, and societal well-being, exponentially increasing the threat exposure and negative consequences of compounding and/or cascading impacts.

Critical infrastructure sectors are increasingly interdependent, meaning that the functioning of one sector (such as energy, water, transportation, telecommunications, or financial services) both depends on, as well as has an impact on the functioning of other sectors. Therefore, disruptions in one infrastructure can cascade and impact multiple other sectors, highlighting the need for comprehensive cross-sector risk management and coordinated protection strategies to ensure overall resilience and stability in the provision of services.

Vulnerabilities in critical infrastructure are also growing due to a greater dependency on digital technologies, with sophisticated cyber threats targeting the very technologies designed to improve efficiency and climate resilience. This increased digitization exposes more devices and systems to potential cyber-attacks, which are further exacerbated by climate-related disruptions such as extreme weather events. Cyber-attacks can disrupt vital climate monitoring services and undermine efforts to mitigate and adapt to climate change, destabilizing systems already weakened by environmental stressors. This creates a vicious cycle of escalating risks, where the impacts of climate change and cyber threats are mutually reinforcing and amplifying the overall threat landscape.

Due to the interconnected nature of modern infrastructure, the impacts of combined environmental and cyber threats extend far beyond immediate physical damage. An attack on one system can trigger cascading failures across multiple sectors, resulting in widespread economic disruption, as well as social and political instability [1]. From an economic perspective, such threats may jeopardize global trade, financial stability, and corporate operations. Socially, they may compromise public health, safety, and community resilience, disproportionately affecting the most vulnerable populations. Politically, these disruptions can undermine public trust in government and institutions, incite social unrest, and contribute to geopolitical tensions. Political systems, including voting, dissemination of information to the public or to state actors, can also be directly impacted by these hazards leading to immediate geopolitical consequences. As a result, the convergence of climate and cyber hazards, compounded by critical infrastructure interdependencies, creates complex vulnerabilities and cascading risks. Addressing these challenges requires coordinated, cross-sectoral, and holistic strategies to enhance the protection and resilience of essential systems.

This chapter explores the vulnerabilities of cyber-physical systems, examining how climate change and cybersecurity threats amplify each other in an interconnected world and the broader implications for infrastructure and societal resilience. The chapter also emphasizes the importance of developing integrated strategies to enhance the resilience of critical infrastructures, improve cybersecurity measures, and foster cooperation across sectors and countries. By understanding and addressing the intersection of climate change and cybersecurity, we can better protect our societies and navigate the complexities of this evolving threat landscape.

1.2 Climate Hazards and Their Direct Impacts

The accelerating impacts of climate change, including more frequent and severe weather events, rising sea levels, and shifting climate patterns, magnify vulnerabilities in critical infrastructure. Climate hazards threatening critical infrastructure can compound to create a larger effect, with climate and cyber risks exacerbated by the degree and extent of interdependence between and across critical infrastructure sectors with the potential for cascading impacts. The direct impacts of climate hazards on cyber, physical, and social infrastructure highlight the urgent need for strategies to build resilience. Understanding these threats and interdependencies to prepare for various scenarios is crucial to mitigate the risks and protect human welfare. Addressing these compounded threats requires comprehensive planning, investment, and collaboration among stakeholders as well as integrated strategies, robust infrastructure, and coordinated global efforts to enhance resilience and security.

Due to the interconnectedness of critical infrastructure systems, and specifically the direct link between critical infrastructure operations and climate hazards, there is a need for improved systems to better understand socio-technical-economic interdependencies in the design, operation, and adaptation of critical infrastructure [2]. Understanding interdependencies between sectors is critical to improving the resilience of the system as a whole and avoiding mal-adaptation. For instance, any disruptions to energy, water and transport infrastructure would likely produce cascading impacts on supply chains and access and/or availability to other critical infrastructure such as health, medical and emergency services, provision and supply of food and the functioning of defence forces [3]. Therefore, understanding the impact of climate hazards such as floods requires an understanding of the individual and combined impacts on multiple sectors. The impacts on energy, water and transport sectors and their cascading effects on the supply chain (e.g., food and medical supplies storage and distribution), health system (e.g., access to emergency and medical services and risk of increased infectious disease) and financial system (e.g., impact on health and insurance costs) needs to be better understood.

It is also critical to understand how a climate and cyber event could compound. Consider how a cyber-attack on water treatment facilities during a flood event could further exacerbate water disruption, leading to a larger health crisis; or how a cyber-attack on a power grid during a heatwave could result in widespread blackouts, leaving millions without air conditioning and heightening the risk of heat-related illnesses. Due to the increased reliance on digitized systems and technology by many critical infrastructures, a cyber event affecting these technologies could have far-reaching impacts due to the interconnectedness of sectors, hazards and risks, and technology dependence. Similarly, it is critical to consider the interconnectedness of critical infrastructure systems when planning for new critical infrastructure. For example, investments that aim to mitigate carbon emissions (such as clean hydrogen) could require additional water investments that could be jeopardised during a drought. Thorough risk assessments should be conducted to ensure that all potential cascading effects and interconnected risks, as well as non-intuitive interdependencies, are identified in advance to ensure resilience and minimize risk for new investments. Ideally, all new infrastructure investments should consider the adoption of emergent technology and where practical, adopt modular platform-agnostic solutions that will enable an easier transition with shorter disruptions when upgrades are required.

The convergence of climate hazards and cyber threats significantly impacts a wide range of critical sectors essential to societal functioning and economic stability, including transportation, healthcare, financial systems, telecommunications, emergency services, and urban development. These sectors are increasingly reliant on digital infrastructure, making them vulnerable to cyberattacks that can be exacerbated by climate-related disruptions such as extreme weather events. For instance, a cyberattack on a hospital's digital systems during a heatwave or flood could paralyze emergency response and patient care. Similarly, disruptions to financial networks or telecommunications during climate emergencies can hinder crisis coordination, economic continuity, and key functions of participatory democracy.

Despite these broad vulnerabilities, current research often prioritizes water, energy, and food security as focal sectors. This is because they form the foundational triad of human survival and societal stability with interdependencies that can trigger cascading failures across all other sectors if compromised. Water, energy, and food systems are not only deeply interconnected but also highly exposed to both environmental stressors and cyber vulnerabilities, making them critical leverage points for building resilience against compounding hazards and threats with cascading impacts.

1.3 Water, Energy and Food Security

Both climate hazards and cyber threats can individually and collectively jeopardize the safety and security of critical infrastructure. As infrastructure systems become increasingly digitized, they face heightened vulnerabilities from both climate hazards and cyber threats. In addition, the interconnectedness of critical infrastructure sectors allows risks to cascade between sectors and systems resulting in greater impacts. In this section, we explore how climate hazards and cyber threats may compound and/or cascade to impact water, energy and food security, drawing from interdisciplinary research and real-world case studies.

The intersection of climate change and cybersecurity presents a complex and evolving threat landscape for critical infrastructure systems that underpin water, energy, and food security. As digital technologies become more embedded in climate-sensitive sectors, the potential for cascading failures that are triggered by simultaneous environmental and cyber disruptions amplifies. We consider infrastructure vulnerability, digital dependency and system complexity, emergency response disruption, economic and market instability, as well as governance and policy gaps as key areas of concern. Addressing these risks requires integrated resilience strategies that bridge environmental science, cybersecurity, and infrastructure governance to ensure the stability and sustainability of essential services in an increasingly volatile world.

1.3.1 Water Security

Water security is critical to both ecosystems and human society and is increasingly being threatened by the convergence of two global challenges: climate change and cybersecurity risks. While climate change disrupts hydrological cycles and intensifies water scarcity and contamination risks [4], cybersecurity threats target the digital systems that manage water infrastructure. Together, these forces create a complex and evolving risk landscape that demands integrated, forward-looking solutions.

Climate change threatens water security through impacts on the water cycle by influencing when, where, and how much precipitation falls [5]. It also leads to more severe weather events over time, impacting the availability and quality of freshwater supplies. In addition, the availability of water impacts local topography, affects animal migration and influences drinking water and sanitation needs for both animal and human health. For instance, an increase in flood events increases the runoff of contaminants such as fertilizers, polluting the water supply and limiting access to safe water for humans and the ecosystem. Rising sea levels also contaminate underground freshwater with salt water, while droughts decrease the availability of water for human consumption and energy and food production. Water availability and quality also have cascading impacts on several other infrastructure sectors such as energy, food and health sectors [6].

Cyber-attacks on water infrastructure can compromise the availability, quality, and safety of water supplies, posing a direct threat to public health and essential services. This is because modern water systems rely heavily on digital control technologies, such as supervisory control and data acquisition (SCADA) and Internet of Things (IoT) sensors for monitoring and control, which can disrupt treatment systems, contaminate supplies and halt or compromise water distribution. A real-world example occurred in 2001, when a cyber-attack on the Australian Maroochy Shire's sewage control resulted in 265,000 gallons of untreated sewage flowing into local parks and rivers, causing significant damage to the local environment [7].

When climate and cyber threats occur at the same time, such as during climate-induced crises such as floods or droughts [8], cyberattacks can exacerbate disruptions by disabling water treatment or distribution systems. An illustration of this is the cyber-attack in 2021 targeting the water treatment system in Oldsmar, United States, involving an attempt to increase the levels of sodium hydroxide (lye) in the water supply to dangerous levels. Although the attack was thwarted before any harm was done, it highlighted the potential for cyber breaches to cause serious environmental and public health consequences by contaminating water supplies [9]. Another realization of these combined threats happened during a period of extreme heat and drought in the United Kingdom when South Staffordshire Water experienced a cyber-attack that disrupted internal systems. Although the company stated that water supplies were not directly affected, the timing of the attack – during a climate-related water stress event – raised serious concerns about the vulnerability of water infrastructure to compound threats [10].

In the long term, the compounded effects of climate hazards and cyber threats could undermine the reliability, safety, and governance of water systems worldwide. Challenges in attributing such attacks can lead to a decrease in public trust in institutions, enhancing conspiratorial thinking and reducing the ability of governments and societies to meaningfully respond. Building resilience to joint threats will require not only technological innovation but also cross-sector collaboration, adaptive policy frameworks, and sustained investment in both environmental and digital infrastructure.

1.3.2 Energy Security

The resilience of energy infrastructure is increasingly challenged by the dual pressures of climate change and cybersecurity threats. Climate-related disruptions such as extreme heat and intense storms can damage physical energy systems, while cyber-attacks can exploit digital vulnerabilities in grid management, fuel supply chains, and renewable energy platforms. Together, these converging risks threaten not only the reliability of energy supply but also the broader stability of critical infrastructure systems that depend on secure and continuous power delivery.

Energy security is critical for the functioning of society, and climate-related disruptions can pose significant risks. On the supply side, extreme weather events can impact the generation of renewable energy, as well as energy transmission and distribution. Consider how extreme heat can lower solar power cell efficiency; wildfires and cyclones can damage wind and solar energy infrastructure; drought can reduce water availability for hydropower; wildfires, floods and cyclones can damage transmission infrastructure such as transmission towers and power poles. Extreme conditions such as high winds and ice storms could also strain energy grids. On the demand side, extreme hot and cold temperatures are major drivers of customer electricity demand, increasing the risks of power outages [11]. A real-world example of where climate-related events disrupted energy security includes the flooding in India and Bangladesh between 2020 and 2022, where monsoon-related flooding disrupted coal transport and submerged power plants, straining the energy supply chains [12].

Another example is the 2019/20 Australian wildfires, which damaged transmission lines and substations, leading to blackouts and energy supply disruptions, and an estimated AUD$100 billion in direct and indirect economic losses to the Australian economy [13].

The transition to cleaner energy sources is also altering the vulnerability profile of the energy system due to increased electrification and the growing reliance on digital technologies. Digital technologies enhance efficiency and control but also open new avenues for cyber threats as illustrated by the cyber-attacks that disabled thousands of digitally managed wind turbines in Germany [14] and the cyber-attack that left around 230,000 people without power in western Ukraine in 2015 [15–17]. Incidents such as these highlight the vulnerability of critical infrastructure to sophisticated cyber-attacks, demonstrating the need for robust cybersecurity measures in the face of increasing digitization.

A combination of extreme weather events and cyber-attacks on electricity infrastructure can lead to widespread blackouts with cascading impacts on other critical infrastructure such as telecommunications (e.g., mobile phone coverage), healthcare (e.g., access to safe water for consumption, wastewater treatment, appropriate indoor temperatures, life support equipment and/or storage of fresh food and medication) and transportation (e.g. access to fuel, public transport) [1]. A cyber-attack during extreme heat or cold conditions could lead to prolonged power outages, leaving communities without cooling or heating and light, posing serious health risks, while a cyber-attack specifically on the liquid fuels sector could have cascading impacts on the transport and financial sectors. An example of this occurred in May 2021, when nearly half of the US East Coast's fuel supply was shut down after a cyber-attack forced Colonial Pipeline to shut its entire network, causing fuel prices to spike ahead of peak US summer driving season, negatively impacting consumers and the economy [18, 19].

In the long term, the compounded effects of climate and cyber threats could destabilize energy markets, hinder decarbonization efforts, and erode public trust in critical infrastructure and the socio-political institutions that manage the infrastructure. Building resilience will require a holistic approach that integrates climate adaptation, cybersecurity protocols, and cross-sector collaboration to safeguard the future of energy security.

1.3.3 Food Security

Food security is increasingly vulnerable to the compounded risks posed by climate change and cybersecurity threats. Climate-related disruptions such as droughts, floods and heatwaves can damage agricultural production, supply chains, and storage infrastructure, while cyberattacks can target the digital systems that manage everything from irrigation to logistics. Together, these threats create a volatile environment where both physical and digital vulnerabilities converge, threatening the availability, accessibility, and stability of food systems [20].

Climate change poses a significant threat to food security by affecting agricultural and fishery productivity, compromising biosecurity and food safety, and impacting the health and wellbeing of animals and humans. Additionally, it jeopardizes the livelihood of those working in farming [21]. In particular, climate change impacts the production of animal feedstock such as corn. A study forecasts a 20–40% decline in US corn belt production from 1991–2000 levels by 2046–2055 [22]. Increased extreme high temperatures will also lower livestock productivity due to heat stress, increasing the likelihood of disease, and lowering animal feed efficiency and reproductive health. Higher temperatures will also demand more water to maintain animal health and wellbeing, putting additional pressure on both water and food security.

The threat of cyber-attacks on food security is growing. This is because the increased use of digital-based technologies (e.g., robotics, aerial imaging, digital mapping of seeds, and GPS) to improve resource efficiency and access to climate forecasting to minimize the impact of climate change on food production also increases exposure to cyber-attacks [23]. For example, in June 2021, the world's largest meat processing company, JBS Foods, paid US$11 million in ransom to resolve a cyber-attack that halted their US cattle slaughtering for a day [24]. Cyber-attacks targeting agricultural data and automated systems can exacerbate climate risks, further impacting food security.

Food security can also be threatened through cascading effects originating from the disruption to other sectors such as electricity, communication and transport which can impact both on-farm and off-farm supply chain infrastructure. Consider how food security was threatened in the East Kimberley, a remote region of Western Australia, by the disruption of transport routes that prevented trucks from delivering food supplies in 2023. Such disruption, which originated in the transport system, had cascading effects on the food systems and the defence forces, which were called upon to assist in supplying food during the disruption [25]. In countries with a high reliance on food imports, such as the UK, any climate disasters or cyber threats on the food producing country or on global food supply chains can have a significant impact on food security. A reduction in the availability and/or quality of food tends to lead to rising prices and an increase in food inequity with cascading impacts on the financial and health systems, which can increase the risk of social unrest [26].

In the long term, the convergence of climate hazards and cyber threats could erode the resilience of global food systems, particularly in regions already facing environmental and economic stress. Addressing these challenges requires a holistic approach that integrates climate adaptation, digital risk management, and cross-sector collaboration to ensure food security in an increasingly interconnected world.

1.4 The Economic, Social and Political Ramifications

Climate change and cybersecurity threats have profound economic, political and social impacts. In addition to the direct impact on the environment, people and infrastructure, threats to water, energy and food security might impact global trade and financial stability as well as public health and safety, community resilience and social cohesion. It may also destabilize governments, alter geopolitical landscapes, and impact international relations. Integrated strategies and coordinated policies are critical to safeguard, economic, social and political stability as well as international cooperation. This section highlights some of these impacts.

1.4.1 Economic Impacts

Increasingly, climate-related risks such as property damage from extreme weather and investment losses in fossil fuel industries can affect asset values and increase the volatility of financial markets [27]. In addition, financial institutions are prime targets for cyber-attacks, which can result in direct financial loss, data breaches, and erosion of customer trust. The integration of robust cybersecurity measures is essential to protect financial stability. An example of this is when financial institution JPMorgan Chase implemented comprehensive cybersecurity measures to protect its operations. This followed a major data breach in 2014, where the personal information of 76 million households and 7 million small businesses was compromised. In response, the bank allocated US$600 million annually to cybersecurity and employed a team of over 3,000 IT security professionals. By adopting advanced security protocols, establishing a robust Security Operations Centre, and engaging in continuous employee training, JPMorgan Chase has significantly enhanced its ability to defend against cyber threats, ensuring the security and resilience of its data assets and critical infrastructure [28].

Climate-related weather and cyber-attacks on critical infrastructure can have significant economic impacts due to disruption to global supply chains. Rising sea levels, more frequent and severe hurricanes, changes in precipitation patterns as well as cyber-attacks on logistics and transportation infrastructure can disrupt global trade routes and port operations [29]. Major port cities are particularly vulnerable to these disruptions, which can lead to significant economic losses [30]. For example, a 2017 ransomware attack on Maersk, a leading shipping company, caused severe disruptions [31], halting operations across 76 ports worldwide. This incident resulted in significant delays in shipping schedules and supply chain disruptions, with Maersk reporting losses of up to USD$300 million due to the attack. Similarly, the NotPetya malware attack, which affected several industries including logistics, resulted in global economic damages estimated at USD$10 billion. Companies such as FedEx and pharmaceutical giant Merck also reported losses in the hundreds of millions due to disrupted operations and recovery costs [32].

Due to cascading effects, disruption to one critical infrastructure sector, such as telecommunications, can generate large economic impacts. To illustrate, in November 2023, Australia's second-largest telecommunications company, Optus, suffered an operational failure which resulted in mobile phone and internet services outages, directly affecting 10 million customers. The outage also impacted 400,000 businesses, as well as government departments, health and transport systems in Melbourne, and health and water services in South Australia. It is estimated this event cost the Australian economy AUD$1 billion [33].

Cyber threats may also impact the accuracy and reliability of climate and weather forecasting and prediction services, which are key to the functioning of critical infrastructure services and modern society. Disruption to these services has the potential to disrupt global trade, affect critical industries such as energy and agriculture, and pose risks to financial system stability. In terms of food production and distribution, the forecasting of crop yield is fundamental to the agricultural industry to better mitigate and manage climate risks. The success of rain-fed crops is highly volatile and uncertain due to climate variability and change, which impacts both costs and market prices [34]. Forecasts of the impact of seasonal climate on crop yield and production is critical for decisions across the supply chain.

Climate and weather forecasting is also important to other sectors of the economy, such as the energy sector, where it is used to estimate peak energy demands (e.g., due to heatwaves or cold spells) and forecast renewable energy generation. The increasing penetration of variable renewable energy generation in the grid will place even greater importance on climate and weather forecasting to ensure the reliability of the energy supply. Forecasting whether key transport routes might be disrupted by extreme weather events is critical to minimize supply chain disruptions. Forecasting might also be critical to trade. For example, a study found that losses from market street vendors in Ghana could be reduced by 50% if extreme weather is anticipated [35].

In addition, cyber-attacks can disrupt critical climate monitoring and forecasting services and organizations. Consider the instance in 2014, when the National Oceanic and Atmospheric Administration (NOAA) was the target of a cyber-attack that compromised its weather and satellite systems. The attack temporarily disrupted the agency's ability to provide accurate weather forecasts and climate data, which are crucial for monitoring environmental changes and preparing for natural disasters. Another significant example is the 2017 WannaCry ransomware attack, which affected various critical infrastructure sectors worldwide, including some meteorological services. Such attacks can delay the dissemination of vital climate information, affecting everything from disaster response to long-term climate research. These disruptions can have cascading effects, hindering scientific research, delaying climate action, and compromising the reliability of data used to track global climate patterns [36].

1.4.2 Social Impacts

Both climate change and cyber threats disproportionately affect vulnerable populations, among which are low-income communities, the elderly, and those with preexisting health conditions [37]. These groups often lack the resources and capabilities to effectively respond to and recover from such threats. The increasing reliance on digital solutions for climate adaptation and response can also exacerbate the effects due to the digital divide. That is, communities without adequate access to technology and internet services tend to be at a disadvantage, exacerbating existing social and economic inequalities [35].

The disruption to water, energy and food security due to climate and cyber threats can have wide implications for public health and safety. For instance, extreme weather events and cyber-attacks on water supply systems can lead to the contamination or disruption of water services, increase the spread of infectious diseases and compromise access to clean water and food, posing significant health risks to communities [38]. Extreme weather events such as heatwaves, bushfires, floods, and hurricanes can also have direct health impacts, causing injuries, heat-related illnesses, and stress on healthcare systems.

Extreme weather events can also force communities to relocate, leading to social displacement and the loss of homes and livelihoods [39]. It is estimated that every year, 20 million people worldwide are displaced from their homes due to climate disasters, with 80% of human displacement in the last decade occurring in Asia. Moreover, people in low- and lower-middle-income countries are four times more likely to be displaced by extreme weather disasters than people in high-income countries [40]. Cyber-attacks that disrupt essential services during such crises can further aggravate the situation, making recovery and adaptation more challenging. An example of this occurred in 2017, when Hurricane Maria devastated Puerto Rico, causing widespread displacement [41]. The island's power grid was severely damaged, and a subsequent cyber-attack on the Puerto Rican Electric Power Authority (PREPA) further delayed power restoration. This compounded the crisis, making recovery more challenging for affected communities. Disruption to forecasting and prediction services can also have implications for public health and safety, as well as disaster prevention and recovery, with wide social implications. For example, in terms of public health and safety, weather forecasts in high-income countries are more than five times as accurate as they were 40 years ago [35]. Studies show that providing longer lead times for severe winter weather forecasts reduces traffic accidents, saving lives and generating broader economic benefits. These include lower insurance costs and fewer disruptions to sectors affected by transport system disruptions. Meteorological data in urban spaces is also key for monitoring urban air quality, planning the capacity of combined storm and sewage systems, determining the location of new developments to avoid flood risk, and for managing urban heat islands [42].

In terms of disaster prevention and recovery, forecasting allows governments and organizations to alert individuals about extreme weather events ahead of time, allowing them time to adjust their actions and reduce exposure or vulnerability to the weather event [43]. For instance, the provision of advance warning of events such as bushfires, flooding and hurricanes provides the opportunity for individuals living or traveling in the area at risk, their families, the communities in which they live and the agencies with responsibilities for ensuring their safety, to make appropriate preparations to mitigate the adverse impacts from such events. Advance warning ahead of extreme events also benefits primary producers and those operating businesses, critical utilities and infrastructure (power, water, roads, etc.) directly by providing time to protect their assets and operations. Reducing potential damages and losses saves lives and assists insurance companies and governments by reducing the level of compensation and costs of funding relief and recovery operations. For example, it is estimated that forecasting the impact of flooding hurricanes in the US has already improved by three times in the last four decades, reducing damages by 5% [35].

1.4.3 Political Impacts

Climate change intensifies competition for natural resources, such as water and arable land, exacerbating existing vulnerabilities. This can further increase resource scarcity and forced migration, as well as social tensions and conflict within and between communities, impacting social cohesion [44]. This can lead to geopolitical conflicts, particularly in regions already prone to instability. Consider the effects of climate change in the Lake Chad Basin, having significantly reduced the size of Lake Chad which is a critical water source for Nigeria, Chad, Cameroon, and Niger. This reduction has led to heightened competition over water and arable land, exacerbating existing conflicts and contributing to the rise of militant groups like Boko Haram. The resulting social instability highlights the need for integrated resource management and conflict resolution strategies to enhance community resilience in the face of climate-related resource scarcity [45].

Increased competition for natural resources may also create opportunities for cyber-attacks, further destabilizing regions [46]. An illustration of this is increased activism both offline and online, which has risen as a method to draw attention to climate issues. Activists may use cyber-attacks such as 'hacktivism', or hacking for political or social causes, to further their agendas. Nations may also use cyber capabilities to gain strategic advantages in resource competition and political influence. Cyber-attacks on critical infrastructure highlight the potential to disrupt essential services, undermine public trust, and create political instability as well as pose significant national security threats, with anticipated adverse impacts on military assets, infrastructure, and operations. For example, a cyber-attack on the Ukrainian power grid [47] temporarily disrupted power to Kiev, affecting hundreds of thousands of residents. This disruption was attributed to a state-sponsored hacking group, which used the cyber-attack as tools of geopolitical strategy.

Cyber espionage is another threat that involves the covert acquisition of confidential information from other nations, including intellectual property, government secrets, and strategic plans. Consider the cyber-espionage group APT29 [48], believed to be linked to a nation-state, which conducted extensive campaigns targeting government agencies and industries to steal sensitive data, enhancing its competitive position in global markets. A more recent example includes Volt Typhoon, a state-sponsored actor that typically focuses on cyber disruption and information gathering, with the intent to disrupt communications infrastructure in the Pacific [49]. Cyber warfare is another attack vector that involves aggressive tactics such as disrupting critical infrastructure, manipulating data, and spreading misinformation to destabilize opponents. A notable example is the Stuxnet worm, which targeted Iran's nuclear program in 2010, allegedly developed by the United States and Israel [50]. This sophisticated cyber weapon disrupted Iran's uranium enrichment process, setting back its nuclear capabilities.

1.5 Integrated Strategies for Resilience

To date, several approaches have been used to develop both preventive and reactive resilience strategies within critical infrastructure sectors. These include risk-based approaches to identify and prioritize critical assets and systems based on their vulnerability to both cyber threats and climate hazards. Such approaches consider the likelihood, potential consequences, and cascading effects of disruptions [51].

For investment in resilience to be effective, both preventive and reactive risk management responses are required to assess and manage climate-related risks across the systemic value chain. For example, in the aftermath of Hurricane Sandy in 2012, New York City undertook significant infrastructure investments to enhance resilience against future extreme weather events. This included the construction of the East Side Coastal Resiliency Project [52], a $1.45 billion initiative aimed at protecting vulnerable neighborhoods from flooding and sea-level rise. Additionally, NYC Health + Hospitals implemented robust cybersecurity measures to safeguard its digital infrastructure, ensuring continued healthcare services during emergencies. However, greater focus is still needed to invest in infrastructure design and upgrades to withstand the simultaneous threat of both cyber-attacks and climate-related weather events [53].

Preventive strategies to address cybersecurity threats may include the implementation of advanced security protocols and continuous monitoring to defend against potential cyber intrusions. In parallel, resilience strategies often incorporate redundancy, analogue controls, and manual overrides where necessary. Investments are also being directed toward novel technologies and renewable energy sources to enhance overall system resilience. Diversifying and decentralizing renewable energy generation can improve the resilience of the energy system, as different sources are vulnerable to different types of extreme weather events. Distributed generation also helps dilute the impact of localized disruptions [54]. Other strategies might include developing systems that combine climate modeling with cyber threat intelligence to anticipate and mitigate cascading risks to critical infrastructure and communities, and investing in climate-resilient and cyber-secure infrastructure, such as smart grids, water systems, and transport networks, with built-in redundancies, encryption, and adaptive capacity to withstand both environmental and digital disruptions [55]. Other strategies might involve establishing clearly defined severity thresholds that trigger specific response protocols to ensure timely and effective action during cyber-climate incidents.

Reactive strategies may include coordinated emergency response controls and cyber-climate national frameworks. Currently, countries such as the UK [56] and Japan [57] have national frameworks, with critical infrastructure protection and resilience policies generally driven by a singular hazard focus, such as climate change. Canada [58] requires sector-related information sharing to promote communication and hazard identification, but this is also focused only on the impact of climate on physical disturbances. Some countries have already recognized and begun to address the need for a multi-hazard focus. The USA [59], Switzerland [60], France [61] and Germany [62] have developed policies based on an all-hazards approach that provides holistic management and response to critical infrastructure threats and hazards. Even countries not advanced in the protection and resilience of their critical infrastructure acknowledge the importance of information sharing for system resilience to address the mega trends of climate change, digitization, cyberattacks and rising geopolitical tensions. Although critical infrastructure policy frameworks are yet to be developed in the Oceanic region, countries such as New Zealand [63] have identified threats to cybersecurity, climate change and the criticality of data storage and processing infrastructure, touching on the cyber-climate nexus. More work needs to be done in this space to mature reactive cyber-climate strategies.

Developing and implementing cohesive legal frameworks that integrate cybersecurity and environmental governance is critical for effectively dealing with this dual threat. Current national and international legal frameworks often treat cyber threats and climate impacts separately, lacking a unified approach to address the complex challenges posed by their intersection. This separation creates gaps where cyber incidents can cause environmental harm, or climate events can expose cybersecurity vulnerabilities. For example, the European Union's General Data Protection Regulation (GDPR) works in tandem with the Network and Information Security (NIS) Directive to ensure that personal data protection and cybersecurity measures are aligned, creating a more resilient digital environment [64]. However, these directives do not address climate risks. Expanding such frameworks to integrate environmental considerations can help safeguard essential systems and promote a more holistic, resilient approach. In this context, the Cyber Resilience Act serves as a model for evolving international agreements to tackle interconnected cybersecurity and environmental challenges, ensuring nations are better prepared to handle dual threats. Strengthening international agreements to include provisions for cybersecurity in climate policies can help nations prepare for and respond to interconnected threats.

While proactive and reactive strategies offer a strong foundation for enhancing resilience to the cyber-climate nexus, their integration remains a work in progress. Continued refinement, cross-sector collaboration, and adaptive governance are essential to ensure these strategies evolve in step with emerging threats and complex interdependencies.

1.6 Current Gaps in Resilience Measures

As climate change and cybersecurity threats increasingly converge, critical infrastructure systems face complex, compounding risks. Despite growing awareness of this cyber-climate nexus, current resilience measures remain fragmented, reactive, and often sector-specific. This section examines the key gaps in existing frameworks, highlighting the need for integrated, forward-looking strategies that address both environmental and digital vulnerabilities in a cohesive manner.

1.6.1 Infrastructure Vulnerability

Climate change and cybersecurity pose significant challenges to the safety and functionality of critical infrastructure. Critical infrastructure remains highly exposed to the dual threats of climate-related damage and cyberattacks, yet resilience planning often fails to account for how these risks can interact and amplify one another. This gap can leave systems increasingly vulnerable to cascading failures [65], where a disruption in one domain can rapidly compromise others, compounding the overall impact [66].

A key component of strengthening this resilience lies in sustained investment in research and innovation. Supporting future research towards the development of advanced technologies such as AI-driven monitoring systems, IoT-enabled infrastructure, and predictive analytics can significantly enhance the ability to detect, respond to, and recover from compound threats. Related to this, investing in data availability, quality and transparency might also be a good avenue to improve climate and weather forecasting and the usability of such data to mitigate any identified vulnerabilities. For example, developing a data inventory and making it easily accessible to research, non-profit, and private-sector modeling communities could help test and further improve the usability of data models [36]. These innovations not only improve situational awareness and operational flexibility but also help future-proof infrastructure against evolving climate and cyber risks.

1.6.2 Digital Dependency and System Complexity

Critical infrastructure systems, including water, energy, and food security, are threatened by both climate change and cyber-attacks. Although rapid digitization can mitigate climate risks, it also increases susceptibility to cyber threats, particularly during climate-related disruptions. This dual threat underscores the need to address both climate resilience and cybersecurity to protect essential services and infrastructure and the socio-political institutions that provide them. The interdependencies of these critical infrastructure systems mean that disruptions in one sector can cascade across others, triggering widespread failures.

As critical infrastructure systems, and specifically energy, water, and food systems, become increasingly digitized, their growing complexity and interconnectivity introduce new points of failure that are not adequately addressed in current resilience frameworks. While these innovations offer efficiency and responsiveness, they also introduce complex interdependencies and novel cyber-physical vulnerabilities. Current resilience measures often fail to capture the full scope of these digital transformations, lacking the systemic perspective needed to understand how disruptions in one digital node can propagate across interconnected networks.

Due to the increased digitalization and digital dependencies, the integration of cybersecurity into infrastructure planning is vital. Some of the current inhibitors include limited cross-sector data sharing, and a lack of dynamic modeling tools that reflect real-time system behavior. Developing a holistic, system-wide approach that integrates digital risk mapping, real-time monitoring, and cross-sector coordination is essential to ensure infrastructure resilience keeps pace with technological advancement, and ensuring that critical infrastructure systems can achieve optimal performance, leveraging technological advances whilst considering all system interfaces.

1.6.3 Emergency Response Disruption

Emergency response systems are increasingly strained by the convergence of climate-related hazards and cyber threats, yet current resilience measures often overlook how these risks can compound and disrupt critical response capabilities. For instance, extreme weather events like floods or bushfires can damage physical infrastructure while simultaneously overwhelming digital communication networks, leaving emergency services unable to coordinate effectively. Key gaps include the lack of redundant and secure communication channels for first responders, insufficient integration of cyber contingency planning into climate disaster protocols, and limited scenario-based training that accounts for simultaneous physical and digital disruptions. Bridging these gaps requires a unified resilience strategy that embeds cybersecurity into climate adaptation planning, ensures interoperability across emergency systems, and invests in real-time threat detection and response coordination technologies.

In addition, emergency response systems are frequently designed to handle either physical or digital crises, but not both simultaneously. This emergency response disruption can leave significant gaps in preparedness for compound cyber-climate events. These gaps are further exacerbated by limited public understanding of how climate change and cyber threats can interact to disrupt essential services. Increasing public awareness and education around these interconnected risks is critical. Educational initiatives such as community workshops, school curricula, and public information campaigns can empower individuals and local organizations to take proactive measures in disaster preparedness and response. By fostering a more informed and engaged public, emergency systems can benefit from stronger community-level resilience and more coordinated responses during complex crises. As demonstrated by responses to events such as Hurricane Katrina in New Orleans and the September 11 attacks in New York City, the long-term political stakes of success or failure in this area are significant, and the consequences enduring.

1.6.4 Economic and Market Instability

The economic consequences of cyber-climate disruptions are often underestimated, with limited mechanisms in place to mitigate cascading effects on markets, supply chains, and consumer access to essential services. Cascading failures across sectors can disrupt supply chains, inflate prices, and limit consumer access to essential goods and services. Yet, current economic planning and risk management frameworks often fail to account for the compounding nature of cyber and climate threats. Further research initiatives are therefore needed to develop integrated forecasting tools that combine climate hazard prediction with cyber risk modeling, and improve the translation of scientific outputs into actionable insights for decision-makers.

Economic and market instability is another unforeseen implication of the cyber-climate nexus. Conventional approaches to risk assessment tend to be based on a retrospective survey of historic hazards and impacts. As extreme events become more severe, more frequent, and more correlated, the true costs of climate-related damages tend to be underestimated. Under a changing climate, forecasting and predictive models that heavily rely on observational weather-related data are becoming less reliable, which may harm economic activity, public health and safety as well as the capacity of governments and communities to prepare for extreme climate events. Research focused on novel ways to improve data models, such as the research led by the University of Utrecht which aims to investigate whether Paleo data can improve climate models, is critical to further scientific understanding [67].

It has been suggested that human and capital investment is urgently needed to enhance forecasting systems that predict not only the frequency and severity of extreme weather events, but also the economic and human losses associated with them [36]. However, a persistent challenge is that climate science outputs are often not user-friendly, and many end-users – especially in the private sector – lack the technical capacity to interpret them effectively. To address this, climate information services must become more user-driven, tailored to the specific needs and technical capabilities of different sectors. The goal should be to make climate information 'useful, usable, and used', enabling more climate-sensitive and cyber-aware economic decisions that support both mitigation and adaptation goals [68].

1.6.5 Governance and Policy Gaps

The impacts of cyberattacks and natural disasters are anticipated to increase significantly in scale and complexity over the coming decades. However, current governance frameworks often treat cybersecurity and climate change risks separately, leading to fragmented and ineffective responses. Fragmented governance across environmental, cybersecurity, and infrastructure domains hinders the development of cohesive resilience strategies, resulting in policy blind spots at the intersection of climate and digital risk [65]. For example, environmental agencies may focus on physical climate adaptation, while cybersecurity regulators prioritize digital threat mitigation, with little coordination between the two. These governance silos can delay response times, complicate accountability, and leave critical infrastructure systems exposed to cascading failures. To address these gaps, it is essential to establish integrated policy frameworks that promote cross-sector collaboration and shared responsibility.

There is a critical need for integrated policy frameworks that combine climate resilience and cybersecurity measures that can evolve with emerging threats and changing environmental conditions. This might include harmonizing national security strategies with environmental policies to create a cohesive response to these dual threats, integrating climate resilience into economic and urban planning, developing regulations and strategies that encourage businesses to invest in climate resilience and allowing for timely and effective responses to new challenges. It could include creating standards for data sharing, incident reporting, and compliance with cybersecurity protocols, as well as research to explore the effectiveness of providing financial incentives for businesses to invest in climate adaptation and cybersecurity measures such as tax breaks, grants, and low-interest loans or establishing minimum security and resilience requirements for critical infrastructure projects receiving federal funding, ensuring compliance through robust accountability mechanisms. The development of comprehensive risk assessment frameworks that account for both climate and cyber threats would support organizations in developing and implementing their strategic planning to enhance resilience. Additionally, embedding cyber-physical risk assessments into infrastructure planning and emergency preparedness policies could help ensure that resilience strategies are both comprehensive and future-proof.

While awareness of the cyber-climate nexus is growing, significant gaps remain in the resilience of critical infrastructure systems. These include vulnerabilities in infrastructure vulnerability, digital dependency and system complexity, emergency response disruption, economic and market instability, and fragmented governance and policy gaps. Ultimately trust in security systems and the institutions that wield them is a vital component of their continued functioning. Addressing these challenges requires a coordinated, cross-sectoral approach that integrates technological advancement, policy reform, and community engagement to build truly adaptive and future-ready systems.

1.7 Collaborative Approaches to Address the Nexus of Climate and Cybersecurity

Addressing the complex and interconnected challenges at the nexus of climate change and cybersecurity requires collaborative approaches that transcend traditional boundaries. Governments, private industry, academia, and civil society must work together to develop integrated strategies that leverage technological innovation, shared knowledge, and coordinated governance to build resilience across critical systems. We consider some of these collaborative approaches.

1.7.1 Governmental Collaboration

Governments play a critical role in managing the dual threats of climate hazards and cyber risks, but their effectiveness depends on inclusive collaboration with the public. By engaging communities in the design and implementation of early warning systems, climate-resilient infrastructure, and digital risk education, governments can ensure that adaptation strategies are equitable, locally informed, and responsive to the needs of the most vulnerable populations [69].

Strengthening collaboration and information sharing between stakeholders in all levels of government (local, national and international), as well as the public and private sectors and civil society is critical to identify shared responsibilities for developing and implementing effective risk management practices to safeguard critical infrastructure. Collaboration is key to developing comprehensive climate and cybersecurity governance models to improve resilience and response capabilities, leading to more robust security practices and quicker responses to cyber incidents. For instance, regional cooperation and initiatives can facilitate targeted strategies and resource sharing as well as address specific geopolitical challenges posed by climate change and cyber threats.

Despite growing recognition of the interconnected risks posed by climate change and cybersecurity threats, effective governmental collaboration remains hindered by several key barriers. These include fragmented institutional mandates [39], lack of interoperable data systems [65], differing national priorities, and limited trust or information-sharing mechanisms between agencies and across borders [69]. Additionally, the absence of integrated policy frameworks and funding streams often results in siloed responses that fail to address the systemic nature of climate-cyber risks.

1.7.2 Cross-Sector Collaboration

Similarly, strengthening cross-sector collaboration within the private sector would assist businesses to incorporate climate change risks into cybersecurity strategies and help build more resilient infrastructures capable of withstanding both cyber and climate threats. Such cross-sector collaboration would be key in several stages including risk assessment and planning, resilient design and infrastructure, integrated incident response plans, as well as training and awareness activities.

Data and information exchange among stakeholders helps to enhance situational awareness and coordinated responses to threats. For example, the Climate Prediction Center (CPC) and National Centers for Environmental Information (NCEI) which operate within NOAA share climate data and forecasts with stakeholders, helping to coordinate responses to climate-related threats and integrate this information into broader risk management frameworks [70]. Strengthening public-private partnerships may also help governments leverage innovative technologies and practices to improve the security and sustainability of critical infrastructure to enhance resilience against climate and cyber threats. Engaging communities in the decision-making process can lead to more effective and inclusive solutions and ensure that strategies developed address the community's specific needs and vulnerabilities [71]. This is because the extent of impacts and capacity to respond to both climate change and cyber threats to critical infrastructure would be dependent on the regions' socio-demographics and geographical characteristics. Communities characterized by resource restrictions, inequity, poverty and conflict are more vulnerable.

Collaboration and data sharing would help develop response strategies that consider the interdependencies of critical infrastructure sectors, which are largely lacking. To reduce the vulnerability to climate change and cyber-attacks it is critical that risk management approaches focus on integrated approaches that consider the various types of hazards and threats, how events compound and cascade as well as cross-sectoral and cross-national interdependencies. As illustrated in the Stimson Center's analysis of the Caribbean, addressing these converging risks requires breaking down national policy siloes and adopting 'polyresponse' strategies that build resilience across both digital and environmental systems simultaneously [72].

Cross-sector collaboration in the climate-cyber nexus faces significant barriers due to institutional silos, misaligned incentives, and differing risk perceptions between sectors. Climate and cybersecurity domains often operate under distinct regulatory frameworks and governance structures, making coordination complex. Additionally, the lack of shared language and metrics for assessing interconnected risks hinders mutual understanding and joint action. Trust deficits, especially between public and private sectors, further complicate data sharing and joint planning. These challenges are exacerbated by geopolitical tensions and concerns over digital sovereignty, which can limit international cooperation on shared climate-cyber threats [73, 74].

1.7.3 International Collaboration

International cooperation is also essential to address the global nature of cyber threats and climate impacts. Initiatives such as the United Nations Framework Convention on Climate Change (UNFCCC) [75], Kyoto Protocol [76], and Paris Agreement [77] on climate change and international cyber norms aim to foster collaborative efforts, but more cohesive policies are needed. Australia and the US cooperate on climate change with both countries engaging in bilateral climate change partnerships and working together in international settings like the Quadrilateral Security Dialogue [78], the Group of Twenty [53], and the Conference of Parties [79].

International climate change actions are primarily organized under the UNFCCC, which expands climate change actions through the Kyoto Protocol in 1997 and the Paris Agreement in 2015. Under the 1987 Montreal Protocol, which was negotiated to ban substances harmful to the environment, the United Nations adopted the Kigali Amendment to ban potent greenhouse gases. Most United Nations members are parties to these treaties. Of specific relevance is the NATO Science for Peace and

Security (SPS) Programme that fosters collaboration between NATO member and partner countries through science and innovation, particularly in addressing emerging security challenges. It supports multi-sectoral projects that integrate government, academia, private industry, and civil society, with a strong emphasis on technology-driven solutions such as cyber defense, environmental monitoring, and infrastructure resilience [80].

International collaboration on the climate-cyber nexus is often hindered by divergent national priorities, inconsistent regulatory standards, and geopolitical tensions. These challenges are compounded by gaps in trust [69], limited interoperability of digital systems, and unequal access to climate and cybersecurity technologies [27], all of which slow coordinated global responses [73].

Increasing risk complexity and the need for new, more sophisticated mitigation responses are driving uncertainty and impacting action and investment decisions across government and industry. There is also an emerging loss of community confidence regarding the ability of government and businesses to ensure essential services continue during and after hazard events. Climate change and cybersecurity threats to critical infrastructure are likely to increase. The protection of critical infrastructure against the dual threats of climate change and cyber-attacks is a multifaceted challenge with profound and far-reaching implications for society. While a range of strategies are being developed or underway to ensure the security and resilience of critical infrastructure systems, much still needs to be done.

1.8 Conclusion

Understanding the intersection of climate change and cybersecurity is key to protecting societies and ensuring a secure, sustainable future. This chapter explored the dual threats of climate events and cybersecurity to critical infrastructure, society, and the economy, looking specifically at the convergence of climate risks and cyber threats to energy, water and food security. To combat these challenges, the chapter considered the importance of integrated strategies to unravel the cyber-physical-social infrastructure climate change nexus and to enhance infrastructure resilience, improve cybersecurity, and foster global cooperation. Table 1.1 presents an overview of the key challenges identified in the cyber-climate nexus, and ten actionable recommendations that can address these challenges. These recommendations are intended not only as immediate guidance but also as a strategic foundation for future research and policy development. They can inform national and international policy frameworks, including initiatives such as NATO's SPS Programme, by offering direction for collaborative resilience-building and innovation in dual-use technologies.

Response strategies to build system resilience and mitigate climate or cyber threats tend to be well developed within sectors, but not yet between sectors. Cross-sectoral, national and international collaboration is key to continue developing knowledge and strategies to improve our understanding of such risks, reduce the vulnerability of critical infrastructure and protect the functioning of society. Systemic risk governance is

Table 1.1 Key challenges and actionable recommendations to advance the cyber-climate nexus

Identified challenge	Actionable recommendation
Destabilized financial markets, eroding asset values and reducing investor confidence	Integrate climate and cyber risk assessments into financial sector regulations, including stress testing, cybersecurity investment, and public-private threat intelligence sharing
Disruption of global supply chains	Strengthen digital and physical resilience of critical infrastructure through cross-sector collaboration to ensure coordinated emergency response planning in supply chain disruption
Disproportionate impact on vulnerable populations	Bridge the social divide by investing in equitable access to both physical and digital infrastructure, as well as essential services
Public health and safety risks from disrupted critical services	Enhance resilience of public health and utility systems through integrated climate-cyber risk planning, including backup systems, early warning technologies, and community-based emergency preparedness
Climate-driven resource scarcity amplified by cyber-enabled political destabilization	Establish integrated climate-cyber risk monitoring and conflict prevention mechanisms through regional cooperation, combining environmental data with cyber threat intelligence to anticipate and mitigate politically destabilizing events
Cascading infrastructure failures across interdependent systems	Invest in interdisciplinary research and innovation to develop advanced and emerging technology-driven systems that can model, predict, and mitigate compound cyber-climate threats to critical infrastructure
Heightened risk of systemic failures due to cascading disruptions	Conduct cross-domain risk assessments and multidisciplinary simulation exercises to build a systemic understanding of interdependencies and failure points across digital, environmental, and infrastructure domains
Disrupted coordination and communication during cyber-climate events	Embed cybersecurity into climate adaptation planning by ensuring interoperability across emergency systems and conducting joint scenario planning that considers both cyber and climate risks during crisis triage and response
Forecasting and planning tools fail to capture combined cyber-climate market impacts	Incentivize integrated cyber-climate risk planning through financial support for interoperable systems and joint forecasting tools
Fragmented policy approaches to climate resilience and cybersecurity	Develop integrated, cross-sectoral policy frameworks that align environmental, digital, and national security strategies

essential to develop and implement comprehensive, inclusive and integrated policy strategies that strengthen the resilience of both physical infrastructure and communities and enhance community preparedness. A coordinated effort nationally and internationally across various sectors and governance levels that spans policy, technology innovation, and business practice is key to implementing effective responses to manage climate and cyber-threats to critical infrastructure and reduce the risk to significant national security threats.

Looking ahead, future efforts should prioritize integrated, cross-sectoral approaches that bridge the current silos between climate and cybersecurity domains. This includes establishing dedicated task forces or working groups that bring together experts from infrastructure, cybersecurity, climate science, and emergency management to co-develop joint risk frameworks. Governments and industry could invest in pilot programs that test adaptive technologies, such as AI-driven threat detection systems and climate-resilient infrastructure designs, under real-world conditions. Standardized protocols for climate-cyber risk assessments could be developed and adopted across sectors to ensure consistency and comparability. Additionally, regular cross-sector simulation exercises and scenario planning can help identify cascading vulnerabilities and improve coordinated response capabilities. Knowledge-sharing platforms, such as open-access databases and regional resilience hubs, could be expanded to facilitate the dissemination of best practices and lessons learned. Embedding these strategies into infrastructure planning, regulatory frameworks, and investment criteria will be essential to ensure long-term resilience. As the risk landscape continues to evolve, proactive, science-informed, and inclusive approaches will be key to safeguarding essential services and maintaining public trust.

As this chapter has shown, climate-related disruptions and cyber-attacks increasingly intersect, threatening the stability of critical infrastructure systems. In an era defined by infrastructure vulnerability, digital dependency, and growing system complexity, the convergence of climate change and cybersecurity presents an urgent and multifaceted challenge. This chapter specifically considered the systems underpinning energy, water, and food security. These dual threats not only compromise emergency response capabilities but also ripple across economies, destabilizing markets and undermining public trust and the ability of institutions to steer a path toward security.

Addressing this nexus requires more than isolated technical fixes; it demands coordinated efforts across policy, technology, and business practices. Cyber-attacks on critical infrastructure and climate-related instability now constitute significant national and global security threats. To safeguard societies and ensure a secure, sustainable future, it is imperative to develop integrated strategies that enhance infrastructure resilience, strengthen cybersecurity, and foster international cooperation. Only through such holistic approaches can we begin to unravel the complex cyber-physical-social dynamics at play and build systems capable of withstanding the shocks of an increasingly volatile world.

References

1. McKenna D (2024) Double trouble: When weather emergencies meet malicious hackers. https://engineering.jhu.edu/news/double-trouble-when-weather-emergencies-meet-malicious-hackers/. Accessed 6 Jan 2025
2. Hou Y, Brewer T (2013) System reliability and its metrics. In: Handbook of networked and embedded control systems. Springer, pp 229–257
3. Lawrence J, Blackett P, Cradock-Henry NA (2020) Cascading climate change impacts and implications. Clim Risk Manage 29:100234
4. Gelfan A (2023) Climate change and water security: Global and regional hydrological changes. Water Resour 50(5):725–735
5. National Geographic Society (2023) How climate change impacts water access. https://education.nationalgeographic.org/resource/how-climate-change-impacts-water-access/. Accessed 6 Jan 2025
6. du Plessis A (2017) Global water availability, distribution and use. In: Freshwater challenges of South Africa and its Upper Vaal River. Springer, pp 3–11
7. Cohen G (2025) Throwback attack: an insider releases 265,000 gallons of sewage on the Maroochy Shire. https://www.industrialcybersecuritypulse.com/facilities/throwback-attack-an-insider-releases-265000-gallons-of-sewage-on-the-maroochy-shire/. Accessed 6 Jan 2025
8. Tuptuk N, Hazell P, Watson J, Hailes S (2021) A systematic review of the state of cyber-security in water systems. Water 13(1)
9. Vera A, Lynch J, Carrega C (2025) Oldsmar, Florida, water poisoning hack: What happened?. https://edition.cnn.com/2021/02/08/us/oldsmar-florida-hack-water-poison/index.html. Accessed 6 Jan 2025
10. Yang YCE (2024) Dual threat: climate-induced flooding and cyberattacks. Cris Response J. Accessed 27 May 2025
11. ESCI (2021) ESCI project final report – climate change in Australia. https://climatechangeinaustralia.gov.au/en/projects/esci/esci-publications/esci-project-reports/. Accessed 6 Jan 2025
12. International Energy Agency (2022) Climate resilience for energy security. https://www.iea.org/reports/climate-resilience-for-energy-security. Accessed 27 May 2025
13. Ahmed I, Ledger K (2023) Lessons from the 2019/2020 'Black summer bushfires' in Australia. Int J Disaster Risk Reduct 96:103947
14. Arghire I (2025) German wind turbine firm discloses targeted professional cyberattack. https://www.securityweek.com/german-wind-turbine-firm-discloses-targeted-professional-cyberattack. Accessed 6 Jan 2025
15. Lee RM, Assante MJ, Conway T (2016) Analysis of the cyber attack on the Ukrainian power grid: Defense use case. https://connections-qj.org/article/analysis-cyber-attack-ukrainian-power-grid-defense-use-case. Accessed 6 Jan 2025
16. McElfresh M (2016) Cyberattack on Ukraine grid: Here's how it worked and perhaps why it was done. https://theconversation.com/cyberattack-on-ukraine-grid-heres-how-it-worked-and-perhaps-why-it-was-done-52802. Accessed 6 Jan 2025
17. Council on Foreign Relations (CFR) (2015) Compromise of a power grid in Eastern Ukraine. https://www.cfr.org/cyber-operations/compromise-power-grid-eastern-ukraine. Accessed 6 Jan 2025
18. World Economic Forum (2021) Cyber-attack on the US major oil and gas pipeline: What it means for cybersecurity. https://www.weforum.org/agenda/2021/05/cyber-attack-on-the-us-major-oil-and-gas-pipeline-what-it-means-for-cybersecurity/. Accessed 6 Jan 2025
19. Russon M (2025) US fuel pipeline hackers 'didn't mean to create problems'. https://www.bbc.com/news/business-57050690. Accessed 6 Jan 2025
20. Cassotta S, Pettersson M (2019) Climate change, environmental threats and cyber-threats to critical infrastructures in multi-regulatory sustainable global approach with Sweden as an example. Beijing Law Rev 10(3)

21. Zhang Y, Wang H, Liu Y, Chen X, Zhao Y et al (2024) Climate change and the global burden of disease: A comprehensive assessment. Nature Clim Change 14(5):492–500
22. World Economic Forum (2024) How climate change impacts food supply chains. https://www.weforum.org/agenda/2024/01/food-system-impact-of-climate-change/. Accessed 6 Jan 2025
23. US Agency for International Development (2023) Cybersecurity briefer: Agriculture and food security. https://www.usaid.gov/digital-development/cybersecurity/agriculture-food-security-briefer. Accessed 6 Jan 2025
24. Council on Foreign Relations (2024) Digital agriculture gap: Critical infrastructure protection. https://www.cfr.org/blog/digital-agriculture-gap-critical-infrastructure-protection. Accessed 6 Jan 2025
25. Hayes J, O'Connor T (2023) Australian defence force to help re-supply East Kimberley amid food shortage due to NT flooding. https://www.abc.net.au/news/2023-03-08/defence-force-to-aid-food-shortage-in-flooded-east-kimberley/102068096. Accessed 6 Jan 2025
26. Bridle S, Jones A (2023) Climate change could lead to food-related civil unrest in UK within 50 years, say experts. https://theconversation.com/climate-change-could-lead-to-food-related-civil-unrest-in-uk-within-50-years-say-experts-214754. Accessed 6 Jan 2025
27. Vikas N (2025) Navigating climate risks: 3 strategies for building resilient financial institutions. https://www.weforum.org/stories/2024/07/navigating-climate-risks-key-strategies-for-resilient-financial-institutions/. Accessed June 2025
28. Financial Times (2023) Cyberattack on critical infrastructure: Russia and the Ukrainian power grid attacks. https://www.ft.com/content/cd287352-cb3b-48d8-a85b-668713b80962. Accessed 6 Jan 2025
29. Economist Impact (2025) Climate change's disruptive impact on global supply chains and the urgent call for resilience. https://impact.economist.com/projects/trade-in-transition/climate_change/. Accessed June 2025
30. NASA Science (2023) The effects of climate change. https://science.nasa.gov/climate-change/effects/. Accessed 6 Jan 2025
31. BBC News (2017) Hacker hijacks car computer system to make vehicle speed up and brake suddenly. https://www.bbc.com/news/technology-40416611. Accessed 6 Jan 2025
32. Greenberg A (2018) Lessons to learn from devastating NotPetya cyberattack: Wired investigation. https://www.cbsnews.com/news/lessons-to-learn-from-devastating-notpetya-cyberattack-wired-investigation. Accessed 6 Jan 2025
33. Williams T (2023) Optus outage disrupts mobile phones and internet: What happened?. https://www.abc.net.au/news/2023-11-08/optus-outage-mobile-phones-internet-what-happened/103077180. Accessed 6 Jan 2025
34. Potgieter AB, Schepen A, Brider J, Hammer GL (2022) Lead time and skill of Australian wheat yield forecasts based on ENSO-analogue or GCM-derived seasonal climate forecasts – a comparative analysis. Agric For Meteorol 324:109116
35. Behrer P (2023) The economic benefits of weather forecasting. https://blogs.worldbank.org/en/impactevaluations/economic-benefits-weather-forecasting. Accessed 6 Jan 2025
36. George S (2023) A higher standard for climate risk modeling? https://www.wtwco.com/en-au/insights/2023/11/a-higher-standard-for-climate-risk-modeling. Accessed 6 Jan 2025
37. Benevolenza MA, DeRigne L (2018) The impact of climate change and natural disasters on vulnerable populations: A systematic review of literature. J Hum Behav Soc Environ 29(2):266–281
38. World Health Organization (2023) Climate change and health. https://www.who.int/news-room/fact-sheets/detail/climate-change-and-health. Accessed 6 Jan 2025
39. United Nations Security Council (2021) Climate change 'Biggest threat modern humans have ever faced', world-renowned naturalist tells security council. https://press.un.org/en/2021/sc14445.doc.htm. Accessed June 2025
40. Oxfam International (2019) Forced from home: Climate-fueled displacement. Oxfam Briefing Paper, published for COP25, Madrid
41. Lawrence J, Hossain N, Jaradat R, Hamilton M (2020) Leveraging a Bayesian network approach to model and analyze supplier vulnerability to severe weather risk: A case study of the U.S. pharmaceutical supply chain following Hurricane Maria. Int J Disaster Risk Reduct 49:101607

42. National Academy of Sciences (2012) Urban meteorology: Forecasting, monitoring, and meeting users' needs. https://nap.nationalacademies.org/resource/13328/urban-meteorology-brief.pdf. Accessed 6 Jan 2025
43. Bureau of Meteorology (n.d.) National arrangements for flood forecasting and warning. http://www.bom.gov.au/water/floods/document/National_Arrangements_V4.pdf. Accessed 6 Jan 2025
44. International Security Advisory Board (2024) Report on new security challenges. Tech. Rep., U.S. Department of State. Final Report
45. Frontières MS (n.d.) Lake Chad crisis in depth. https://www.msf.org/lake-chad-crisis-depth. Accessed: 6 January 2025
46. World Economic Forum (2025) Why we need to power cyber resilience in the energy sector. https://www.weforum.org/stories/2025/05/powering-cyber-resilience-energy-sector/. Accessed June 2025
47. Walstrom M, Park D (2017) Cyberattack on critical infrastructure: Russia and the Ukrainian power grid attacks. https://jsis.washington.edu/news/cyberattack-critical-infrastructure-russia-ukrainian-power-grid-attacks/. Accessed 6 Jan 2025
48. MITRE (n.d.) Group G0016. https://attack.mitre.org/groups/G0016/. Accessed 6 Jan 2025
49. Tsioutis A (2023) Volt typhoon: Critical infrastructure an ongoing cyber target. https://arpc.gov.au/resources/volt-typhoon-critical-infrastructure-an-ongoing-cyber-target/. Accessed 6 Jan 2025
50. Fruhlinger J (2023) Stuxnet explained: The first known cyberweapon. https://www.csoonline.com/article/562691/stuxnet-explained-the-first-known-cyberweapon.html. Accessed 6 Jan 2025
51. National Institute of Standards and Technology (NIST) (2012) Guide for conducting risk assessments: Special Publication 800-30 Rev. 1. Accessed: 6 Jan 2025
52. New York City Department of Parks & Recreation (2019) East Side coastal resiliency: final scope of work. https://www.nyc.gov/assets/escr/downloads/pdf/ESCR-Final-Scope-of-Work_040519.pdf. Accessed 6 Jan 2025
53. OECD (2020) Building back better: A sustainable, resilient recovery after COVID-19. OECD Policy Responses to Coronavirus (COVID-19), OECD Publishing, Paris. https://doi.org/10.1787/52b869f5-en
54. Wang C, Ju P, Wu F, Pan X, Wang Z (2022) A systematic review on power system resilience from the perspective of generation, network, and load. Renew Sustain Energy Rev 167:112567
55. Stanik J, Napiórkowski J (2023) Cyber resilience as a new strategy to reduce the impact of cyber threats. In: Borzemski L, Świątek J, Wilimowska Z (eds) Information systems architecture and technology: proceedings of 44th international conference on information systems architecture and technology—ISAT 2023, vol 790 of Lecture Notes in Networks and Systems. Springer, pp 57–66
56. National Preparedness and Security Authority (n.d.) Critical national infrastructure. https://www.npsa.gov.uk/critical-national-infrastructure-0. Accessed 6 Jan 2025
57. National Information Security Center (2024) Cybersecurity policy. https://www.nisc.go.jp/eng/pdf/cip_policy_2024_eng.pdf. Accessed 6 Jan 2025
58. Public Safety Canada (n.d.) Strengthening critical infrastructure. https://www.publicsafety.gc.ca/cnt/rsrcs/pblctns/srtg-crtcl-nfrstrctr/index-en.aspx. Accessed 6 Jan 2025
59. Cybersecurity and Infrastructure Security Agency (n.d.) Critical infrastructure security and resilience. https://www.cisa.gov/topics/critical-infrastructure-security-and-resilience. Accessed 6 Jan 2025
60. Center for Security Studies (CSS) (2023) Focal report 1: Critical infrastructure protection. https://css.ethz.ch/content/dam/ethz/special-interest/gess/cis/center-for-securities-studies/pdfs/Focal-Report-1-CIP.pdf. Accessed 6 Jan 2025
61. European Commission, European programme for critical infrastructure protection [summary]. https://eur-lex.europa.eu/EN/legal-content/summary/european-programme-for-critical-infrastructure-protection.html. Accessed 6 Jan 2025

62. Federal Office for Information Security (BSI) (2024) KRITIS-Dachgesetz: Sicherheitsgesetz für Kritische Infrastrukturen [Website]. https://www.openkritis.de/it-sicherheitsgesetz/kritis-dachgesetz-sicherheitsgesetz-3-0.htm. Accessed 6 Jan 2025
63. Department of the Prime Minister and Cabinet (DPMC) (2023) Strengthening the resilience of Aotearoa New Zealand's critical infrastructure system: Discussion document. https://consultation.dpmc.govt.nz/national-security-group/critical-infrastucture-phase-1-public-consultation/user_uploads/discussion-document--strengthening-the-resilience-of-nzs-ci-system.pdf. Accessed 6 Jan 2025
64. General Data Protection Regulation (GDPR) Info, "GDPR Info." https://gdpr-info.eu/. Accessed 6 Jan 2025
65. Romanach L, Boulaire F, Fleming A, Capon T, Bluhm S, Lin BB (2024) Australia's national climate risk assessment: Identifying climate risk interdependencies within the infrastructure and built environment system for effective climate adaptation. Clim Risk Manag 46:100670
66. Ibraheem IO, Bello-Ahmed L (2025) Green tech, safe networks: the role of cybersecurity in combating climate change. Discov Internet of Things 5(58)
67. Utrecht University (2025) Major project launched: improving climate models with historical data. https://www.uu.nl/en/news/major-project-launched-improving-climate-models-with-historical-data. Accessed 25 June 2025
68. Findlater K, Webber S, Kandlikar M, Donner S (2021) Climate services promise better decisions but mainly focus on better data. Nature Clim Change 11:731–737
69. Pacillo G, Medina L, Liebig T, Carneiro B, Schapendonk F, Craparo A, Ramirez-Villegas J, Basel A, Minoarivelo HO, Achicanoy Estrella H, Villa V, Belli A, Caroli G, Madurga-Lopez I, Scartozzi C, Dutta Gupta T, Mendez A, Kenduiywo B, Resce G, Ruscica G, Sax N, Mastrorillo M, Läderach P (2024) Measuring the climate security nexus: The integrated climate security framework. PLOS Clim 3(10):e0000280. https://www.noaa.gov/. Accessed June 2025
70. National Oceanic and Atmospheric Administration (NOAA) (n.d.) NOAA. Accessed 6 Jan 2025
71. United Nations Development Programme (n.d.) Communities and local development. https://www.undp.org/tag/communities-and-local-development. Accessed 6 Jan 2025
72. Gruber C, Pytlak A (2025) Cyber and climate threats: Shared risks, resilience, and response strategies. https://www.stimson.org/2025/cyber-and-climate-threats-shared-risks-resilience-and-response-strategies/. Accessed June 2025
73. von Luepke H, Neuhoff K, Marchewitz C (2024) Bridges over troubled waters: Climate clubs, alliances, and partnerships as safeguards for effective international cooperation? Int Environ Agreements: Polit, Law Econ 24:289–308
74. Aaheim A, Orlov A, Sillmann J (2022) Cross-sectoral challenges for adaptation modelling. In: Climate adaptation modelling. Springer, pp 11–18
75. United Nations (n.d.) United Nations framework convention on climate change. https://unfccc.int. Accessed 6 Jan 2025
76. United Nations Framework Convention on Climate Change (n.d.) Kyoto Protocol. https://unfccc.int/kyoto_protocol. Accessed 6 Jan 2025
77. Australian Government Department of Foreign Affairs and Trade (DFAT) (2023) International cooperation on climate change. https://www.dfat.gov.au/international-relations/themes/climate-change/international-cooperation-on-climate-change. Accessed 6 Jan 2025
78. Australian Government Department of Foreign Affairs and Trade (DFAT) (2023) The Quad. https://www.dfat.gov.au/international-relations/regional-architecture/quad. Accessed 6 Jan 2025
79. United Nations Convention to Combat Desertification (n.d.) Governance—COP. https://www.unccd.int/convention/governance/cop. Accessed 6 Jan 2025
80. NATO (2023) Science for Peace and Security (SPS) Programme. https://www.nato.int/cps/en/natohq/topics_85373.htm. Accessed June 2025

Chapter 2
CPSICC Nexus Information Gathering

Jason Reinhardt, Rossitza Homan, Danielle N. Sanchez, Lynn Yang, and Nick Winstead

Abstract Investigating the intersecting challenges of climate impacts across cyber, physical, and social infrastructure is crucial due to their intricate interdependencies and heightened vulnerability to extreme weather events. In July 2024, DHS, NATO, Sandia National Laboratories, Purdue University, and CSIRO collaborated to host a workshop aimed at identifying R&D gaps within the Cyber-, Physical-, Social-Systems Infrastructure and Climate Change nexus. The workshop was thoughtfully designed and executed to gather input from a diverse array of participants representing various countries, organizations, and scientific disciplines. Additionally, Sandia conducted preparatory activities for the workshop to collect insights on the nexus, identify key topics, and a review of wargaming literature to develop engaging interactive activities for the event. This chapter highlights the information gathered from the preparatory activities for the workshop and the wargaming review, as well as during the execution of the workshop. Given that climate impacts are a global challenge, understanding these nexus issues promotes international collaboration and knowledge sharing, informs policy development, drives technological innovation, underscores the importance of social equity in addressing climate impacts, and ultimately aids organizations in achieving long-term sustainability objectives.

Keywords Cyber-, Physical-, Social-systems infrastructure and climate nexus · National security · Critical infrastructure · Resilience · Climate impacts · Interdependencies · Wargaming

J. Reinhardt · R. Homan · D. N. Sanchez · L. Yang (✉)
Complex Systems Risk and Resilience, Sandia National Laboratories, Albuquerque, NM, USA
e-mail: liyang@sandia.gov

J. Reinhardt
e-mail: jcreinh@sandia.gov

N. Winstead
Zeichner Risk Analytics, Washington, DC, USA

M. Huber et al. (eds.), *Critical Infrastructure Systems and Extreme Threat Nexus: Preparing for Resilience and Security*, NATO Science for Peace and Security Series C: Environmental Security, https://doi.org/10.1007/978-3-032-13170-6_2

2.1 Introduction

The escalating climate impacts present multifaceted challenges that extend beyond environmental concerns, affecting the intricate web of cyber, physical, and social infrastructures. These systems are deeply interdependent, and their vulnerabilities to extreme weather events are becoming increasingly pronounced. As climate-related disruptions intensify, it is essential to investigate how these infrastructures intersect and influence one another, as well as to identify the gaps in research and development (R&D) that must be addressed to enhance resilience.

In July 2024, a collaborative workshop was convened by a consortium of organizations, including the Department of Homeland Security (DHS), the North Atlantic Treaty Organization (NATO), Sandia National Laboratories (Sandia), Purdue University, and the Commonwealth Scientific and Industrial Research Organisation (CSIRO). The primary objective of this workshop was to explore the Cyber-, Physical-, Social-Systems Infrastructure and Climate Change (CPSICC) nexus, with a focus on identifying critical R&D gaps. The workshop was meticulously designed to foster an inclusive environment, drawing participants from a diverse range of countries, organizations, and scientific disciplines. This diversity was instrumental in generating a rich tapestry of insights and perspectives, which are vital for understanding the complexities of the nexus.

To complement the workshop, Sandia undertook an internal preparatory activity for the workshop aimed at gathering insights on the nexus. This initiative not only helped identify key topics of interest but also provided a foundation for reviewing existing literature on wargaming—a strategic tool that can facilitate interactive discussions and scenario planning. The insights gleaned from the preparatory activity for the workshop and the wargaming literature review were pivotal in shaping engaging activities for the workshop, ensuring that participants could actively contribute to the discourse.

The findings from the workshop and the internal initiatives underscore the importance of addressing climate impacts globally that necessitates international collaboration and knowledge sharing. By fostering dialogue among stakeholders, we can inform policy development, drive technological innovation, and emphasize the significance of social equity in mitigating climate impacts. Ultimately, understanding the complexities of the CPSICC nexus will aid organizations in achieving their long-term sustainability objectives, ensuring that they are better equipped to navigate the challenges posed by a changing climate.

2.2 Background and Information Gathering Approach

Studying the intersecting challenges across CPSICC is crucial due to their complex interdependencies. Climate exacerbates extreme weather events, increasing the vulnerability of infrastructure systems. By examining these intersections, researchers can develop holistic solutions that address technological, environmental, and social dimensions. Understanding the impacts of climate on the CPSICC nexus aids organizations, such as NATO and DHS, as well as governments, such as the governments of NATO member countries, in managing risks and preparing for potential disruptions.

The intersecting challenges of climate across cyber, physical, and social infrastructure pose a significant global challenge due to their interconnected nature. Modern societies depend on these interdependent systems, where disruptions in one area can lead to failures in another, particularly as climate impacts cascade through them. As a global phenomenon, climate affects multiple countries simultaneously, necessitating coordinated international responses, as local efforts alone may be inadequate. Infrastructure vulnerabilities are evident, with physical structures like roads and bridges susceptible to extreme weather, while cyber systems can be compromised during disasters. This strain can lead to social unrest and exacerbate inequalities, particularly affecting vulnerable populations. The economic implications are profound, as disruptions can ripple through global markets, impacting supply chains and trade.

Addressing challenges across the CPSICC nexus requires comprehensive strategies that encompass technological innovations and policy frameworks at both local and global levels. Collaboration among nations, industries, and communities is crucial to develop effective responses. Additionally, as reliance on digital technologies increases, cybersecurity risks grow, with potential threats exploiting vulnerabilities in already stressed systems. Overall, a holistic approach is essential to build resilience against the multifaceted impacts of climate.

Sandia's team employed a two-pronged approach to information gathering in preparation for the workshop. First, they conducted a preparatory activity for the workshop with internal subject matter experts (SMEs), alongside a review of wargaming literature. These efforts aimed to identify R&D gaps at the intersection of the CPSICC nexus. The information was compiled and categorized, ultimately distilling the R&D gaps into a focused set of research questions and challenges. The insights gained from the wargaming literature review informed the design of the workshop's gaming elicitation approach. Second, Sandia partnered with Purdue University and CSIRO to develop three Gameplay scenarios for the workshop (Johnson et al., 2025, Chap. 3).

2.3 Pre-workshop Activities

In preparation for the workshop, Sandia collected input from internal SMEs and analyzed the results to enhance the workshop organization. This activity consisted of three main components: First, it asked participants about their technical backgrounds and prior research related to the CPSICC nexus. Second, it inquired about which areas of the CPSICC nexus are well-researched and which present open R&D gaps. Finally, participants were asked to prioritize recommended R&D across the CPSICC nexus.

Additionally, Sandia conducted a literature review of the current state of the art of wargaming relevant to the CPSICC nexus. This involved reviewing past examples of wargames and lessons learned across government and the private sector. A full account of the wargaming literature was outside the scope of this review, which focused on exercises incorporating multiple elements of the CPSICC nexus.

2.3.1 Internal SME Preparatory Activities

All Sandia SMEs who participated in the pre-workshop activities had relevant backgrounds related to the CPSICC nexus. Some individuals specialized in systems analysis and engineering, focusing on climate-physical and social infrastructure, with expertise in energy security and critical infrastructures. Others concentrated on climate impact modeling, specifically erosion's effects on transportation and future predictions across various spatial scales. Additionally, some respondents had expertise in policy and social science, particularly political science. They worked on bridging the gap between policymakers and technology, addressing decision-making challenges, and developing risk analysis frameworks. Others explored the intersection of technology and policy, including critical infrastructure analysis, interdependencies, cascading consequences, and the impact of climate on national security. Finally, some SMEs specialized in data science and analytics, covering areas such as AI/ML, physical infrastructure modeling, risk and resilience of complex systems, system security, disruption consequence analysis, and scenario development.

Identifying and addressing R&D gaps across the intersections of the CPSICC nexus is essential for fostering resilience, informing decision-making, promoting innovation, and ensuring effective resource allocation in an increasingly interconnected world. Section 2.4.2 contains a detailed analysis of the intersectional gaps for the CPSICC nexus.

2.3.2 Wargaming Literature Review

To prepare for the workshop, Sandia conducted a literature review of the current state of the art of wargaming relevant to the CPSICC nexus. This involved reviewing past examples of wargames and lessons learned across government and the private sector. For the purposes of this chapter, "wargames" is used loosely to define experiential, game-oriented exercises that may be used for a variety of purposes beyond military planning to include policy, organizational, and other goals. A full account of the wargaming literature was outside the scope of this review, which focused instead on exercises incorporating multiple elements of the CPSICC nexus. The review had two objectives: (1) validate the need and gaps in the literature; and (2) identify lessons learned to inform game design for the CPSICC workshop. This section summarizes the findings of the literature review. An annotated bibliography is included in the appendix section of this report.

The literature review largely validated the need for more deliberate work around the CPSICC nexus. While there are many wargames that focus on cyber-, physical-, and social infrastructures, and climate, there are few if any that incorporate all these elements in a deliberate and structured way. Of the wargames surveyed, the most relevant fall into two categories: exercises focused on cyber-physical threats to critical infrastructure, and exercises centered around climate.

There are several examples of wargames centered on the nexus of cyber and physical infrastructure risk. These wargames are often to facilitate operational planning and coordination within and between government agencies and the private sector. For example, the Naval War College's Defend Forward: Critical Infrastructure Wargame involved cyber attacks against electrical grids, bringing together private sector and government participants to explore response options and strategies [1]. CISA's Cyber-Physical Convergence Scenarios focus mostly on cyber-physical threats to critical infrastructure, though they are customizable to include other elements of the CPSICC nexus such natural disasters (i.e., climate) and civil disturbances (i.e., social) [2]. Similarly, the Cyber Guard Exercise hosted by CYBERCOM, the DHS, and the Federal Bureau of Investigations focused on threats to critical infrastructure from cyber attacks, with the goal of identifying new response methodologies and whole-of-government approaches [3].

The second relevant category is climate-focused wargames. The review identified several examples of climate wargames. Many of these have are tabletop exercises designed to consider the long-term impacts of climate on national security. These games are generally designed to facilitate issue spotting, education, and brainstorming. For example, the Center for a New American Security's (CNAS's) Clout and Climate Change exercise gave participants tailored simulations and climate projections for 2050, and broke participants into different country teams that explored national security consequences and potential strategies for addressing them [4]. Other climate exercises focus mainly on operational planning and disaster response in relation to climate. This may include military planning, such as the Department of Defense's (DoD's) Elliptic Thunder exercise [5]. It may also include community response planning, such as the Federal Emergency Management Agency's (FEMA's) National Level Exercise where extreme weather impacts Hawaii's critical infrastructure, businesses, and residents [6]. FEMA's exercise is notable for including climate, cyber, and physical risks—though relatively little attention is given to social infrastructure risks. Relatively few climate-related exercises incorporated cyber infrastructure as a central scenario point.

One notable example of a climate exercise that combined both goals of operational planning and longer-term horizon scanning was the Research ANd Development's (RAND's) Corporation Ho'okele MUA II: Navigating the Future (RAND Corporation, Ho'okele MUA II Navigating the Future, 2023). The first portion of the exercise focuses on mitigating operational risk in the Indo-Pacific region, followed by a second portion where players invest in different adaptation and resilience measures over the next few decades.

Taken together, the examples cited here demonstrate the need for additional research into the design of games that can accommodate the complexities of the intersection of cyber, physical, and social infrastructures and climate.

2.3.3 Gameplay Scenario Development

The workshop was designed around interactive scenario-based Gameplay, near-term (2023) and long-term (2050). Three future scenarios were developed including,

1. **Arctic Shipping Lanes**—focused on addressing geopolitical tensions, developing technologies and strategies to counter vulnerabilities, and exploring international cooperation to mitigate cybersecurity threats.
2. **Power Grid Failures**—focused on creating contingency plans for smart grid infrastructure, assessing impacts from climate events, and prioritizing national responses during extreme weather combined with cyber incidents.
3. **Solar Radiation Modification**—focused on discussing international governance for regulating R&D and deployment of solar radiation modification technologies while protecting these technologies from adversarial interference; and managing transboundary impacts.

Sandia led the development of the Power Grid Failures scenario including the background, role assignments, and gameplay objectives. During the workshop, participants were subdivided into three Gameplay groups, comprised of a mix of SMEs and non-SMEs on the scenario topic. Each participant was assigned a role to play during their group's brainstorm activities. The role-based approach was designed to foster comprehensive discussions and innovative solutions for each scenario (Johnson et al., 2025, Chap. 3). The Scenario Gameplay executed during the workshop informed the R&D gaps and recommendation that emerged from the workshop discussions (Reinhardt et al., 2025, Chap. 4).

2.4 Key Takeaways from Pre-workshop Activities

This section summarizes the key takeaways from the three pre-workshop activities: the internal preparatory activities, the wargaming literature review, and the collaborative scenario development activities.

2.4.1 Internal Preparatory Activities Takeaways

The key takeaways from the internal preparatory activities are summarized below:

1. **Research Focus Areas**: Respondents indicated that physical infrastructure is regarded as the most thoroughly researched area, closely followed by cyber infrastructure. This perception is primarily due to the tangible nature of physical infrastructure, which has led to a longer history of research. Additionally, there is a substantial body of work focused on understanding cyber vulnerabilities and their mitigations.
2. **Research Gaps**: Social infrastructure was identified as the area with the most urgent need for further research, followed by climate, cyber infrastructure, and physical infrastructure. The prioritization of social infrastructure as a critical R&D gap within the CPSICC nexus arises from its higher degree of uncertainty and challenges in definition. Climate is ranked second because of its significant impacts on social, physical, and cyber infrastructures.
3. **Interdependencies**: Most respondents emphasized the importance of examining all combinations of interdependencies to identify emerging vulnerabilities. Among those who suggested specific interdependencies for investigation, the impacts of climate on social and physical infrastructures were particularly highlighted, along with the effects of social infrastructure on both physical and cyber infrastructures. Respondents noted that while climate affects many systems, there remains a lack of clarity regarding the extent and timing of these impacts. Additionally, since cyber and physical infrastructures are operated by humans, understanding social infrastructures is crucial for identifying potential vulnerabilities and developing effective mitigation strategies.

Additionally, respondents emphasized several key considerations for the CPSICC nexus:

1. **Leverage Interdisciplinary Techniques**: Respondents suggested that future research should capitalize on interdisciplinary strengths, incorporating both qualitative and quantitative research methods, as well as focusing on "systems-level solutions."
2. **Enhance Communication**: It was noted that effective communication with policymakers and mission-focused SMEs should be prioritized. This could involve creating "systems-informed white papers" for policymakers and establishing forums where mission-focused SMEs can pose critical questions to address existing gaps.
3. **Address Funding and Adoption Challenges**: Respondents highlighted ongoing difficulties in securing funding and encouraging partners to adopt existing solutions. These challenges should be considered as additional opportunities for improvement are pursued.

2.4.2 Intersectional Gaps in the CPSICC Nexus

Considering R&D gaps at the intersections of the CPSICC nexus is crucial for several reasons. It fosters a holistic understanding of the complex interactions among cyber, physical, social, and climate systems, which enhances resilience against disruptions like natural disasters and cyber-attacks. This understanding aids policymakers in making informed decisions and developing effective strategies. Additionally, identifying these gaps can uncover innovation opportunities, allowing researchers to create new technologies and methodologies that address pressing challenges. It also facilitates more effective resource allocation by prioritizing research in high-impact areas and encourages interdisciplinary collaboration among experts from various fields. Finally, recognizing R&D gaps helps stakeholders adapt to the evolving nature of the CPSICC nexus, ensuring that responses remain relevant and effective in the face of change.

These preparatory activities indicated R&D gaps at the intersections of the CPSICC nexus in the following areas:

- Cyber-Physical Security and Climate Resilience
- Cyber Systems and Energy Integration
- Data Privacy and Ethical Considerations in Climate Monitoring
- Economic and Financial Implications of Climate-Resilient Infrastructure
- Human Behavior, Decision-Making, and Technology Adoption.

The relationship between climate and social infrastructure is frequently highlighted, yet it remains poorly understood. While there is an acknowledgment that climate will impact social systems, the degree and timeline of these impacts are still unclear. This uncertainty is largely attributed to the complexities inherent in social infrastructure rather than a lack of understanding of climate itself.

Similarly, the relationship between climate and physical infrastructure presents challenges. A primary gap lies in the need to update outdated models to reflect changing atmospheric conditions, temperature, and moisture parameters. Many infrastructure requirements are based on old data and models, which must be revised to account for current and future conditions.

In the cyber-social domain, the dynamics of misinformation and disinformation in cyberspace, along with their social consequences, represent a significant security issue. There is currently no clear framework for assessing, prioritizing, and mitigating these risks.

While climate stresses on physical infrastructure have garnered considerable attention and research, the intersection of these areas—particularly the propagation of risk to social infrastructure—requires more focus. The ongoing impacts of these risks highlight the need for a dedicated R&D agenda to assess, prioritize, and mitigate intersectional risks.

To effectively address emerging vulnerabilities, it is essential to explore all four domains—cyber, physical, social, and climate—together. This approach is akin to Earth System Modeling (ESM), where separate components require their own models with distinct exogenous inputs (such as economics) that must be coupled effectively. The challenge lies in how to couple architectures, time scales, and spatial scales. In ESMs, partial differential equations (PDEs) of physical processes facilitate the development of coupling schemes, but this is more complex in the CPSICC context. It is crucial to determine the outputs of one component that need to be passed to another and how to model these interactions in the absence of PDEs.

The intersections of social infrastructure with other elements of the CPSICC nexus warrant further R&D, particularly since neither physical nor cyber infrastructures can function without human interaction, which inherently connects them to social infrastructure. Additionally, the relationships between climate and each infrastructure type should be more clearly articulated, as the causal pathways are often indirect and complex.

Several key intersections within the CPSICC nexus require more R&D to enhance understanding of complex interdependencies and to develop integrated solutions that improve resilience, mitigate risks, and adapt to climate. These intersections include:

- **Cyber-Physical Security and Climate Resilience**: Developing climate-resilient cybersecurity strategies is crucial for addressing vulnerabilities introduced by climate, such as the increased frequency of natural disasters that can disrupt cyber-physical systems. It is important to create adaptive security measures capable of responding to dynamic threats exacerbated by climate disruptions. Research needs include conducting integrated risk assessments that consider both cyber threats and climate risks to critical infrastructure, as well as developing resilient design principles for cyber-physical systems. As climate introduces new vulnerabilities, establishing adaptive cybersecurity measures is essential for protecting critical infrastructure during climate-induced disruptions.
- **Social Systems and Climate Adaptation**: Investigating community-based adaptation strategies that leverage local knowledge and resources is essential for enhancing resilience to climate. It is also important to examine the implications of climate and Cyber-Physical-Social Systems (CPSS) interventions to ensure that vulnerable populations are not disproportionately affected. Research needs include developing inclusive adaptation strategies and understanding how human behavior influences the effectiveness of these measures. Given that climate disproportionately impacts vulnerable populations, it is crucial to create equitable adaptation strategies that strengthen community resilience.
- **Cyber Systems and Energy Integration**: Developing smart grid technologies is essential for monitoring and managing electricity flows, integrating energy sources, and enhancing grid resilience. Implementing energy management systems that optimize the use of energy and improve overall efficiency is also crucial. There is a pressing need for research focused on establishing interoperability standards for smart grids and ensuring their cybersecurity. The ongoing transition to energy sources is driving significant research in smart grid technologies, making this a vital and active area of development.

- **Data Privacy and Ethical Considerations in Climate Monitoring**: Exploring the ethical implications of using data from cyber-physical systems for climate mitigation is crucial, particularly concerning privacy and consent issues. Developing regulatory frameworks that balance data-driven decision-making with individual privacy protections is essential. There is a need for research focused on creating models for ethical data governance and fostering public trust through transparent practices. Addressing these ethical concerns is vital for gaining public confidence in data-driven solutions to climate challenges.
- **Economic and Financial Implications of Climate-Resilient Infrastructure**: Conducting climate risk assessments is essential for evaluating the economic risks that climate poses to Cyber-Physical-Social Systems, particularly regarding impacts on insurance and investments. To support resilience initiatives, it is important to explore innovative funding mechanisms and financial instruments. Research needs include performing cost–benefit analyses of investments in climate-resilient infrastructure and developing models for public–private partnerships. Understanding these economic risks is vital for maintaining financial stability and promoting sustainable development in the face of climate challenges.
- **Human Behavior, Decision-Making, and Technology Adoption**: Investigating behavioral insights is crucial for understanding how human behavior affects the adoption and effectiveness of CPSICC interventions. Developing strategies for public engagement in climate action is essential, as it fosters community involvement and support. Research needs include applying behavioral economics to comprehend decision-making processes related to climate and creating effective communication strategies to convey important information. Recognizing that human behavior plays a significant role in the success of climate interventions underscores the necessity of actively engaging the public in these efforts to enhance their effectiveness.

The intersections within the CPSICC nexus that require further R&D are critical for addressing the complex challenges posed by climate challenges. By focusing on these intersections, researchers, policymakers, and organizational leaders can develop integrated solutions that enhance resilience, security, and sustainability. Addressing these under-researched areas will necessitate interdisciplinary collaboration, innovative approaches, and a commitment to understanding and mitigating the multifaceted impacts of climate on cyber, physical, and social systems.

2.4.3 Wargaming Takeaways

The wargaming review highlighted important lessons learned from previous wargaming that can be applied to game design for the CPSICC nexus. Four key wargaming challenges were identified related to (1) scoping and goals, (2) complexity, (3) game structure, and (4) interdisciplinary participation.

The review underscored the need for clearly defining the scope and goals of the wargame. This may be particularly challenging at the CPSICC nexus, where there is inherently a large scope with many different contingencies. As described by one expert commentary: "Climate is the ultimate systems-level challenge, so it is easy to overreach. Games that try to be about everything can end up being about nothing. Too many sponsors or stakeholders with different agendas can make for incoherent outcomes" [7]. In general, there are several different goals that wargames could advance: concept development, capability analysis and development, strategic discussion and senior leader engagement, coordination, international exchanges, and education and training [8, 9]. Selecting from these goals and then designing against this scope will help ensure game success.

Second and relatedly, wargaming best practices require adequate attention to the balance between complexity/realism and tractability. As respected wargaming expert Jacquelyn Schneider summarizes, "the balance between abstraction and reality is one of the big challenges of gaming… A strategic level game… doesn't need a lot of detailed technical knowledge about how cyber exploits occur but it does need realistic strategic considerations," but in contrast a more tactical, operationally focused game "needs less realism about strategic policy and greater attention on technical realism" [10]. This may be especially difficult when combining elements of the CPSICC nexus, where a game scenario may involve many different technical issues that can distract from the overall objectives of the exercise.

Third, game structure should facilitate participation toward the desired outcomes of the exercise. Key considerations include the scenario design (time scale of the scenario, precipitating events, etc.), the use of injects or red teams to move the scenario forward, and the roles of participants. Here again, designing games for the CPSICC nexus may pose unique challenges. For one, the CPSICC nexus incorporates risks that happen at very different timescales and levels of abstraction (e.g., cyber scenarios are often more tactical and short-term, whereas climate risks are long-term systemic risks). Indeed, some experts have pointed to difficulties incorporating climate into game design and begun experimenting with solutions. The RAND Corporation highlights several aspects of game play that could be used to inject climate aspects into scenarios, including future state narrative, road-to-war/starting conditions, crisis injects, and game moves [9].

Finally, game designers must ensure that the participant groups are both appropriate for the game's goals and diverse enough to capture a breadth of perspectives and expertise. A lack of diversity may result in biased results or missed opportunities, while a lack of expert participants or critical stakeholders may produce misinformed results. Lessons learned from previous games have underscored the difficulty in assembling interdisciplinary participants [11]. This is very relevant for the CPSICC nexus, which is inherently highly multidisciplinary and involves practitioners that are not always familiar with wargaming exercises.

In summary, the review of best practices and lessons learned highlights some of the challenges with designing wargames at the CPSICC nexus. This provides a solid basis for game design for the CPSICC workshop.

2.4.4 Scenario Development Takeaways

Developing gaming scenarios for the CPSICC workshop requires careful planning and consideration to ensure engagement, relevance, and effectiveness. The following best practices were identified by the scenario development team:

1. **Define Clear Objectives**: Establish specific goals for the gaming scenario, such as enhancing understanding of the CPSICC nexus, fostering collaboration, and identifying R&D gaps. Clear objectives will guide the design and facilitate meaningful outcomes.
2. **Involve Stakeholders Early**: Engage a diverse group of stakeholders in the scenario development process. This includes representatives from various sectors (government, industry, academia, and civil society) to ensure that multiple perspectives are considered.
3. **Create Realistic Scenarios**: Develop scenarios that reflect plausible future situations based on current trends and challenges within the CPSICC nexus. Use data and expert insights to ground scenarios, making them relatable and engaging for participants.
4. **Encourage Interdisciplinary Collaboration**: Design scenarios that require participants to collaborate across disciplines. This fosters a holistic understanding of the CPSICC nexus and encourages innovative problem-solving.
5. **Incorporate Role-Playing**: Assign participants specific roles that reflect different stakeholders within the CPSICC nexus. Role-playing can encourage diverse viewpoints and stimulate dynamic discussions.
6. **Facilitate Open Communication**: Create an environment that encourages open dialogue and respectful debate. Establish ground rules for communication to ensure that all voices are heard and valued.
7. **Provide Context and Background**: Offer participants background information on the CPSICC nexus and the specific scenario being explored. This ensures that all participants have a foundational understanding, regardless of their prior knowledge.
8. **Incorporate Feedback Mechanisms**: Build in opportunities for participants to provide feedback throughout the gaming process. This can help refine the scenario in real-time and ensure that it remains relevant and engaging.
9. **Debrief and Reflect**: After the gaming scenario, conduct a debriefing session to discuss outcomes, insights, and lessons learned. Encourage participants to reflect on their experiences and how they can apply the knowledge gained to real-world challenges.
10. **Document Outcomes**: Capture key findings, discussions, and recommendations from the gaming scenario. This documentation can serve as a valuable resource for future workshops and inform ongoing R&D efforts.

By following these best practices during the CPSICC workshop execution, the gaming scenario facilitators created engaging and effective gameplay that promote collaboration and innovation among the diverse set of SME and non-SME participants.

2.5 Designing the CPSICC Nexus Workshop

The key takeaways from the pre-workshop activities played a key role in designing the CPSICC workshop. This section summarizes how the outcomes of the SME solicitation, wargaming review, and scenario development activities informed the CPSICC workshop development process.

2.5.1 Workshop Development

The internal preparatory activities highlighted leveraging interdisciplinary strengths, incorporating both qualitative and quantitative research, and focusing on "systems-level solutions." Additionally, they highlighted the importance of effective communication with policymakers and mission-focused SMEs, suggesting the creation of "systems-informed white papers" and forums for critical discussions. Finally, respondents pointed out existing challenges related to securing funding and encouraging partners to adopt existing solutions, indicating that these challenges should be addressed as new opportunities are pursued. These recommendations informed the wargaming scenarios developed for the workshop.

The wargaming literature review identified four important lessons from previous wargaming experiences that informed the design of games for the CPSICC nexus workshop. Four key wargaming challenges were identified:

1. **Scoping and Goals**: It is crucial to clearly define the scope and goals of the wargame, which can be particularly challenging in the CPSICC nexus due to its broad scope and numerous contingencies. Experts noted that attempting to cover too many aspects can lead to incoherent outcomes, emphasizing the importance of selecting specific goals such as concept development, capability analysis, strategic discussions, coordination, international exchanges, or education and training.
2. **Complexity**: Wargaming best practices require a careful balance between complexity and tractability. The challenge lies in finding the right mix of abstraction and realism. Strategic-level games may not need detailed technical knowledge but must consider realistic strategic factors, while tactical games require more technical realism. This balance can be especially difficult when integrating various elements of the CPSICC nexus, as different technical issues may distract from the overall objectives.
3. **Game Structure**: The structure of the game should facilitate participation toward achieving the desired outcomes. Key considerations include scenario design, the use of injects or red teams to advance the scenario, and the roles of participants. The CPSICC nexus presents unique challenges due to the varying timescales and levels of abstraction involved, with cyber scenarios often being tactical and short-term, while climate risks are long-term and systemic. Experts have begun

experimenting with ways to incorporate climate aspects into game design, such as using future state narratives and crisis injects.

4. **Interdisciplinary Participation**: It is essential to ensure that participant groups are appropriate for the game's goals and diverse enough to capture a wide range of perspectives and expertise. A lack of diversity can lead to biased results or missed opportunities, while insufficient expert participation may result in misinformed outcomes. Previous experiences have highlighted the difficulties in assembling interdisciplinary participants, which is particularly relevant for the CPSICC nexus, given its multidisciplinary nature and the involvement of practitioners who may not be familiar with wargaming exercises.

These lessons informed the design of the workshop gaming activities.

2.5.2 Workshop Approach

The workshop was a collaborative effort involving Sandia, Purdue University, CSIRO, the DHS Science and Technology Directorate, and NATO's Science for Peace and Security (SPS) program, in partnership with the workshop's host, the Institute for a Sustainable Future at Purdue University. For more information about the workshop, please visit the workshop website [12].

The workshop aimed to identify R&D gaps that, when addressed, could enhance the resilience of interdependent critical infrastructure and institutions against the growing threats within the CPSICC nexus. To achieve this objective, the workshop convened a diverse, interdisciplinary, and international group of subject matter experts from various institutions, including academia (e.g., Lund University, Auckland University), national laboratories (e.g., Sandia, Oak Ridge National Laboratory), industry (e.g., Fannie Mae, Viridien), and security organizations (e.g., NATO, DHS). Participants came from a balanced range of backgrounds, including the finance sector, water and energy utilities, multidisciplinary organizations, and think tanks.

2.5.3 Workshop Discussions

The speakers and participants engaged in various structured activities to identify R&D gaps within the CPSICC nexus. These activities included a mix of presentations, including TED-style talks, scenario-based games, and an R&D proposal contest.

Discussions at the workshop focused on climate challenges within and across CPSICC systems, domains, and stakeholders. Topics ranged from artificial intelligence and data governance to specialized interdependent models, risk metrics, stakeholder engagement, and public–private partnerships. The R&D gaps identified during the workshop reflect participants' insights into critical knowledge, process, and technological deficiencies that impede our ability to assess, prevent, or respond to

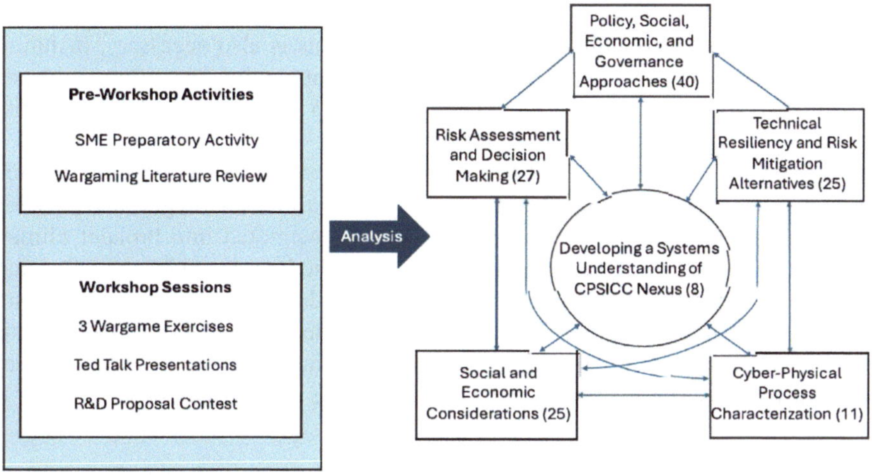

Fig. 2.1 Workshop sessions and resulting R&D gaps

climate-driven risks affecting interdependent cyber, physical, social, and infrastructure systems. Six general categories of gaps emerged from the workshop discussions (Reinhardt et al., 2025, Chap. 4). Figure 2.1 lists the pre-workshop activities, the workshop sessions, and the resulting R&D gaps. These R&D categories are adopted as the overarching categorization for the CPSICC book.

2.5.3.1 Developing a Systems Understanding of CPSICC Nexus Discussions

Workshop participants agreed that climate acts as a threat multiplier and that complex interactions exist among various elements of the CPSICC nexus. They also agreed that methodologies, common frameworks, and information are lacking to foster understanding across these elements. Key questions included:

- How can the complex interdependencies among disparate systems within the CPSICC nexus be more effectively characterized?
- In what ways do these interdependencies introduce additional risks or affect resilience?
- How are cyber-physical-social systems interconnected across borders, sectors, and nations?

The discussions included several key themes. First, research is essential to investigate the trade-offs and risks associated with various architectures and topologies in rapidly evolving energy infrastructure. Second, enhancing the modeling of complex systems is crucial for understanding the dynamics and emergent behaviors that arise over different time horizons, particularly regarding interactions among AI systems, human behaviors, and long-term, non-linear climate challenges. Additionally, a comprehensive understanding of compounding events is also necessary, including multivariate, spatially compounded, temporally compounding/cascading, and pre-conditioned compounds. This knowledge will help in assessing the cumulative impacts of these events on infrastructure and society.

Furthermore, research should evaluate the effectiveness of SRM as a climate intervention, especially when considered alongside other climate strategies and policies. This will provide insights into how SRM can be integrated into broader climate action frameworks. Lastly, it is vital to understand the interdependencies of critical infrastructure systems that extend across national borders, as current mapping efforts tend to focus primarily on national contexts. This understanding will facilitate better planning and resilience strategies in an increasingly interconnected world, ensuring that infrastructure can withstand and adapt to climate-related challenges.

2.5.3.2 Policy, Social, Economic and Government Approaches Discussions

The workshop discussions and activities highlighted substantial challenges in developing and enforcing international agreements, legal frameworks, and governance models essential for effectively managing and regulating activities within the CPSICC nexus. The discussions underscored the need for policy and governance mechanisms that address long-term climate and infrastructure risks, including:

- How to ensure coordination amid the complexity of decisions across various scales, sectors, and partners within the CPSICC nexus.
- The roles of different institutions in governing and managing a distributed array of risks.
- Identifying specific technologies or mechanisms that could alleviate coordination challenges across sectors.

The discussions covered many topics. First, engaging citizens in climate resilience initiatives is crucial for fostering community involvement and promoting behavior change. Innovative strategies should be developed to encourage citizen participation in data collection and other resilience-building activities. Second, as the use of AI in climate data increases, establishing improved practices and enforcement mechanisms for climate data security is essential to protect sensitive information and models. Additionally, researching community-based adaptation strategies that leverage local knowledge, and resources can significantly enhance resilience against climate events. Furthermore, enhancing best practices and enforcement for Cyber-Physical-Social Systems data security is vital, particularly considering proprietary and security concerns. Moreover, investigating deterrence strategies for climate interventions, including preemptive actions and security measures, will help mitigate potential threats. Creating educational resources and promoting reliable data sharing, especially regarding supply chain cyber challenges, will empower communities and stakeholders.

Furthermore, collaboration between the public and private sectors is necessary to effectively understand and respond to threats in the shipping sector. Formulating international agreements to address potential environmental degradation in the Arctic is also critical, identifying key countries involved in maintaining shipping lanes. Finally, developing standards and protocols for microgrids, including procedures for automatically transitioning to "island mode" during cyberattacks, will enhance infrastructure resilience. Additionally, creating research agreements among Arctic states will facilitate the sharing of advancements in shipping technologies. Finally, understanding government strategies for responding to threats against climate interventions, including investigation, negotiation, and disaster response approaches, is essential for effective risk management and resilience planning.

2.5.3.3 Technical Resiliency and Risk Mitigation Alternatives Discussions

Various technologies are being developed to mitigate risks associated with data, monitoring, and analysis, particularly in the context of climate. However, while some of these technologies target climate issues, their potential to exacerbate other vulnerabilities has not been thoroughly examined. Additionally, several mitigation gaps remain unaddressed. The discussions focused on how emerging technologies may either reduce risks or introduce new threat vectors within the CPSICC nexus. Key questions included:

- What unintended consequences could arise from emerging climate and non-climate solutions?
- Can enhanced monitoring requirements be established to identify unauthorized or undesirable mitigation actions?
- What opportunities are available for community-based sensors to improve early-warning forecasts?
- How might new technological advancements positively or negatively disrupt risks associated with increased digitization?

A variety of topics were discussed. First, research is needed to explore the dual role of AI in combating misinformation while also recognizing its potential to amplify misinformation threats, particularly through audience segmentation and deepfakes. Second, it is crucial to assess the requirements and timelines for deploying alternative SRM strategies if primary interventions are disrupted. Additionally, developing redundant systems is essential to reduce reliance on technological advancements, including backup communication and GPS technologies. Furthermore, advanced energy management systems should be designed to ensure reliability during significant events, enhancing overall resilience.

Moreover, investigating microgrid technologies can help decrease the dependence of infrastructure and architectural grids on communication systems, promoting greater autonomy. In the shipping industry, advancing technologies to integrate energy can mitigate environmental impacts and support sustainability efforts.

Furthermore, applying systems engineering concepts will facilitate the effective implementation of promising R&D initiatives. Lastly, establishing capabilities for timely threat impact assessments is vital for evaluating climate impacts, especially during crises. These assessments will guide response decisions related to threats, mitigation efforts, and disaster responses, ultimately enhancing preparedness and resilience in the face of climate challenges.

2.5.3.4 Cyber-Physical Process Characterization Discussions

There was a clear acknowledgment of the increasingly cyber-physical nature of most infrastructure systems, particularly as more systems transition to digital formats. Various gaps were identified regarding the intersection of these cyber-physical processes with climate challenges and other aspects of the CPSICC nexus. The discussions focused on questions concerning the characterization of associated threats and vulnerabilities:

- Does climate exacerbate existing threat vectors within cyber-physical systems?
- What types of cyber-physical data should be collected to support impact modeling and analyses of both acute (e.g., extreme weather) and chronic (e.g., aridification) events?
- How can data sharing and analysis processes enhance resilience, particularly considering challenges posed by misinformation, disinformation, and malinformation?

The workshop discussions encompassed several critical areas. First, there is a pressing need for comprehensive environmental research and data, particularly to enhance understanding of ecological impacts in the Arctic and improve forecasting through early environmental assessments. Second, the sustainability of cyber infrastructure requires better data and comparative studies, including guidance on integrating climate hazards into digital risk management. Additionally, understanding the effects of rapid warming on cyber infrastructure is vital, as such events can have cascading impacts on data centers and other systems reliant on cool temperatures. Research should focus on developing effective cooling methods and backup options, such as distributed infrastructure.

Moreover, more research is needed on SRM to understand its impacts and interactions with Earth systems, necessitating small-scale experiments to advance knowledge. Investment in high-resolution climate modeling, remote sensing, and impact analysis is also essential. Lastly, supply chain risk management must address risks associated with emerging energy technologies, as the current focus on material provenance is insufficient, particularly given the long-term nature of CPSICC risks influenced by geopolitical changes.

2.5.3.5 Social and Economic Considerations Discussions

Workshop participants discussed the challenges of effectively accounting for relevant geopolitical and technological impacts to socioeconomic systems, considering the following questions:

- How might climate challenges intensify geopolitical tensions, such as those related to migration, supply chain constraints, and the formation of new alliances?
- In what ways could AI techniques either mitigate or exacerbate risks to socioeconomic systems, including the potential for adversarial interference through climate disinformation campaigns?
- What types of investments could help reduce the impact of critical infrastructure disruptions, particularly for vulnerable populations during crises?

The workshop discussions encompassed several critical topics. First, applying principles of behavioral economics can enhance understanding and influence decision-making related to climate challenges. Second, it is essential to evaluate the costs associated with SRM interventions, considering their scale and duration to assess feasibility as a large-scale climate solution. Additionally, effective crisis communication strategies are also needed to determine what information should be disclosed to the public while combating misinformation and disinformation. Understanding the varying definitions and datasets for critical infrastructure across nations is crucial, particularly among NATO members, who may interpret these concepts differently.

Moreover, research should also focus on global inequities, examining how climate crises impact social systems and interstate relations, leading to disparities in population health and economic stability. Furthermore, assessing NATO's political feasibility for governing SRM is necessary, along with fostering trust among member nations. Enhancing forecasting and understanding of population movement decisions in response to climate and infrastructure changes is vital. Lastly, improving insights into public trust is essential for the successful deployment of climate interventions and effective crisis management.

2.5.3.6 Risk Assessment and Decision-Making Discussions

The management of complex risk profiles was a central theme at the workshop, with stakeholders emphasizing a significant gap in comprehensive frameworks, methodologies, and governance structures for addressing risks within the CPSICC nexus. The discussions focused on the convergence of multiple systems characterized by high interdependencies, emergent behaviors, and diverse time horizons. Key questions raised included:

- What techniques currently exist or need to be developed to effectively characterize cascading risks across various scales and sectors?

- What risks arise from making decisions based on outdated, invalid, or incomplete data and models?
- How might adversarial responses to climate challenges, from both supportive and opposing communities, alter system trajectories?
- In what ways can improved attribution of climate impacts enhance requirements, performance metrics, and data collection needs for cyber-physical systems?

A variety of risk assessment and decision-making topics were discussed. First, there is a need for enhanced analytical gaming techniques to better understand the risks associated with the CPSICC nexus. This includes developing new risk metrics that can accurately measure and model climate-related risks, as traditional cost–benefit analyses often fall short in this context. Second, creating a catalog of potential bad actors and their responses in supply chains, particularly concerning Arctic shipping lanes, is essential for improving resilience. Research should also explore the implications of SRM as a potential vector for chemical, biological, radiological, and nuclear (CBRN) threats. Additionally, understanding the dynamics of climate misinformation is critical, as it can exacerbate existing threats. This necessitates the development of effective strategies to mitigate misinformation's impact.

Furthermore, it is vital to characterize the direct and indirect effects of climate on critical national and international security missions. Expanding existing deterrence and game theory frameworks to include scenarios involving climate modification and related conflicts is also necessary. Understanding the economic impacts of threats to climate interventions will help inform decision-making. R&D efforts should prioritize the needs of policymakers, ensuring that they have the tools necessary for effective planning. Lastly, a method for conducting scenario analysis that considers multiple interacting factors across domains is needed, along with tailored risk management strategies that address local energy usage, risk tolerance, and demographic factors. These efforts will enhance resilience and preparedness in the face of complex, evolving climate challenges.

2.6 Conclusion

The pre-workshop activities and the CPSICC workshop sessions resulted in eight initial priority recommendations for consideration by NATO and member countries:

1. **Bridge Engineering Lifecycle Gaps**: The CPSICC nexus poses R&D challenges across the systems engineering lifecycle, requiring interdisciplinary systems thinking to effectively align people, processes, and technologies. To bridge the gap between promising research and operational deployment, it is recommended to connect research entities with operational divisions within NATO and the government organizations of individual countries.
2. **Integrate Existing Frameworks and Models**: Although numerous frameworks and models exist for cyber, physical, social, and climate systems, they often operate in silos. Integrating these capabilities is both crucial and challenging.

It is recommended to combine various existing platforms and models with data from government, industry, and academia to capture key CPSICC dependencies.

3. **Understand CPSICC Behaviors**: The CPSICC nexus encompasses multiple complex systems exhibiting emergent behaviors across various time horizons. It is recommended to investigate AI applications to model these dynamics and enhance the current understanding of CPSICC system interactions and behaviors.
4. **Map Climate-Infrastructure Impacts**: There is a lack of established frameworks to systematically identify how climate impacts critical infrastructures. It is recommended to leverage U.S. national critical function (NCF) and asset data from DHS CISA to map climate impact mechanisms and prioritize the most vulnerable NCFs. Furthermore, if other NATO member countries assess national-level infrastructure resilience using NCFs, the U.S. can share valuable lessons learned and best practices.
5. **Map Technology-Infrastructure Dependencies**: Understanding the impact of emerging technologies and environmental changes on dependencies in critical infrastructure is essential. It is recommended to conduct research into scalable and shareable dependency mapping and analytic methods among stakeholders.
6. **Map Cross-country Infrastructure Dependencies**: Both the CPSICC nexus and the NATO alliance extend beyond national borders, yet there is a limited understanding of cross-country dependencies in critical infrastructures. It is recommended to collaborate with NATO allies to evaluate existing frameworks and data, standardizing and improving the understanding of cross-border dependencies, thereby enhancing risk mitigation across NATO.
7. **Improve Climate Data Security**: There is insufficient focus on the cybersecurity of climate data and models, which presents potential threats. Collaborating with climate science stakeholders to improve data security and implement secure-by-design initiatives could lead to beneficial outcomes.
8. **Map Mis/Dis/Mal Information Impacts**: Public trust is essential for the resilience of critical infrastructure, and misinformation can intensify threats. It is recommended to address the impacts of misinformation and coordinate R&D efforts to develop effective strategies for tackling this challenge.

It is recommended for NATO and its member nations consider investing in these recommendations to enhance the resilience of the CPSICC nexus. This can be achieved by promoting collaboration, improving the understanding of complex systems, and addressing vulnerabilities across interdependent critical infrastructures. If implemented, these strategies could also positively impact other areas of national and global security, including water security, food security, and migration.

Acknowledgements The authors would like to express their gratitude to our colleagues at the DHS Science and Technology (S&T) Directorate, NATO's Science for Peace and Security (SPS) program, Purdue University, CSIRO, Zeichner Risk Analytics (ZRA), and Sandia for their excellent collaboration across multiple time zones in organizing an effective and informative workshop. We extend our thanks to our program managers and workshop sponsors, Eyup Turmus (NATO SPS), David Alexander (DHS S&T), and Bradley Dean (DHS S&T), for their leadership and guidance. We are particularly grateful to David Alexander for funding Sandia's contributions to the workshop and the follow-on report. DHS's strong participation in the workshop yielded valuable insights, and we appreciate the contributions of Alfred Bender, Julie Brewer, Merideth Secor, Zachary Smith, Sunny Wescott, and Kristin Wykoff. Our partners at Purdue University, including workshop lead Matthew Huber, David Johnson, Lynne Dahmen, Tara Greene, and Karan Hustedt-Warren, along with Marthie Grobler and Alex Blauensteiner from CSIRO, were engaging technical collaborators and teammates. Within the Sandia team, we thank Leslie Proctor for her leadership and project management, as well as Rob Leland and Amanda Dodd for their important contributions and inputs during the workshop. Diana Hackenburg provided strong communications support, and we appreciate the technical contributions from Thushara Gunda, Stephanie Kuzio, Ben Bonin, Nicole Jackson, and Mika Armenta. Finally, we would like to thank Shannon Walsh and Gavril Torrijos from ZRA for their invaluable technical support for the workshop and this chapter, as well as Kelli Jorstad from DHS and Danny Webber and Julie Thomas from Sandia for their business support.

References

1. U.S. Naval War College (2019) Defend forward: critical infrastructure war game
2. CISA, Cyber-Physical Convergence Scenarios (2023) https://www.cisa.gov/resources-tools/resources/cyber-physical-convergence-scenarios
3. US Cyber Command, Cyber Guard Exercise (2024)
4. Center for New American Security (2008) War game: clout and climate change
5. U.S. Department of Defense, Elliptic Thunder TTX (2021)
6. FEMA, National Level Exercise (2024)
7. Cameron A, Burke S (2022) War on the rocks, "wargaming climate change: who plays for the red team?"
8. RAND, Wargaming Climate Change (2024)
9. RAND, Wargaming Climate Change (2024) A structure for incorporating physical and social effects into strategic military planning
10. Atlantic Council (2020) The 5 × 5—Is it a game or is it real? Simulations and wargaming in cyber
11. FEMA Long-Term Community Resilience Exercise Resource (2021)
12. https://cpsiccnexusworkshop2024.org

Chapter 3
A Participatory Risk Elicitation Exercise to Identify Knowledge and Policy Gaps at the Intersection of Climate Change and Cyber-Physical-Social Infrastructure

David R. Johnson, Lynne Dahmen, Courtney Falk, Carly Frank, Thushara Gunda, Matthew Huber, Pragathi Jha, and Jason Reinhardt

Abstract The necessity of adapting to and mitigating the impacts of climate change is driving global efforts to develop and deploy new technologies for critical infrastructures. These new systems must simultaneously advance the safety, security, and prosperity of individuals and societies while progressing towards fulfilling the United Nations Sustainable Development Goals. However, the rapid deployment of new climate-smart technologies and critical infrastructure may increase the risk of cyber-attacks by introducing novel and unanticipated vulnerabilities. So how does one effectively assess the vulnerabilities associated with these infrastructure advancements? This chapter describes a novel risk elicitation methodology, deployed at the CPSICC workshop in July 2024, to identify research gaps and policy gaps associated with risks at the nexus of climate change and cyber-physical-social systems. The workshop attendees, drawn from a diverse range of countries, organizations, and scientific disciplines, engaged in a combination of scenario mapping with tabletop role-playing exercises. Participants were assigned varied roles and asked to react to scenarios in 2030 and 2050 wherein climate change would exacerbate the vulnerabilities or consequences of an attack on cyberphysical infrastructure. The novel risk elicitation methods are generalizable to other contexts, and the key research and policy gaps identified by workshop attendees will inform future funding opportunities and research agendas for NATO member states and NATO partners.

Keywords Cyber- · Physical- · Social-systems infrastructure and climate change nexus · National security · Critical infrastructure resilience · Climate change · Strategic foresight analysis

D. R. Johnson (✉) · L. Dahmen · C. Falk · C. Frank · M. Huber · P. Jha
Purdue University, West Lafayette, IN, USA
e-mail: davidjohnson@purdue.edu

T. Gunda · J. Reinhardt
Sandia National Laboratories, Albuquerque, NM, USA

M. Huber et al. (eds.), *Critical Infrastructure Systems and Extreme Threat Nexus: Preparing for Resilience and Security*, NATO Science for Peace and Security Series C: Environmental Security, https://doi.org/10.1007/978-3-032-13170-6_3

3.1 Introduction

Security organizations like the North Atlantic Treaty Organization (NATO) are charged to anticipate the future security environment and establish a common preparedness for long-term trends, threats, and challenges [12]. Today's world represents an unprecedented confluence of economic, environmental, and social systems at scale, with traditional sources of risk interacting across networks in ways that can produce unimagined cascading failures. Rapidly advancing technologies such as artificial intelligence, robots, and quantum computing threaten to transform systems more quickly than analysts can detect or plan for, even when anticipating their importance and monitoring innovations [5, 13, 17]. The weakening of multilateral governance institutions, as evidenced by withdrawals from bodies such as the European Union and World Health Organization, and ongoing wars threaten to limit the capacity for transnational coordination. This is not only the case for responding to rapidly evolving threats like the COVID-19 pandemic, but also for countering persistent threats like climate change, which will require voluntary commitments to the Paris Agreement that are vulnerable to shifting political headwinds.

These kinds of dynamics may invalidate risk models based on historical conditions that are increasingly detached from current realities. As systems become increasingly interconnected and complex, future conditions become impossible to predict due to deep uncertainty, where system parameters cannot be assigned probability distributions with confidence [11]. This limits the credibility of probabilistic risk assessments, except perhaps within narrowly tailored scenarios that make assumptions about fundamentally unknowable factors. It also increases the value of techniques from future studies and strategic foresight analysis [6], which focus less on predicting likely future conditions and more on imagining plausible ones. This is particularly important for organizations concerned with low-probability, high-consequence events, in this context, knowing the precise probability of an adverse event can be less critical than being prepared to respond effectively should it come to pass.

Unlike traditional planning exercises based on a "predict, then act" paradigm, strategic foresight acknowledges the fundamental uncertainties about future conditions while providing structured approaches to elicit and explore the range of plausible system trajectories. Drawing on systems thinking and complexity theory, strategic foresight analysis grapples with the multiplicity of futures that can emerge from the interaction of multiple dynamic systems operating at different temporal and spatial scales. Physical infrastructure systems are commonly designed with multi-decadal useful lifetimes and large spatial extents (e.g., interconnected power grids) or zones of influence (e.g., dams affecting larger watersheds). Strategic foresight is a theoretical lens particularly well-suited, then, to address the compound risks associated with the intersection of climate change with infrastructure systems, given climate change's gradual progression and spatial heterogeneity of impacts.

In this chapter, we provide guidance on best practices for risk elicitation exercises inspired by strategic foresight analysis. We then describe a series of group exercises undertaken at the 2024 "Unravelling the Cyber-Physical-Social Infrastructure Climate Change (CPSICC) Nexus" workshop, a NATO Advanced Research Workshop supported by NATO's Science for Peace and Security Programme and the U.S. Department of Homeland Security's Science and Technology Directorate. The exercises were inspired by other risk elicitation and wargaming methods from strategic foresight and futures studies [18], but tailored to fit the need of identifying as broad a collection of knowledge and policy gaps as possible. This chapter primarily describes the methodological advances used to elicit research and policy gaps at the intersection of climate change and infrastructure systems. We provide some detail on the identified gaps themselves but primarily leave that discussion to Chap. 4 [15]. The analysis of gaps was based on both the exercises described in this chapter and data collection efforts described in Chap. 2 [16].

3.2 Strategic Foresight Analysis to Identify Research and Policy Gaps

Strategic foresight analysis is still an emerging field, and research in the area is "quite heterogeneous when it comes to theoretical frameworks and reference theories" [6]. As applied to defense and security, it has largely evolved from techniques such as scenario planning, developed by Herman Kahn at RAND Corporation for the U.S. military [7], horizon scanning to identify emerging technologies or threats in various domains [2, 8], and assumption-based planning [3]. The latter is organized around the following steps for exploring scenarios, which are especially useful for identifying research and policy gaps:

1. **Identify important assumptions** on which current operations or plans are founded
2. **Identify vulnerabilities**, in terms of plausible changes that could occur within a relevant planning horizon which would violate assumptions or cause them to be wrong
3. **Define signposts**, events or thresholds that give an unambiguous indication about the vulnerability of an assumption
4. **Define shaping actions** that are designed to avert or cause the failure of a vulnerability
5. **Define hedging actions** that involve rethinking an organization's plans as if an important assumption had failed

Each of these steps represents a place where knowledge and policy gaps might be hiding. For example, new research may be needed to identify when a particular signpost has occurred or to achieve the capability of taking a specific shaping action. New policies may be needed to strengthen institutions and formalize compliance with normative actions that have heretofore only been implicitly assumed.

Collaborative scenario analysis techniques address individual cognitive biases and disciplinary limitations by involving diverse perspectives and domains of expertise, with research showing that utilizing diverse group compositions yields scenarios that individual experts would not generate independently [1]. This is particularly valuable when dealing with intersectional risks that span across multiple systems or knowledge domains. Prior foresight exercises have been used to develop concepts for risk governance of nexus-related threats, such as solar radiation modification (SRM) and geoengineering [10]. In the context of this chapter, identifying threats in the CPSICC nexus requires expertise in atmospheric sciences, infrastructure engineering, cybersecurity, policy analysis, and other domains.

Scenario development is likewise enhanced by participatory methods that explicitly include stakeholders who would be affected by future developments or drive policy responses. Participatory methods ensure that scenarios acknowledge real-world constraints and that responses are feasible, while collaboration builds shared understanding and generates buy-in towards shaping and hedging actions.

Modern approaches to scenario mapping consider group dynamics to avoid common problems such as anchoring on early ideas or failing to adequately explore possibilities considered to be unlikely. They also blend quantitative approaches with qualitative narrative development and leverage technology to enhance collaboration, run models in conjunction with discussions, or provide real-time group feedback to steer exercises.

3.3 Eliciting Gaps at the CPSICC Nexus

The "Unravelling The Cyber-Physical-Social Infrastructure Climate Change (CPSICC) Nexus" workshop, held 29 July–1 August 2024 in Washington, DC, had the following purpose (emphasis added):

> This interactive workshop aims to identify the building blocks of risk that lie at the nexus of climate change (e.g., direct impacts, tipping points, geoengineering, mitigation and adaptation efforts), cybersecurity (e.g., AI/deep learning/quantum computing), critical physical infrastructure with an emphasis on food-energy-water (FEW) systems, and social systems with an emphasis on policy, legal frameworks, institutions, and migration. In bringing together experts from various fields of academia, government, and industry from NATO member and partner countries, we **will identify gaps in our collective knowledge of risk that lay at the nexus of climate change, cybersecurity, and the critical infrastructure that supports society through FEW services, national security, and defense systems**.
>
> —Purdue University Institute for a Sustainable Future [14]

Gaps were conceived as a lack of knowledge or a need for research and development (R&D) or policies to effectively address risks at the CPSICC nexus.

3.3.1 Exercise Concepts and Logistical Considerations

Workshop attendees included 61 in-person speakers and participants, along with 10 support staff. Over half of the speakers and participants were from the government sector, coming from 12 NATO member and partner countries. About a quarter of attendees were from academia, with the remainder from industry and non-profit organizations. Subject-matter expertise included economics, climate science, wargaming, cybersecurity, critical infrastructure, cognitive sciences, national and/or global security, systems engineering, and decision analysis. We obtained consent from participants for recording/transcription and data collection during the workshop.

The workshop spanned three and a half days, with the group elicitation exercises being held on the final two days. The first day and a half of the workshop included keynote presentations and a variety of technical talks designed to level-set on perspectives from different countries and disciplinary backgrounds, as well as to build a common vocabulary around terms like "security," "risk," and "resilience."

Design considerations for the group elicitation exercises were guided by the workshop purpose of identifying a collective set of R&D and policy gaps spanning the nexus between climate change, cybersecurity, and critical infrastructure. As such, we wanted to ensure that discussions covered a range of climate risks and infrastructure systems. This was accomplished by splitting participants into multiple groups, each of which was assigned a different scenario to explore, simultaneously and independently of the other groups. In addition to the goal of spanning multiple risks and infrastructure systems, we crafted each scenario for its relevance to NATO operations and member state priorities.

3.3.2 Participant Roles

To increase engagement and familiarity, we also decided to leverage many attendees' experience with wargaming or tabletop simulation exercises by incorporating role-playing elements, wherein participants would be assigned a generic bio and perspective to take. However, a standard wargaming exercise was deemed to be unfit for purpose, given its nature of only playing through one realization of a scenario rather than exploring the full breadth of how a situation might play out. Consequently, we decided to incorporate role-play into scenario mapping, a technique designed to build out a branching tree of plausible actions and their consequences that then lead to further choices.

In order to leverage subject-matter expertise and to encourage creativity unconstrained by familiarity with stakeholder jurisdiction and regulatory constraints, we assigned up to three participants to play the same role as a group within each scenario, with each role comprising a mix of subject-matter experts and "outsiders." Roles varied by scenario but included positions with responsibilities related to cybersecurity, defense, economic development, etc. Assigning some participants to roles outside of their expertise was intended to aid in generation of novel ideas unconstrained by the role's typical authority or responsibilities; this was also a practical consideration in acknowledgement that the response to real scenarios would involve other roles not represented in the workshop activities. The total number of scenarios was thus constrained by this consideration, resulting in a total of three base storylines.

3.3.3 Scenario Timelines

Finally, we chose to include some continuity between the gaming sessions on the first and second day of the group exercises. This was done by situating the scenarios of the first day in a hypothetical year 2030, with the second day prompting participants to consider a hypothetical year 2050 following a particular branch of the scenario map developed from the 2030 vignette. Examining multiple time periods enabled exploration of differing degrees of climate change severity, technological change and evolution beyond current global conditions. The year 2030 was chosen as being sufficiently far from 2024 to allow for imaginative storytelling about "various short-term climate, technological, geopolitical, and economic developments" (i.e., political cycles and cultural shifts), while still being situated in participants' current experiences of climate and infrastructure risks. The year 2050 was chosen to allow for plausible differences in climate conditions and for relevance to current planning horizons.

Retrospective analyses of scenario planning exercises in the defense and security contexts have found that short-term scenarios (1–2 years) may "reveal gaps and measures that can mitigate threats at the margin to improve operational performance but are unlikely to substantially affect force structure," while medium-term scenarios (5–7 years) can be used to "identify new joint operational concepts… and force structure changes that might affect up to 10 percent to 20 percent of the force." Longer-term scenarios (7–20 years) are "far more likely to be useful" in guiding "research and development and underwrit[ing] more futuristic operational concepts and formations" [9]. Playing through related threats in both 2030 and 2050 thus allowed us to generate insights relevant to short-, medium- and long-term agenda setting.

3.3.4 Operational Details

The three storylines developed for the scenario mapping exercises included (1) a threat of attacking the U.S. power grid during an extreme heatwave, (2) jamming navigational systems to disrupt shipping operations in newly-opened Arctic shipping lanes, and (3) sabotage of pilot deployments of solar radiation modification (SRM) via stratospheric aerosol injection (i.e., spraying sulfuric aerosol compounds in the upper atmosphere to reflect solar radiation back into space). Each story consisted of a two to three page description of the hypothetical situation in 2030, which included a set of roles that participants were assigned to and references to relevant scientific literature informing the plausibility of the scenario and provided context.

Roles were defined by stakeholder categories, in terms of playing positions related to economic development, security, public affairs, cybersecurity, diplomatic relations, etc. Roles included specific detail about, for example, being a steering committee member for the NATO Cooperative Cyber Defence Centre of Excellence, along with context about the types of resources available to them and example actions they could take. However, participants were told not to feel overly constrained about jurisdiction and authority, but instead to focus on embodying the priorities and perspectives of the assigned stakeholder roles. For example, one security role included direction to "focus your decision making on those options available to military personnel. You are advocating for military positions while identifying where there exist synergies with other roles."

One member of the workshop organizing team was assigned to play the role of the bad actor in each scenario. The bad actors were empowered to respond to actions proposed by the participants to ensure that they did not assume that all of their actions succeeded in preventing or resolving the attack. Scenarios varied in the information provided about the bad actor (e.g., attribution, motives), and they were not required to answer questions from the other players.

Participants were also sent briefing notes ahead of the workshop to give additional information about the climate hazards and infrastructure technologies that would be the focus of the group exercises. These notes provided a baseline level of understanding and terminology to all attendees, not all of whom were familiar with the topics and technologies involved in each scenario. Participants were not given their assignments to specific groups in advance, but the role and scenario descriptions were generally shorter and more digestible in real time than the read-ahead materials. Example materials of the 2030 vignette, roles, and briefing context for the SRM scenario are provided as appendices to this chapter.

The initial gameplay session started with a brief recap of the scenario, roles, and logistics for the exercise, followed by breaking the room into small-group discussion among participants assigned the same role. Players for each role were asked to provide possible actions that they would take in response to the scenario, and to brainstorm consequences or impacts of what would happen as the scenario unfolded. These were written on Post-it notes color-coded by role. A facilitator helped to organize the notes into a hierarchical tree or "scenario map" of how the various stakeholders could navigate the scenario, shown in part for the 2030 SRM scenario in Fig. 3.1. In some cases, tiers in the tree represented cause-and-effect relationships (e.g., Action A could produce Effect X, which would then provoke Action B); tiers could also reflect chronology if one action required a longer lead time to execute. Participants were encouraged to think of multiple alternative actions or outcomes that were distinct in nature, rather than simple negations; for example, an alternative to "flights are grounded" might be "flights are re-routed" instead of just "flights are not grounded."

After an initial round of scenario map development, participants had the opportunity to vote on the importance of further expanding different branches of the map.

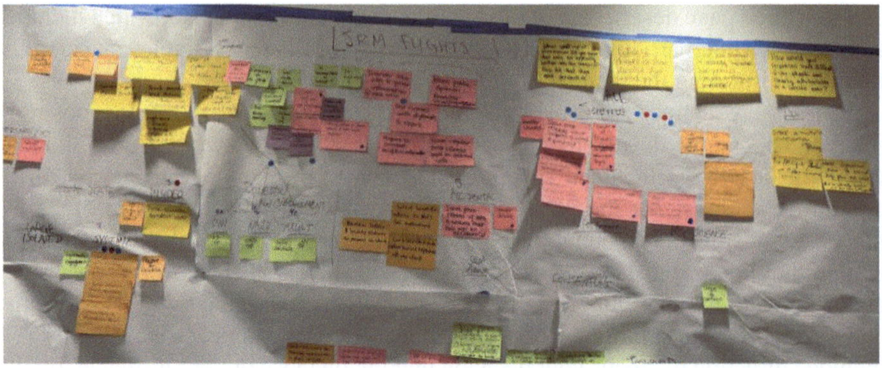

Fig. 3.1 Partial scenario map for the 2030 solar radiation modification attack scenario

Facilitators managed the remaining time of the 2030 sessions to explore branches, roughly in order of the voted importance but with some subjective judgement, until the group reached consensus that they have elicited the most plausible eventualities and contingencies. Branches varied widely in their success at countering the attack scenarios.

During the scenario mapping procedure, a recorder entered the scenario maps into an online spreadsheet template, capturing them in digital format. The spreadsheet template included forms with questions about each item of the map. Participants were asked to rate, from 0 to 5, the likelihood of taking an action or observing an outcome in the map contingent upon reaching the preceding item above it (0 = Not Possible, 1 = Very Unlikely, 2 = Plausible, 3 = Probable, 4 = Almost Certain, 5 = Certain). They also rated the positive or negative impact of each item from − 3 to + 3 along various dimensions (economic, geopolitical, and overall), where − 3 was considered the most negative outcome, + 3 the most positive outcome, and 0 a neutral outcome with respect to each dimension.

A brief synthesis session identified actions or outcomes that were contentious, in terms of having a high standard deviation of scores with respect to their likelihood or the connotation of impacts. This allowed groups to discuss disagreements, begin thinking about commonalities and differences across the three scenarios, and identify salient branches judged to be higher probability and strongly negative. In this manner, it was also helpful to organizers in crafting the 2050 attack scenarios to be played through on the next day of the workshop.

The 2050 vignettes presented scenarios based on the long-term consequences of particularly salient branches of the 2030 scenario maps (i.e., fast-forward to 2050 and explore what the world looks like after that branch occurred in 2030, or consider how a branch could have turned out differently after two decades of subsequent climate, economic, sociodemographic and technological changes). Due to the need for rapid creation of scenarios informed by the 2030 scenario maps, the 2050 scenarios were presented as single PowerPoint slides with bullet points detailing necessary world-building in addition to the attack or threat of attack.

Participants then spent an hour producing a 2050 scenario map using a similar procedure as the 2030 session. After this was accomplished, we facilitated a hind-casting exercise looking backwards from the items in the 2050 scenario maps. We asked questions designed to probe more deeply at the knowledge and policy gaps relevant to the 2050 scenarios, such as

- What could have been done to prevent this outcome/impact?
- When would action have been required?
- What data or indicators would have foreshadowed this outcome's likelihood?
- Would new technologies, sensors, or other intelligence capabilities need to be developed in order to collect that data or detect those indicators?

Note that these questions relate directly to the steps of an assumption-based planning process as described in Sect. 3.2, with respect to identifying hedging and shaping actions, signposts, and triggers for action, etc.

Finally, the three groups were reconvened in plenary to discuss the lessons learned and gaps identified in each scenario. Participants identified common responses in how various stakeholder roles approached the three different attack circumstances, differences where particular climate impacts or infrastructure systems shaped unique responses, and needs for near-term investments in capabilities (i.e., assumptions) that would have prevented the 2050 scenarios from coming to pass.

3.4 Results

Collectively across the three groups, participants identified over 140 potential R&D gaps related to the CPSICC nexus. More detail on specific gaps is provided in [4, 15], but we grouped them into six primary themes described below. Many gaps were cross-cutting, reflecting connections or overlaps across the themes.

1. **Developing a Systems Understanding of CPSICC Nexus**: There was strong recognition that climate is a threat multiplier and the existence of complex interactions between various elements of the CPSICC nexus. However, methodologies, common frameworks, and information to generate understanding across these elements are lacking.
2. **Policy, Social, Economic, and Governance Approaches**: The workshop games and discussions highlighted challenges pertaining to development and enforcement of international agreements, legal frameworks, and governance models to effectively manage and regulate activities at the CPSICC nexus.
3. **Technical Resiliency and Risk Mitigation Alternatives**: Various technologies are being advanced and developed across data, monitoring, and analysis dimensions. While some of these technologies are intended to directly address climate, their potential to exacerbate other vulnerabilities is largely unstudied. Further, some risk mitigation gaps remain unaddressed.

4. **Cyber-Physical Process Characterization**: There was clear recognition of the increasingly cyber-physical nature of most infrastructure systems, especially with more systems becoming digital. Various gaps were identified focusing on how these cyber-physical processes intersect with climate change and other dimensions of the CPSICC nexus.
5. **Social and Economic Considerations**: Multiple gaps were identified specific to understanding the role and impact of socio-economic dimensions (across public and institutional landscapes) at the CPSICC nexus.
6. **Risk Assessment and Decision Making**: Management of complex risk profiles was a common topic of discussion at the workshop. Multiple stakeholders identified that there is a significant gap in the availability of comprehensive frameworks, methodologies, and governance structures to manage and govern risks at the CPSICC nexus.

Reinhardt et al. [15], also discusses key insights around these major themes and how the identified gaps can inform future research and policy development.

3.5 Lessons Learned and Conclusion

The workshop activities highlighted the importance of resilience and deterrence in interconnected cyber-physical-social systems impacted by climate change, and underscored the need for a unified framework to address these challenges. Participants were deeply engaged and enthusiastic about the group elicitation exercises, and the end result was a long list of potential R&D gaps, meeting the stated objective of the workshop.

However, we observed some elements associated with the games that could be improved if similar workshops are convened in future as part of establishing a community of practice around the CPSICC nexus. Throughout the exercises, it became clear that the focal points of many discussions reflected the expertise of those in the room, and that greater presence from social scientists and representatives from industry would have been valuable to inform thinking around behavioral (e.g., risk perception and communication, public opinion) and technological (e.g., data privacy, technical capabilities) issues.

Logistically, participants found the spreadsheet exercise rating the likelihood and desirability of scenario map items to be challenging. Confusion emerged around how the scenario map's branching hierarchies were represented in a flat spreadsheet, making it difficult to interpret. A more bespoke software solution might be able to guide participants through the ratings process more clearly than the spreadsheet template designed for the CPSICC workshop; this could possibly be based on taking a photo of the final scenario maps and annotating each Post-it note digitally with a form containing the rating items.

Evaluating the positive or negative quality of certain items was also a challenge, given that individuals' views about many actions hinged on the potential outcomes that would come to pass lower in the scenario map (i.e., the positivity or negativeness was not inherent in the action itself but in the consequences). This resulted in less useful quantitative data being produced by the exercises than intended, which could have potentially been improved by more specific instructions for facilitators around the format of how items in the scenario map should be structured, or by modifying the spreadsheet template to focus on rating the likelihood only of actions, and rating the connotations only of outcomes.

Overall, the elicitation exercises successfully yielded a large set of potential gaps that can drive development of a research agenda in the CPSICC nexus, with greater detail provided in the following chapter. The games were structured to incorporate best practices for strategic foresight analysis and could be easily adapted to other contexts.

Acknowledgements This work was conducted through support from NATO and the US Department of Homeland Security. Sandia National Laboratories is a multi-mission laboratory managed and operated by National Technology and Engineering Solutions of Sandia LLC, a wholly owned subsidiary of Honeywell International Inc. for the U.S. Department of Energy's National Nuclear Security Administration contract DE-NA0003525. Staff from Australia's Commonwealth Scientific and Industrial Research Organisation produced the advanced briefing notes used for each attack scenario. The authors also acknowledge staff support and resources from the Purdue University Institute for a Sustainable Future in producing the workshop. We also gratefully acknowledge the participants who were an integral component of this activity and the lessons learned from it.

Appendix

The following supplementary information are published at https://doi.org/10.5281/zenodo.17329458:

- Solar Radiation Modification 2030 attack scenario and role definitions
- Solar Radiation Modification advance briefing notes
- Solar Radiation Modification 2050 attack scenario slide.

References

1. Bradfield R, Wright G, Burt G, Cairns G, Van Der Heijden K (2005) The origins and evolution of scenario techniques in long range business planning. Futures 37:795–812
2. Cuhls KE (2020) Horizon scanning in foresight—why horizon scanning is only a part of the game. Futur Foresight Sci 2:e23. https://doi.org/10.1002/ffo2.23
3. Dewar JA, Builder CH, Hix WM, Levin M (1993) Assumption-based planning: a planning tool for very uncertain times
4. Gunda T, Reinhardt J, Sanchez D, Winstead N, Walsh S, Torrijos G (2024) CPSICC workshop findings: executive summary of R&D gaps. Sandia National Laboratories, Albuquerque, New Mexico

5. Haimes YY (2009) On the complex definition of risk: a systems-based approach. Risk Anal 29:1647–1654. https://doi.org/10.1111/j.1539-6924.2009.01310.x
6. Iden J, Methlie LB, Christensen GE (2017) The nature of strategic foresight research: a systematic literature review. Technol Forecast Soc Change 116:87–97. https://doi.org/10.1016/j.techfore.2016.11.002
7. Kahn H (1965) On escalation. Praeger
8. Könnölä T, Salo A, Cagnin C, Carabias V, Vilkkumaa E (2012) Facing the future: scanning, synthesizing and sense-making in horizon scanning. Sci Public Policy 39:222–231. https://doi.org/10.1093/scipol/scs021
9. Larson EV (2019) Force planning scenarios, 1945–2016: their origins and use in defense strategic planning
10. Low S (2017) Engineering imaginaries: anticipatory foresight for solar radiation management governance. Sci Total Environ 580:90–104. https://doi.org/10.1016/j.scitotenv.2016.07.200
11. Marchau VAWJ, Walker WE, Bloemen PJTM, Popper SW (eds) (2019) Decision making under deep uncertainty: from theory to practice. Springer Nature
12. NATO Allied Command Transformation (2025) NATO strategic foresight analysis: future capabilities and security. In: NATOs ACT. https://www.act.nato.int/activities/allied-command-transformation-strategic-foresight-work/. Accessed 7 Oct 2025
13. NATO Allied Command Transformation (2024) First allied foresight conference. North Atlantic Treaty Organization, Helsinki, Finland
14. Purdue University Institute for a Sustainable Future (2024) CPSICC Nexus Workshop 2024. https://cpsiccnexusworkshop2024.org/index.html. Accessed 27 June 2025
15. Reinhardt J, Gunda T, Sanchez D, Homan R, Yang L, Winstead N (2025) Cyber, physical, social systems infrastructure and climate change nexus analysis and key findings. In: The cyber-physical-social infrastructure climate change nexus: preparing for resilience and security. North Atlantic Treaty Organization, Washington, DC
16. Reinhardt J, Gunda T, Sanchez D, Homan R, Yang L, Winstead N (2025) Cyber, physical, social systems infrastructure and climate change nexus information gathering. In: The cyber-physical-social infrastructure climate change nexus: preparing for resilience and security. North Atlantic Treaty Organization, Washington, DC
17. Renn O, Laubichler M, Lucas K, Kröger W, Schanze J, Scholz RW, Schweizer P-J (2022) Systemic risks from different perspectives. Risk Anal 42:1902–1920. https://doi.org/10.1111/risa.13657
18. Wright G, O'Brien F, Meadows M, Tapinos E, Pyper N (2020) Scenario planning and foresight: advancing theory and improving practice. Technol Forecast Soc Change 159:120220. https://doi.org/10.1016/j.techfore.2020.120220

Chapter 4
CPSICC Nexus Analysis and Key Findings

Jason Reinhardt, Rossitza Homan, Danielle N. Sanchez, Lynn Yang, and Nick Winstead

Abstract Identifying and prioritizing the intersecting research and development challenges related to climate impacts across cyber, physical, and social infrastructure is essential due to their complex interdependencies and increased vulnerability to extreme weather events. This chapter presents the analysis and key findings from the information gathered during the preparation and the execution of the Cyber-, Physical-, Social-Systems Infrastructure and Climate Change nexus workshop. We contend that the primary R&D challenge within the CPSICC nexus is to enhance decision-making processes for the NATO alliance and its member nations, enabling them to allocate limited resources effectively to mitigate risks to the nexus system amidst deep uncertainty, competing priorities, and high complexity. Therefore, we utilize a Decision Quality framework to structure the findings, define a cohesive research agenda, and inform future investment decisions. By mapping the identified research and development gaps to this framework, Sandia pinpoints five priority themes along with corresponding recommendations for future resource investments.

Keywords Cyber-, Physical-, Social-systems infrastructure and climate nexus · Decision quality framework · National security · Critical infrastructure resilience · Climate · Interdependencies

J. Reinhardt · R. Homan · D. N. Sanchez · L. Yang (✉)
Complex Systems Risk and Resilience, Sandia National Laboratories, Albuquerque, NM, USA
e-mail: liyang@sandia.gov

J. Reinhardt
e-mail: jcreinh@sandia.gov

N. Winstead
Zeichner Risk Analytics, Washington, DC, USA

M. Huber et al. (eds.), *Critical Infrastructure Systems and Extreme Threat Nexus: Preparing for Resilience and Security*, NATO Science for Peace and Security Series C: Environmental Security, https://doi.org/10.1007/978-3-032-13170-6_4

4.1 Introduction

Identifying and prioritizing the intersecting research and development (R&D) challenges related to climate impacts across cyber, physical, and social infrastructure is essential due to the complex interdependencies among these systems and their increased vulnerability to extreme climate events. As climate affects critical infrastructures, it disrupts essential services and threatens community resilience. The interconnected nature of these systems means that a failure in one area can have cascading effects on others, highlighting the need for a coordinated approach to R&D.

This chapter presents the analysis and key findings from the Cyber-, Physical-, Social-Systems Infrastructure and Climate Change (CPSICC) nexus workshop, which brought together experts from various fields to explore these pressing challenges. Through collaborative discussions, participants highlighted the necessity of enhancing decision-making processes for the North American Treaty Organization (NATO) alliance and its member nations. Effective decision-making is crucial for allocating limited resources in a manner that mitigates risks to these interconnected systems, especially when facing deep uncertainty, competing priorities, and high complexity.

To structure our findings and define a cohesive research agenda, we employed a Decision Quality (DQ) framework. This framework serves as a valuable tool for informing future investment decisions, allowing stakeholders to systematically evaluate and prioritize R&D initiatives. By mapping the identified R&D gaps to the DQ framework, we pinpointed five priority themes, each accompanied by specific recommendations for future resource investments. These recommendations not only aim to enhance the resilience and sustainability of the CPSICC nexus but are also broadly applicable to promoting community resilience and supporting security missions in the context of a changing climate. In addition to enhancing the resilience and sustainability of the nexus, these recommendations are broadly applicable to promoting community resilience and supporting security missions in the context of a changing climate.

4.2 The Decision-Quality Framework

This section describes how Sandia analyzed the information gathered during the workshop. For additional details regarding the information gathering approach before and during the workshop (Reinhardt et al., 2025, Chap. 2). To focus future R&D efforts aimed at addressing gaps in the CPSICC nexus on empowering decision-makers to mitigate risks and enhance resilience against climate threats, Sandia employed the DQ framework. Sandia analyzed the R&D gaps identified by the workshop participants and Sandia SMEs through the lens of DQ and distilled the findings into a series of recommendations and research agenda items. This section offers an overview of the analytical approach and highlights the key findings.

NATO, along with its Member States and partners, is increasingly concerned about the effects of climate on cyber, physical, and social systems [1]. We contend that the primary R&D challenge within the CPSICC nexus is to enhance decision-making processes for the NATO alliance and its member nations, enabling them to allocate limited resources effectively to mitigate risks to these systems amidst deep uncertainty, competing priorities, and high complexity. Therefore, we utilize a DQ framework to structure the findings and define a cohesive research agenda.

We adopt a widely recognized version of a DQ framework that consists of six elements: frame, values, information, alternatives, integration, and commitment (Fig. 4.1). This framework is drawn from decision science and research literature and serves as a well-established rubric for evaluating decision-making processes. The concepts of decision quality were first introduced in 1964, drawing upon advancements in statistical decision theory and game theory by Professor Howard Raiffa of Harvard University, as well as dynamic probabilistic systems developed by Professor Ronald A. Howard of Stanford University. The initial application of decision analysis/decision quality concepts in a professional context is documented in Professor Howard's paper titled "Decision Analysis: Applied Decision Theory" [2]. Each of these six elements represents a critical aspect of a decision problem that must be addressed to ensure a normatively high-quality decision. Importantly, the elements of DQ are not intended to be addressed sequentially, rather, they can be developed in parallel, provided there is effective communication and coordination among them. In this sense, they are not steps in a process but rather pillars of effort that work together to ensure high-quality decisions. We assert that decisions related to policies, investments, analyses, and the implementation of measures to mitigate the impacts of climate on cyber, physical, and social systems share these same requirements. Next, we describe each of the DQ elements through the lens of the CPSICC nexus.

4.2.1 *Frame*

Ensuring that decisions are well-framed is crucial for establishing a shared understanding of scope among stakeholders. The frame specifies the assumptions made, what is included or excluded from consideration, and the specific questions that need to be answered. Researchers highlight three key components for an appropriate frame: clear purpose, defined scope, and conscious perspective. In the context of the CPSICC nexus, this involves formally defining what the nexus entails and the community's objectives, including clarifying the meanings of cyber, physical, and social systems, as well as the relevant aspects of climate.

The framing process aims to identify potential decisions and create consensus among various stakeholders, including governments, system operators, communities, and the R&D community. It requires dedicated efforts to clarify the CPSICC decision space and recognize how the frame may differ among decision-makers. Additionally, identifying gaps that hinder shared understanding is part of this framing process. Ultimately, framing also involves pinpointing scenarios of interest that will

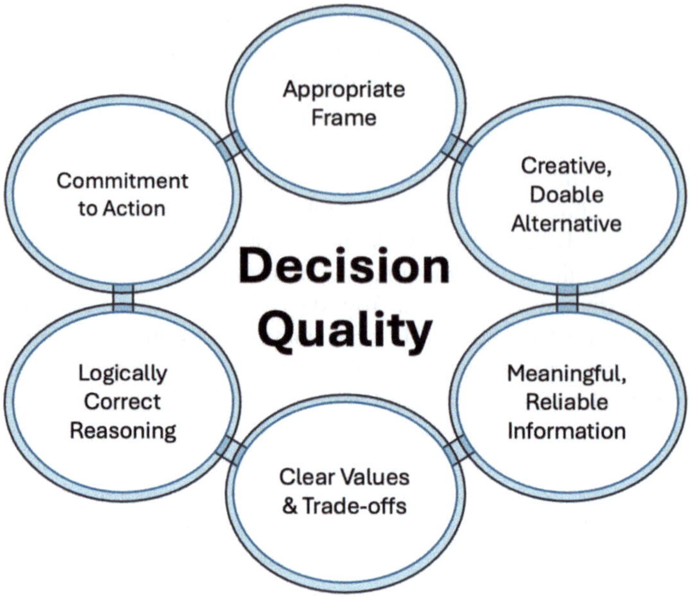

Fig. 4.1 The decision quality framework

foster a common understanding of risks, mitigation strategies, and how the CPSICC community can effectively work towards beneficial outcomes.

4.2.2 Alternatives

Building on the results of the framing element of DQ, the CPSICC nexus community must identify and develop actionable alternatives for its members. This work involves understanding, at a systems level, which options are feasible for implementation and have the potential for risk reduction. It is essential to carefully characterize the expected risk reduction of each option and to outline the trade-offs and costs associated with them. Additionally, this element includes exploring ways to bundle options into comprehensive and robust portfolios and strategies that mitigate risks stemming from individual options.

Developing alternatives also encompasses research aimed at advancing science, technology, and the humanities to create new options. This research includes a spectrum from basic to applied, as well as the development of advanced technologies and systems. It is also important to cultivate a cultural understanding of how different alternatives may be adopted or may fail. Advances in basic sciences can lead to new technological or policy approaches that positively impact CPSICC systems and alleviate burdens on stakeholder communities. Applied research can help the community determine the best implementation strategies for these new approaches and clarify their roles in mitigating risks associated with climate hazards.

4.2.3 Information

The CPSICC community must ensure they have adequate information to make informed decisions about mitigating risks to cyber, physical, and social systems considering climate threats. However, decision-making does not require perfect knowledge; rather, a DQ framework emphasizes research that characterizes and incorporates uncertainty into the community's collective understanding. This approach allows for a clear assessment of known and unknown factors, often revealing areas where assumptions or biases may distort perceptions.

The information element of DQ involves enhancing the community's understanding of the physical processes, systems, technologies, policies, outcomes, and stakeholder communities involved. While there are numerous opportunities to improve information and reduce uncertainty, a principled approach should prioritize investments in information development based on the value they provide. Research that identifies which information gaps are significant for decision-making and which have minimal impact—known as value-of-information research—will be as essential as the research aimed at generating new information.

4.2.4 Values

The stakeholders within the CPSICC nexus must comprehend values at multiple levels. They need to identify the drivers of value, building on the work done to frame the problem space. Research aimed at understanding what is important to various stakeholder communities and codifying those insights into a set of drivers is essential, as it helps refine the scope of scenarios and threats of concern. This exploration may involve applying systems engineering approaches to engage stakeholders and clarify priorities.

Additionally, efforts to establish CPSICC values should focus on developing core metrics and measures for value, both qualitative and quantitative. The community must not only identify and reach consensus on these metrics but also define how to interpret changes in them, as well as the goals and objectives of the CPSICC community as reflected in those metrics. Achieving agreement on these points is as challenging and crucial as developing the metrics themselves. Ultimately, this work will enable clear articulation of scenario outcomes and provide a framework for evaluating potential futures and the impacts of decisions made by the alliance, its members, and partners.

4.2.5 Reasoning (Integration)

The reasoning for integration of the results from the framing, alternatives, information, and values development efforts is essential within the Application of the DQ framework to the CPSICC nexus. This integration aims to create and utilize models that capture the logic of the problem, the interactions among its components, and how system behavior varies across different scenarios and alternatives. Integration efforts should focus on developing a comprehensive understanding of the scenario drivers, as well as the cyber, physical, and social systems, to incorporate them into an integrated model.

Given the complexity of the CPSICC nexus, initial integration efforts should prioritize the creation of a simple, high-level model that effectively addresses the questions identified during the framing process, while also being informed by lower-level assessments. As the CPSICC concerns evolve, additional complexity can be introduced as needed. Moreover, integration involves not only the development of high-level models but also the synthesis of the community's understanding of mitigation options and the value placed on outcomes by the CPSICC community. This process requires gathering, organizing, and safeguarding the information into a comprehensive corpus of data and insights that can guide decision-making. Managing and stewarding this collective understanding presents a significant challenge.

4.2.6 Commitment

The final element of DQ is commitment, which solidifies the motivation and determination to act. A crucial factor in achieving commitment among decision-makers is their involvement in the other five elements of DQ, as this fosters a better understanding and acceptance of the information and alternatives presented. To drive and formalize commitment, specific agreement mechanisms, such as formal agreements and treaties, may be necessary.

Strategies to enhance commitment can include researching cultural factors at the stakeholder level, developing approaches to build consensus within communities, addressing challenges related to collective action, and conducting legal and policy research to ensure alignment with national laws and political efforts. Additionally, establishing protocols and regimes for monitoring and verification is essential to ensure that stakeholders adhere to their commitments and do not act contrary to those agreements.

4.3 The Decision-Quality Framework

The DQ framework is a structured approach aimed at enhancing decision-making processes by ensuring that decisions are informed by comprehensive and relevant information. It includes key components such as defining clear objectives to align with desired outcomes, understanding the context of the decision, generating a range of alternatives, and evaluating the trade-offs of each option by considering risks, benefits, costs, and impacts. The framework also emphasizes the importance of involving stakeholders to gather diverse perspectives, making informed decisions based on evaluations and input, and implementing the chosen alternative while monitoring its effectiveness.

During the post-workshop analysis, Sandia identified 136 separate R&D gaps, which the Sandia team categorized into six groups. Note that this categorization reflects their interpretation of each gap's intent, and alternative interpretations may exist. Additionally, some gaps could fit into multiple categories; in such cases, the team placed the gap in the category that best represented its focus.

The largest group of R&D gaps centered on the need for improvements in policy, social, economic, and governance approaches (40). The categories of risk assessment and decision-making (27), social and economic considerations (25), technical resiliency and risk mitigation alternatives (25) accounted for significant portions of the overall gap set. In contrast, cyber-physical process characterization (11) and developing a systems understanding of the CPSICC nexus (8) had the fewest identified gaps. While the workshop and preparatory activities represented a small yet diverse cross-section of the CPSICC community, it is important to note that definitive conclusions about the overall state of these issues cannot be drawn from this data. On the other hand, these R&D gaps are indicative of broader trends applicable beyond the CPSICC nexus.

4.3.1 Analytical Questions

This subsection outlines the analytical questions developed by Sandia's team for each of the six decision quality elements. It also describes Sandia's method of evaluating each R&D gap against these analytical questions to categorize the gaps into six overarching themes.

1. **Frame**: Does addressing this gap or concept have the potential to enhance our common understanding of terms, purpose, scope, and assumptions? For instance, what exactly is being addressed? How will we know when we have achieved it? What metrics should be used? What is the problem, and do we have a shared understanding of it?
2. **Alternatives**: Does addressing this gap or concept have the potential to provide an improved set of possible mitigations or actions that can be considered to reduce risk or enhance outcome value?

3. **Information**: Does addressing this gap or concept have the potential to improve our collective understanding of underlying processes, impacts, likelihood of outcomes, or the efficacy of mitigation options?
4. **Values**: Does addressing this gap or concept have the potential to foster a better agreement among stakeholders regarding drivers and preferred outcomes? What do we want? What are our preferences?
5. **Integration**: Does addressing this gap or concept have the potential to enhance our ability to collectively understand or compare different scenarios, mitigations, or strategies?
6. **Commitment**: Does addressing this gap or concept have the potential to increase stakeholder willingness or ability to agree on and/or take action to implement solutions?

Each of the identified R&D gaps were evaluated by the Sandia team against these analytic questions, and the most relevant DQ element was identified. For example, a gap such as "the need for advanced energy management systems" was determined to address the alternatives element. The team concluded that addressing this gap would provide additional options for nations seeking to make decisions and allocate resources to reduce risks and improve the resilience of cyber, physical, and social systems. In some instances, a gap could address multiple DQ elements; in these cases, a primary and secondary DQ element were identified. By examining which parts of the DQ framework the gaps covered, the Sandia team could better identify trends from the workshop and pinpoint areas that appeared to be underrepresented.

Tables 4.1 and 4.2 summarize the analysis of gaps identified during the workshop and those from the preparatory activities, respectively. Among the data collected, the most frequently identified primary DQ element in the workshop was Information. Specifically, 62 out of 136 gaps (45.6%) from the workshop and 16 out of 51 gaps (31.4%) from the preparatory activities were classified as addressing an Information DQ. This result indicates that many of the identified R&D gaps are expected to enhance collective understanding of underlying processes, impacts, likelihood of outcomes, and the efficacy of mitigation options.

Table 4.1 Number of primary and secondary DQ element associations for workshop gaps

DQ element	Frame	Values	Alternatives	Information	Integration	Commitment
Primary	16	4	30	62	11	13
Secondary	3	5	24	13	7	10

Table 4.2 Number of primary and secondary DQ element associations for SME questionnaire gaps

DQ element	Frame	Values	Alternatives	Information	Integration	Commitment
Primary	6	7	14	16	7	1
Secondary	5	5	9	13	6	13

The least frequently mentioned DQ element varied between the two data sets. In the workshop data, Values was the least common primary DQ element, with only 4 out of 136 gaps (2.9%) classified as addressing a "Value" DQ. Conversely, in the internal Sandia preparatory activities, "Commitment" was the least frequently cited DQ element, with just 1 out of 51 items (2.0%) classified as addressing the "Commitment" primary DQ. This discrepancy may be attributed to the different populations sampled; the workshop included a diverse range of experts from various fields, while the preparatory activities sample was limited to just seven individuals.

The most frequently observed pairing of primary and secondary DQ elements was Information (primary) linked to Alternatives (secondary), with 22 pairs identified in the workshop data and six from the preparatory activities data. Additionally, Alternatives (primary) and Information (secondary) were paired five times in the preparatory activities data. This underscores the interconnectedness between enhancing collective understanding of underlying processes and impacts and how this can lead to better alternative options.

Overall, the participants identified the greatest need for facilitating CPSICC decisions within the Information category. They may perceive that decision-makers and the community lack a comprehensive understanding of underlying processes, impacts, likelihood of outcomes, and the efficacy of mitigation options. While the problem may be well defined, there appears to be a gap in understanding the underlying mechanisms and variables involved.

Several alternative interpretations of these results should be considered. One possibility is that the workshop and preparatory activity participants were predominantly from research communities focused on developing information and alternatives, leading to a concentration of identified gaps in these areas. Another interpretation is that participants may not have had specific ideas to promote certain DQ elements, despite recognizing significant gaps. In the cases of Alternatives and Information, many gaps centered on improving prescriptive measures while also identifying fundamental shortcomings.

Furthermore, the gaps associated with the less frequently mentioned DQ elements—Frame, Values, Integration, and Commitment—tended to be quite fundamental, often relating to a lack of common definitions for terms, absence of shared metrics for performance assessment, and insufficient governance frameworks. This suggests that additional efforts to directly address these areas could be beneficial for the community. It is recommended that these engagements involve decision-makers and policymakers from across the alliance and its partners whenever possible, to ensure that the identified gaps and proposed solutions can be validated across NATO Member States and partners.

4.3.2 Six R&D Gap Categories

The workshop brought together a diverse, interdisciplinary group of SMEs and designed various activities to effectively elicit R&D gaps. These activities included three analytical games, TED Talk-style presentations, an R&D proposal contest, and SME elicitation sessions (Johnson et al., 2025, Chap. 3).

While the workshop and preparatory activities provided a diverse cross-section of the community, definitive conclusions about the overall state of issues cannot be drawn from this data. However, participants underscored the importance of understanding policy and social dimensions and improving assessments of systemic risk to inform decision-making. Notably, several SMEs highlighted the need for better understanding of trade-offs, interdependencies, and cascading impacts related to climate and proposed mitigation strategies. Additionally, there was a recurring theme regarding the need for improved understanding of direct intervention strategies, such as solar radiation modification (SRM), including their efficacy, unintended consequences, and impacts on different communities.

Sandia categorized the R&D gaps into six groups (Fig. 4.2), focusing on improving knowledge of core physical processes, social and economic impacts, and integrating this understanding into a systems perspective of CPSICC nexus. Other categories included enhancing risk assessment and decision-making approaches, developing technical alternatives for resiliency and risk mitigation, and improving policy, social, economic, and governance options. The largest group of gaps (40) emphasized the need for improvements in policy, social, economic, and governance approaches, followed by risk assessment and decision making (27). Categories related to social and economic considerations (25) and technical resiliency and risk mitigation alternatives (25) represent significant portions of the overall R&D gaps. In contrast, cyber-physical process characterization (11) and developing a system understanding (8) included the fewest number of R&D gaps, (Fig. 4.2). These R&D gap categories were first introduced in Chap. 2 (Reinhardt et al., 2025), and were finalized during the post-workshop analysis.

These gap categories are interconnected and emphasize the need for a systems understanding of the CPSICC nexus. Many identified gaps correlate with multiple categories, illustrating the complexity of the nexus. For instance, a gap related to identifying the types of cyber-physical data needed for impact modeling connects to both cyber-physical process characterization and risk management and governance. Similarly, a gap concerning investments to mitigate the impact of critical infrastructure disruptions for vulnerable populations during crises relates to socioeconomic impact assessments and policy challenges.

4.4 Detailed Analytical Findings

The following subsections outline and describe the specific gaps within their respective R&D categories.

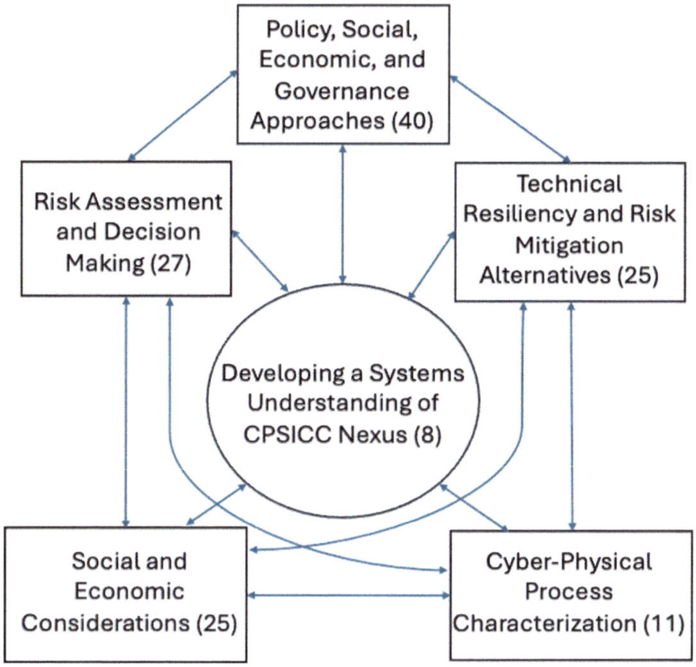

Fig. 4.2 R&D gap categories. Arrows indicate connections between R&D gap categories. Numbers indicate the number of R&D gaps per category

4.4.1 Developing a Systems Understanding (8 Gaps)

Workshop participants and SMEs agreed that the CPSICC nexus comprises multiple systems with significant interdependencies, emergent behaviors, and varying time horizons. Unfortunately, most efforts in this area lack a comprehensive systems understanding, necessitating characterization, modeling, and evaluation of the complex interactions and potential cascading impacts of disruptions within the CPSICC nexus.

1. **Architecture Trade-offs**: Research is needed to explore the trade-offs and risks associated with different architectures and topologies for rapidly evolving energy infrastructure.
2. **Complex Systems**: Improved modeling of complex systems is necessary to understand the dynamics and emergent behaviors occurring at different time horizons, including interactions between AI systems and operational technology, human movements, and long-term climate impacts.
3. **Compounding Events**: Understanding various compounding events, such as multivariate, spatially compounded, temporally compounding/cascading, and pre-conditioned compounds, is essential.

4. **Comprehensive Climate Intervention**: Researching the utility of SRM as part of a mix of climate interventions and policies is needed.
5. **Energy Dependencies**: Understanding energy dependencies as new energy modalities and architectures emerge is crucial, along with identifying ways to make these dependencies more adaptable, modular, and flexible.
6. **More Granular Dependency Mapping**: Precise dependency modeling is needed to understand complex interactions within the CPSICC nexus.
7. **National Critical Infrastructure**: Understanding dependencies between critical infrastructure systems that cross national borders is essential, as current mapping efforts are largely national.
8. **Updating Data Based on Climate**: Existing physical security models need to be updated and refined based on changing climate parameters, such as atmospheric conditions, temperature, and moisture.

4.4.2 Policy, Social, Economic, and Governance Approaches (40 Gaps)

Current policy and governance mechanisms are insufficient to address the risks associated with the CPSICC nexus. Participants emphasized the need for comparative studies on effective governance models to tackle long-term climate and infrastructure risks, necessitating local, national, and international policy research.

1. **Citizen Involvement**: Develop new approaches to engage citizens in climate resilience efforts, such as data collection, behavior change, and outreach initiatives.
2. **Civil Society Strategies**: Strengthen resilience by promoting civil society engagement in the Arctic to leverage local mitigation strategies regarding environmental impacts.
3. **Climate Data Security**: Establish best practices and enforcement mechanisms for climate data and models, particularly considering increasing AI usage.
4. **Climate Resilient Cybersecurity**: Develop cybersecurity strategies that specifically address climate challenges.
5. **Communication Strategies**: Create effective communication strategies to raise awareness and encourage public participation in climate action.
6. **Community-Based Adaptation**: Research community-based adaptation strategies that utilize local knowledge and resources to enhance resilience to climate challenges.
7. **Cyber Attribution**: Develop capabilities for quickly attributing cyberattacks and effectively communicating findings to the public.
8. **Cyber System Climate Resilience**: Improve understanding of best practices for designing cyber systems that are resilient to climate risks, such as data cooling centers vulnerable to heat and storms.

9. **Cyber-Physical-Social Systems (CPSS) Data Security**: Establish best practices and enforcement for CPSS data and models, considering the proprietary nature and security concerns of certain datasets, especially with the rise of AI.
10. **Cybersecurity Standards for the Shipping Sector**: Develop minimum cybersecurity standards for ships.
11. **Data Education**: Create curricula, materials, and tools to better inform stakeholders from diverse backgrounds about data provenance, security, and related issues.
12. **Data Security**: Develop scalable methods and standards to ensure data security across all areas of the CPSICC nexus.
13. **Definitions of Resilience Across Nations**: Establish common definitions and frameworks for resilience concepts and applications to ensure compatibility and alignment of goals across nations and governments.
14. **Deterrence**: Research and develop deterrence options for climate interventions, including security postures and preemptive actions against threats.
15. **Deterrence for Climate Interventions**: Investigate the role and potential effectiveness of deterrence strategies in managing climate interventions and preventing unilateral actions that could harm others.
16. **Develop Educational Resources**: Create educational resources and promote data sharing regarding supply chain cyber challenges, ensuring that data is distributed and shared in a trusted manner.
17. **Develop Regulatory Framework for Cybersecurity Standards in the Shipping Sector**: Establish a framework and monitoring assessments to ensure cybersecurity compliance in the shipping sector.
18. **Educational Resources for Supply Chain Cyber Challenges**: Develop materials and frameworks to enhance education around cyber threats, vulnerabilities, and management challenges.
19. **Enforcement of SRM Governance**: Research and develop mechanisms for enforcing governance related to SRM.
20. **Enumerate the CPSICC Nexus**: Create and integrate frameworks, taxonomies, and lexicons that enable the enumeration of CPSICC risks, as there is currently no overarching framework to understand these complex interactions.
21. **Establish Public/Private Sector Coordination**: Foster coordination between the public and private sectors utilizing shipping routes to share insights on threats or incidents in the shipping sector.
22. **Funding Mechanisms**: Identify innovative funding mechanisms and financial instruments to support climate resilience and adaptation initiatives.
23. **Incident Response Plans**: Develop incident response plans that integrate climate risk assessments with cybersecurity protocols.
24. **Integration of Data**: Integrate data across disciplines, jurisdictions, time horizons, and formats to enhance understanding of the CPSICC nexus.
25. **International Agreements**: Establish international agreements to prepare for potential environmental degradation in the Arctic, focusing on which countries will play significant roles in maintaining shipping lanes.

26. **International Agreements and Laws for AI**: Create international agreements and laws to better govern and regulate the use (and misuse) of AI in the design and operation of CPSS systems.
27. **International Cybersecurity Laws**: Develop enforceable international cybersecurity laws that impact international shipping.
28. **Interoperability Standards**: Establish standards for interoperability between different components of smart grids and renewable energy systems.
29. **Legislative Frameworks for Electronic Waste**: Improve frameworks and standards for the management and disposition of electronic waste streams.
30. **Microgrid Standards/Protocols**: Develop standards and protocols for microgrids, such as protocols for automatically entering "island mode" before a cyberattack occurs.
31. **Military Response Options**: Investigate military options for responding to threats or attacks against climate interventions, considering proportionality and its relation to climate-related deterrence.
32. **Monitoring**: Identify the mechanisms, regimes, infrastructure, technology, and data needed to monitor SRM deployment.
33. **Multilateral Exercise for the Shipping Sector**: Initiate a series of multilateral exercises for shipping emergencies to prepare for potentially degraded environments in the Arctic or other regions.
34. **Partnerships with Local Communities**: Establish partnerships with local and indigenous communities to strategize potential mitigation strategies.
35. **Public Engagement**: Develop strategies to engage the public in climate mitigation and adaptation efforts, leveraging behavioral insights to promote sustainable practices.
36. **Public/Private Sector Coordination**: Create frameworks and methods to enhance coordination between the public and private sectors, including programs, legislation, and regulations that align incentives and promote action.
37. **Research Agreements and Coordination**: Develop research agreements between Arctic states to share advancements in shipping technologies, such as GPS and ship technologies.
38. **Research Community**: Establish a community of practice that incorporates SMEs from the CPSICC nexus.
39. **Trusted AI**: Research and develop methods and frameworks for trusted AI to effectively manage and secure CPSS systems.
40. **Understanding Options for Threat Response**: Investigate government options for responding to threats against climate interventions, considering investigations, negotiations, threat actor demands, disaster preparation, multilateral actions, and recovery operations.

4.4.3 Technical Resiliency and Risk Mitigation Alternatives (25 Gaps)

There are significant R&D gaps related to resiliency and mitigation technologies within the CPSICC nexus. Research is needed in areas such as new energy technologies, cyber-climate resilience, emerging technologies like AI and quantum computing, and secure and sustainable data usage.

1. **AI and Disinformation**: Explore the applications of AI to mitigate misinformation, including identifying and countering extremism, as well as understanding how AI may enhance misinformation threats (e.g., audience segmentation, deepfakes, and generative AI for tailored messaging).
2. **AI Applications in Climate Solutions**: Investigate how AI and other emerging technologies can be utilized in climate solutions for critical infrastructure.
3. **Alternative Energies:** Prioritize R&D for carbon-free energy production, such as inertial confinement fusion and other alternatives, as an effective solution for climate challenges.
4. **Alternative Interventions**: Continuously research and develop diverse climate interventions to provide multiple options, especially in scenarios where the deployment of one or more interventions is disrupted.
5. **Alternative Supply Chain and Shipping Models**: Research alternative supply chain designs that reduce global shipping volumes, potentially increasing domestic production and introducing new shipping methods.
6. **Cybersecurity for Emerging Technologies:** Develop methods and technologies to enhance the cybersecurity of emerging technologies, ensuring that vulnerabilities are addressed as new solutions are implemented to tackle CPSS challenges.
7. **Deployment Alternatives**: Understand the requirements and timelines for deploying alternative solar radiation management (SRM) strategies if primary interventions are disrupted.
8. **Develop Redundant Systems**: Create redundant systems to avoid complete reliance on technological advancements, including backup communications and GPS shipping technologies.
9. **Electric Vehicle Cybersecurity**: Invest in cybersecurity capabilities and standards for electric vehicles.
10. **Electric Vehicles for Power**: Develop infrastructure that enables electric vehicles to supply power to residences and other infrastructure during grid failures.
11. **Energy Management Systems**: Invest in advanced energy management systems.
12. **Energy Storage Methods**: Identify innovative energy storage methods, such as water-based storage solutions.
13. **Grid Resilience Technologies**: Conduct R&D on self-healing and automatic repairing grid technologies.

14. **Improve Physical Infrastructure**: Enhance physical infrastructure along shipping routes to accommodate increased traffic and improve port facilities.
15. **Low-Carbon Solutions Across Critical Infrastructure**: Research ways to apply existing low-carbon solutions from transportation to other critical infrastructure sectors.
16. **Microgrid Technologies**: Investigate methods to make infrastructure and architecture grids less dependent on communications.
17. **Nature-Based Solutions**: Develop nature-based solutions and integrate them with man-made approaches to address climate challenges.
18. **New Shipping Technologies**: Create a reliable tracking system for shipping in the Arctic, such as a resource that effectively schedules container ships in real time.
19. **Quantum Technology Applications**: Research potential applications of quantum technology, such as quantum cryptography, to secure climate data.
20. **Renewable Energy Integration into Shipping**: Advance technologies that integrate renewable energy into the shipping industry to minimize environmental impacts.
21. **Resilient Digital Currency**: Investigate the development of digital currency solutions for austere environments, such as during massive cyberattacks or solar flares.
22. **Space Contingency**: Research the feasibility of evacuations to space.
23. **Systems Engineering Concepts**: Implement systems engineering processes to effectively deploy promising R&D.
24. **Technologies and Methods for SRM**: Research and develop various SRM methods and technologies to better understand options and inform further R&D and deployment decisions. Stratospheric aerosol injection (SAI) is one method, but it is highly controversial, and its long-term consequences are debated.
25. **Threat Impact Assessments**: Develop capabilities for timely assessments of climate impacts, especially during crises, to inform response decisions related to threats, mitigation efforts, and disaster responses (e.g., rapid warming, flooding, acid rain, extreme weather).

4.4.4 Cyber-Physical Process Characterization (11 Gaps)

There is a critical need to enhance the fundamental understanding of the physical properties and behaviors of materials, technologies, and systems within the CPSICC nexus, particularly those that encompass both cyber and physical elements.

1. **Bolster Environmental Research and Data**: Comprehensive studies are needed to enhance data related to ecological impacts in the Arctic, along with earlier development of environmental assessments to improve impact forecasting.

2. **Cyber Infrastructure Sustainability**: Improved data and comparative studies are required to assess the sustainability of cyber infrastructure, including guidance on integrating climate hazards into digital risk management.
3. **Enrich Data**: Existing datasets need to be enriched with a climate focus.
4. **Global Health and Populations**: Understanding the impacts of rapid warming events on global health, including disease, as well as the effects on human and animal populations and their cascading impacts, is essential.
5. **Oceanic/Cryospheric Data**: Enhanced capabilities are needed to observe changes in the ocean and cryosphere.
6. **Rapid Warming and Cyber Infrastructure**: Research is needed to understand the impacts of rapid warming on cyber infrastructure, including cascading effects. This includes developing cooling methods and considering alternatives like distributed infrastructure in rapid-warming scenarios.
7. **Regressive Deployment of SRM**: Research is required to explore how SRM can be deployed and tapered while more sustainable climate interventions are developed.
8. **Science for SRM**: More research is needed to understand SRM, including its impacts on Earth systems and interactions at different timescales, with small-scale experiments to advance knowledge.
9. **Sensors**: Investment in high-resolution climate modeling, remote sensing, and impact analysis is necessary.
10. **Supply Chain Risk Management**: Understanding and preventing supply chain risks for emerging energy technologies is crucial, as current material provenance focuses are inadequate for supply chain risk management.
11. **Understanding the Climate Intervention Spectrum**: A comprehensive understanding of the comparative impacts of different climate interventions is needed to inform decisions regarding controversial interventions like SRM and to shape response options in case of intervention failures.

4.4.5 Social and Economic Considerations (25 Gaps)

There is a significant gap in R&D related to social and economic factors and impacts within the CPSICC nexus. Social infrastructure is poorly defined and challenging to model with current data and methods. More research and data are needed to understand population movements, gather data through sensors, model climate, and analyze different infrastructures and their interdependencies. Additionally, economic impacts and trade-offs have not been adequately addressed.

1. **Behavioral Economics**: Applying principles of behavioral economics is necessary to understand and influence decision-making related to climate and critical infrastructure.
2. **Cost–Benefit Analysis**: Metrics and methods for conducting cost–benefit analyses of climate-resilient infrastructure investments need to be developed to inform decision-making.

3. **Costs of SRM Interventions**: Understanding the costs associated with SRM interventions, considering scale and duration, is essential for evaluating its feasibility as a large-scale climate intervention.
4. **Courses of Action for Crisis Management**: A review of current and alternative crisis management strategies is needed, as existing plans may not be optimal for rapid-warming scenarios.
5. **Crisis Communications**: R&D of effective crisis communication strategies are necessary, including what information to disclose to the public and how to mitigate risks of misinformation.
6. **Definitions of Critical Infrastructure Across Nations**: Understanding the varying definitions and datasets for critical infrastructure among different countries is crucial, as NATO members may have divergent definitions.
7. **Economic Impacts of Climate Interventions**: Models and datasets are needed to help researchers understand the economic impacts of climate interventions, both positive and negative.
8. **Energy Inequalities**: Understanding socioeconomic inequalities in the provision of critical infrastructure services, including energy, is important.
9. **Ethical Data Use**: The ethical implications of using data from cyber-physical systems, including privacy, consent, and data ownership, need to be addressed, along with establishing legal and ethical standards for data use.
10. **Expertise for Governance Structures**: Identifying the necessary expertise for effective governance structures is essential.
11. **Financial Implications**: Assessments of the economic and financial risks posed by climate to critical infrastructure, including impacts on insurance, investments, and economic stability, are needed.
12. **Global Inequity**: Understanding the potential impacts on social systems, including interstate relations, following climate crisis resolutions is important, as some countries may fare better than others.
13. **Governance for SRM**: Research into potential governance structures, laws, treaties, or regimes for employing SRM is necessary, drawing insights from pandemic responses and existing treaties.
14. **Governance Models**: Developing and studying the efficacy of various governance models and policy responses that can mitigate risks at the CPSICC nexus is crucial.
15. **Indigenous Communities Displacement**: Understanding the implications of displacement and ecological degradation on indigenous communities, particularly in the Arctic, is essential.
16. **National Incentives**: Understanding the incentives that may influence a state's support for, opposition to, or attacks on climate interventions is necessary.
17. **National Incentives Regarding SRM**: Investigating the incentives that may affect a state's stance on SRM interventions is important.
18. **NATO Capability for SRM Governance**: Understanding NATO's ability and feasibility to govern SRM, including concerns about member buy-in and trustworthiness, is crucial.

19. **Population Movement**: Improving forecasting and understanding of population movement decisions in response to climate and infrastructure changes is needed.
20. **Public Trust**: Enhancing understanding of public trust and its role in the deployment of climate interventions is essential.
21. **Resilient Communities Initiatives**: Exploring how NATO members can create a resilient communities initiative to optimize alliance resources for preparing populations for climate impacts is important.
22. **Signals for SRM**: Researching and developing signals to support decisions regarding SRM, including indicators of effectiveness and potential disruptions, is necessary.
23. **Social Impacts**: Understanding how public awareness of threats affects social systems is crucial, including risks of unrest, looting, mass migration, hoarding, and loss of government credibility.
24. **Social Infrastructure Definition**: Developing a consensus definition of social infrastructure is needed to promote cooperation and collaboration across communities, nations, and researchers.
25. **Stratification**: Researching how differential behavior among various societal classes may impact climate crises is important, as wealthier individuals may employ unauthorized interventions or isolate themselves.

4.4.6 Risk Assessment and Decision Making (27 Gaps)

A significant challenge in studying the CPSICC nexus is the convergence of complex systems, evolving threats, and challenging risk management problems. There is a pressing need to enhance risk and resiliency analytic methods and metrics to account for the complexities inherent in the CPSICC nexus. Greater interdisciplinary research and approaches are required to define and measure infrastructure, identify new governance strategies, and systematically model complex risks.

1. **Analytical Gaming for CPSICC Nexus**: Additional work on analytical gaming is needed to better understand risks within the CPSICC nexus.
2. **Better Risk Metrics**: New methods for measuring and modeling risk at the CPSICC nexus, particularly for climate risks, need to be identified, as traditional cost–benefit analyses are less effective due to longer time horizons and unpredictable outcome distributions.
3. **Catalog of Potential Bad Actors**: Developing a catalog of potential bad actors and responses to incidents related to supply chains and the shipping sector, especially concerning Arctic shipping lanes, is necessary.
4. **CBRN Threats via SRM**: Research is needed to explore the potential of SRM as a vector for CBRN attacks.
5. **Climate and Mis/Dis/Mal Information**: Understanding the dynamics of climate misinformation and its potential to exacerbate threats is crucial, along

with enumerating potential misinformation scenarios and researching effective mitigation strategies.

6. **Climate Risks to Critical Missions**: Characterizing and understanding the direct and indirect impacts of climate on critical national and international security missions is essential.
7. **Climate-Cyber Impacts**: Frameworks, typologies, and taxonomies are needed to enumerate the impacts of climate on cyber systems.
8. **Data Fidelity**: Determining the appropriate level of data fidelity and precision required for analytical insights is necessary.
9. **Data Gaps**: Methods to identify, prioritize, and fill data gaps that hinder analysis, science, and technical research are needed.
10. **Data Provenance**: Developing methods to verify data provenance is essential to ensure that research and conclusions drawn from datasets can be validated.
11. **Decision Sciences**: Integrating decision sciences into climate and critical infrastructure modeling, including deterrence considerations and game theory, is necessary.
12. **Deterrence and Game Theory**: Investigating and extending existing deterrence and game theory in scenarios involving climate modification and competition or conflict related to CPSS is important.
13. **Economic Impacts**: Understanding the financial and economic impacts of threats and attacks against climate interventions is crucial.
14. **Increase Information Sharing**: Developing mechanisms for information sharing to understand threats or bad actors in the shipping sector is necessary.
15. **Legacy Vulnerabilities to Cyber from Climate**: Identifying and mitigating vulnerabilities in physical and cyber infrastructure that were developed without considering climate is essential.
16. **Non-Linear Emergent Risks**: New methods are needed to measure non-linear, long-term risks within the CPSICC nexus and to understand the changing probability distributions of outcomes resulting from interactions between complex systems.
17. **R&D for Decisions**: R&D efforts should focus more on the needs of policymakers and decision-makers.
18. **Resilience Metrics**: Metrics for measuring the resilience of cyber, physical, and social critical infrastructure against climate and other threats are needed.
19. **Response Capabilities**: Examining the ability of various actors, including "tiger teams," to mount timely responses to crises and emergencies related to the CPSICC nexus is essential.
20. **Risks to Space Assets**: New methods for modeling risks to space assets, including threats from space debris and congestion, are necessary.
21. **Scenario Analysis**: Developing methods for conducting scenario analysis that considers multiple interacting factors across domains is crucial for planning complex, real-world situations.
22. **Strategic Energy Reserve**: Understanding the feasibility, key considerations, and trade-offs of maintaining a strategic energy reserve of electric batteries and other alternative energy sources is important.

23. **Supply Chain Risk Management for the Arctic**: Understanding the various risks and logistical challenges associated with shifting shipping routes to the Arctic is essential.
24. **Supply Chain Risks to Critical Missions**: Characterizing and estimating risks to critical missions, such as national security and defense, resulting from supply chain vulnerabilities is necessary, along with better understanding risks to critical CPSS missions from supply chain disruptions.
25. **Tailored Risk Management Strategies**: Developing tailored risk management strategies based on local needs, such as energy usage, risk tolerance, and demographics, is crucial.
26. **Threat Actor Contingencies**: Understanding how responses may vary depending on the threat actor (e.g., strategic adversaries, allied states, rogue states, non-state actors) is essential.
27. **Unauthorized SRM**: Researching scenarios and response options for situations where an actor unilaterally deploys SRM technology is necessary.

4.5 R&D Recommendations

The pre-workshop activities and the CPSICC workshop sessions resulted in eight initial recommendations for consideration by NATO and member countries (Reinhardt et al., 2025, Chap. 2). The structured and detailed post-workshop analysis yielded revised R&D recommendations aimed at reducing risks and enhancing resiliency across the CPSICC nexus.

1. **Develop Common Definitions, Frameworks, and Metrics**: A fundamental issue identified by participants in both the workshop and preparatory activities was the need for a shared understanding. While there is general agreement on broad concepts, the community lacks coherent definitions of key terms and metrics to compare alternatives across various sectors and consequences. Although this may not be strictly technical R&D, significant collaborative efforts are required to advance these areas, which should be a primary focus for near-term work. Establishing a common foundation will facilitate more effective future R&D.
2. **Develop Systems Models that Describe the Nexus**: While advancements have been made in specific sectors of the CPSICC nexus, participants emphasized the need for integrated systems representations. Systems models provide a comprehensive view of performance and trade-offs within the nexus. Near-term efforts should encourage interdisciplinary collaboration to develop a systems perspective of the CPSICC nexus.
3. **Improve Capabilities to Model and Assess Interdependencies, Cascading Impacts, and Human Consequences**: Participants highlighted the need for enhanced capabilities to conduct risk and resilience assessments that consider complex interdependencies among nexus elements. This approach should help

address cascading impacts that are currently challenging to evaluate. Additionally, there is a gap between technical impact assessments and the actual human consequences, which is essential for informing policymakers and decision-makers in their resource allocation efforts.

4. **Develop Scalable Risk and Resiliency Methodologies**: Existing risk models for individual components of the nexus often focus on detailed representations of complex processes, which may not be sufficient for broader analyses. A "bottom-up" approach that federates complex models is unlikely to succeed due to data limitations and computational complexity. Efforts should be directed toward developing modeling and analytical approaches that can scale with the complexity of the nexus.
5. **Invest in Technologies for Cyber and Data Security**: Data provenance, security, and cybersecurity for both deployed systems and emerging technologies emerged as critical themes during the workshop. Reliable and verified data is essential for informed decision-making, and confidence in the resilience of existing and new systems against cyberattacks is crucial, especially as digital systems like microgrids become more prevalent. The community recognizes these areas as significant vulnerabilities to malicious threats.
6. **Focus on Convening Key Researcher Groups with Practitioners and Policymakers**: Participants expressed a need to better understand the perspectives of practitioners and policymakers. Engaging those who benefit from R&D advancements, particularly in developing foundational knowledge about the CPSICC nexus, is vital for ensuring that R&D is prioritized and scoped effectively for maximum impact.
7. **Enhance and Utilize Analytic Wargaming to Explore and Develop Mutual Understanding**: The workshop employed a serious game to immerse participants in scenarios that revealed the gaps discussed in this report. Further research into analytical wargaming could help both researchers and practitioners understand CPSICC nexus risks. This approach is valuable due to its rigorous, data-driven methodology, flexibility, and applicability in complex problem spaces with limited data. R&D of wargaming methods should be included in near-term efforts.
8. **Include Research in the Social Sciences and Economics**: DHS S&T should invest in research that explores the cultural, social, and economic impacts of climate across diverse stakeholder populations. This research should encompass forecasting population movements, geopolitical responses (e.g., Arctic competition, resource conflicts), governance models, misinformation, and social unrest.

4.6 Conclusion

The CPSICC nexus workshop gathered a diverse group of experts who collaboratively identified key R&D gaps and formulated associated recommendations. The common themes between the pre-workshop and post-workshop recommendations focus on enhancing understanding, collaboration, and resilience in the face of climate impacts on critical infrastructures. Both sets of recommendations emphasize the need for integrated frameworks and models to bridge gaps in the engineering lifecycle and to better understand the complex interdependencies within the CPSICC nexus.

The final priority recommendations are summarized below:

1. **Integration and Collaboration**: Both recommendations stress the importance of integrating existing frameworks and models across disciplines and fostering collaboration among NATO member countries to address cross-country dependencies and vulnerabilities.
2. **Understanding Interdependencies**: There is a shared emphasis on mapping and assessing interdependencies, cascading impacts, and human consequences, highlighting the need for improved capabilities to evaluate risks and resilience in interconnected systems.
3. **Data Security and Cybersecurity**: Both sets of recommendations recognize the critical need for enhancing cybersecurity measures for climate data and models, ensuring reliable data for informed decision-making.
4. **Engagement with Stakeholders**: The importance of involving practitioners and policymakers in the research and development process is highlighted, ensuring that R&D efforts are aligned with real-world needs and challenges.
5. **Investment in Social Sciences**: Both recommendations acknowledge the necessity of incorporating social science and economic research to understand the broader impacts of climate change on communities and infrastructures.

These recommendations represent a comprehensive approach to addressing the challenges posed by climate impacts on critical impacts across cyber, physical, and social infrastructure through interdisciplinary collaboration, enhanced modeling, and stakeholder engagement.

Acknowledgements The authors would like to acknowledge our colleagues at DHS S&T Directorate for funding the post-workshop analysis performed by the Sandia team. Within the Sandia team, we thank Leslie Proctor for her leadership and outstanding project management, Diana Hackenburg for providing excellent communications support, and we appreciate the technical contributions from Thushara Gunda and Mika Armenta. We would like to thank Shannon Walsh and Gavril Torrijos from Zeichner Risk Analytics for their instrumental technical support for the workshop and this chapter.

References

1. NATO (2024) NATO Secretary General welcomes new NATO Headquarters in Brussels
2. Howard RA (1966) Decision analysis: applied decision theory. In: Hertz DB, Melese J (eds) The proceedings of the fourth international conference on operational research. Wiley, New York

Part II
Policy, Social, Economic, and Governance Approaches

Chapter 5
The Cascading Impact of Cyber on Climate

M. A. P. Chamikara, Marthie Grobler, Tooba Aamir, Ejaz Ahmed, Sid Chau, Marcin Niemiec, and Lygia Romanach

Abstract The interplay between cyber technology and climate events presents both challenges and opportunities. Recent climate events highlight how multi-hazard disruptions can impact communities. For example, a community already facing an uncontrolled bushfire can experience catastrophic consequences if it simultaneously suffers from malicious cyber activities targeting critical infrastructure. Conversely, advanced digital solutions, such as AI-driven resource allocation, can enhance emergency responses, limit disaster spread, and save lives. Recent incidents, such as misinformation campaigns during the 2019 Australian wildfires, the 2021 ransomware attack on the Colonial Pipeline, and system overloads on crisis-response platforms during the 2025 Los Angeles wildfires, highlight how cyber vulnerabilities can intensify the impacts of climate-related emergencies. This chapter examines the cascading impact of cyber technology on climate and considers the interconnections between cyber, climate, and the infrastructure that supports our societies. It outlines knowns and unknowns, identifies potential biases, and underscores critical information gaps in the context of climate resilience and security planning. Specifically, it considers cybersecurity challenges that can be used to amplify climate attacks and events, and cyber advancements that can enhance the resilience of our physical and social infrastructures against the impacts of climate. We provide practical insights and recommendations to help policymakers address the interconnected challenges of cyber vulnerabilities and climate-related risks. Consequently, this chapter enhances the understanding of how protecting cyber-physical systems can contribute to a more

M. A. P. Chamikara (✉) · M. Grobler · T. Aamir
CSIRO's Data61, Clayton, VIC, Australia
e-mail: Chamikara.Mahawagaarachchige@data61.csiro.au

M. Grobler
e-mail: Marthie.Grobler@data61.csiro.au

E. Ahmed · S. Chau
CSIRO's Data61, Marsfield, NSW, Australia

M. Niemiec
AGH University of Krakow, Kraków, Poland

L. Romanach
CSIRO's Environment, Dutton Park, QLD, Australia

M. Huber et al. (eds.), *Critical Infrastructure Systems and Extreme Threat Nexus: Preparing for Resilience and Security*, NATO Science for Peace and Security Series C: Environmental Security, https://doi.org/10.1007/978-3-032-13170-6_5

sustainable climate and secure future. The improved capacity for communities to withstand compounded crises, protect critical services, and sustain economic activity underscores the pivotal role that cyber-physical security plays in climate adaptation.

Keywords Cyber risks · Climate threats · Resilience · Interconnectedness · Digital divide · Critical infrastructure · Cascading failures

5.1 Introduction

Cyber technologies are playing an increasingly vital role in forecasting and mitigating the devastating impacts of climate-related hazards. This is evidenced by the World Intellectual Property Organization's 2024 report which shows that approximately 20% of climate-related patents and 60% of climate-related trademarks have a digital component, highlighting the integration of digital technologies in climate solutions [1]. In particular, modern sensors, Internet of Things (IoT) devices, and data analytics applications help capture continuous streams of information on evolving weather conditions, infrastructure integrity, and resource availability [2]. Emerging technologies help to monitor drought, floods, crop and soil health, and advanced historical and real-time data collection and analysis enable climate risk assessments and subsequent crop insurance [1]. These technologies enable advances in climate-related initiatives.

However, these technologies are not foolproof and, in some cases, have introduced unforeseen negative implications [3]. For example, during the 2019/20 bushfires in Australia, a misinformation campaign was spread on social media with the *#arson-emergency* hashtag attributing the widespread devastation to arson,[1] and during the 2025 wildfires in Los Angeles County, an evacuation alert was sent in error.[2] In both instances, the use of technology diverted attention from the emergency response and public safety, escalating any disruptions brought about by the natural hazard. In addition, cybercriminals are increasingly targeting individuals who are either turning to technology for support and help during a crisis or individuals who are trying to support those in need [4]. For example, cybercriminals might set up fake crisis support websites for disaster victims to register for assistance, while the website is actually used to harvest personal information that could be used in future identity theft attempts. Alternatively, cybercriminals might set up fake disaster charities or impersonate authorized government departments, scamming well-intentioned people out of the money they thought they were donating to support disaster victims [4]. In both instances, community social functions would further deteriorate instead of being restored and adapted quickly post-disaster [5].

---

[1] https://theconversation.com/bushfires-bots-and-arson-claims-australia-flung-in-the-global-disinformation-spotlight-129556.

[2] https://techcrunch.com/2025/01/09/every-smartphone-in-la-accidentally-received-a-wildfire-evacuation-alert/.

Beyond these operational challenges, cyber technologies are dependent on reliable data pipelines. This is not a unique application related to environmental factors, as data collection through smart meters and IoT devices has readily been adopted in this space [6], and due care should be taken to ensure that emerging and advanced technologies have adequate security and privacy measures in place. Continuous data collection, reinforced by strong encryption and authentication protocols, enables utilities and government agencies to optimize resource allocation by managing energy distribution during natural hazards [1] (e.g., uninterrupted water distribution during droughts). Conversely, tampered or withheld data undermine future climate defense efforts, resulting in misdirected investments and inadequate preparedness [3]. At the same time, the interdependence among digital, physical, and social infrastructures calls for coordinated action under a unified governance framework. Researchers, technology vendors, policymakers, and local communities have critical roles in developing and disseminating best practices, supporting standardized protocols that protect cyber-physical systems, and ensuring that policy directives keep pace with technological progress [7].

As climate threats intensify, defending cyber infrastructure is no longer a peripheral concern, and it stands at the heart of safeguarding community well-being and economic stability [7]. These risks emerge from the dynamic interactions between climate-related hazards and the exposure and vulnerability of a particular system [8]. Due to the reliance of so many systems on technology, these climate risks can easily be amplified into system risks that can cause significant disruption. The contributions of this chapter include:

- **Exploring the role of cyber technologies in climate mitigation, adaptation, and response**—We demonstrate how digital solutions, ranging from AI-driven modeling to real-time sensor networks, can forecast hazards, streamline emergency procedures, and support longer-term recovery efforts.
- **Investigating the interdependencies between cyber, climate, and the infrastructure that supports our societies**—We highlight how disruptions, whether from climate hazards or malicious cyber activities, can cascade across energy grids, water treatment systems, telecommunications, and other essential services. Such disruptions lead to real-world consequences, including power outages, telecommunication breakdowns, and, consequently, community service failures that undermine economic stability and public trust.
- **Investigating specific cybersecurity challenges that can amplify climate attacks and events**—We analyze how cyberattacks on critical infrastructure can compound the damage of ongoing climate disasters or create new risks. By showcasing targeted scenarios, we highlight the importance of comprehensive cybersecurity strategies in climate-related contexts.
- **Addressing biases, policy gaps, and governance challenges in cyber-climate resilience**—We explore how biases in technology access, policy design, and cybersecurity investments disproportionately affect certain communities and sectors, creating vulnerabilities in climate resilience efforts. We further highlight critical

information gaps that hinder the effective integration of cybersecurity into climate adaptation strategies and propose key policy and regulatory interventions.

- **Evaluating long-term resilience and future outlook for cyber-climate adaptation**—We assess the evolving nature of cyber and climate threats and the long-term resilience of cyber-enhanced infrastructure. We also discuss the role of emerging technologies, such as quantum computing and AI, in addressing future challenges while ensuring sustainability and security.

The intersection of cyber technology and climate presents both opportunities and challenges. By leveraging cyber advancements, we can enhance the resilience of our physical and social infrastructures against the impacts of climate. However, it is crucial to address cybersecurity vulnerabilities to safeguard these advancements. To this end, we investigate the long-term resilience of cyber-enhanced infrastructure in the face of escalating climate events and the potential of emerging technologies, such as quantum computing, in climate mitigation.

This chapter emphasizes the importance of incorporating cybersecurity measures into climate resilience strategies. We provide practical insights and recommendations for policymakers to address the interconnected challenges of cyber vulnerabilities and climate. This chapter enhances the understanding of how protecting cyber-physical systems can contribute to a more sustainable climate and secure future. The intersection of cyber technology and climate presents both opportunities and challenges. By leveraging cyber advancements, we can enhance the resilience of our physical and social infrastructures against the impacts of climate. However, it is crucial to address cybersecurity vulnerabilities to safeguard these advancements. To this end, we investigate the long-term resilience of cyber-enhanced infrastructure in the face of escalating climate events and the potential of emerging technologies, such as quantum computing, in climate mitigation. Table 5.1 summarizes the key interactions and provides relevant real-world examples to illustrate these complex dynamics. To guide the reader through the complex interactions between cyber systems, climate risks, and critical infrastructure, we consider the five core types of cyber-climate-infrastructure nexus interactions. These interaction types form the structural backbone of this chapter:

1. **Cyber-enhanced climate forecasting and mitigation**: Digital technologies such as AI, big data, and IoT enable improved climate risk detection, forecasting, and response.
   (Explored in Sect. 5.2*)*
2. **Climate-driven disruption to cyber-physical infrastructure**: Climate hazards such as floods, heatwaves, and wildfires damage digital infrastructure and disrupt essential services.
   (Explored in Sect. 5.3*)*
3. **Cybersecurity vulnerabilities that amplify climate risks**: Malicious actors exploit climate-related disruptions or target climate technologies to increase impact.
   (Explored in Sect. 5.4*)*

Table 5.1 Summary of cyber-climate-infrastructure nexus types

Nexus type	Description	Real-world examples
Cyber-enhanced climate forecasting and mitigation	Utilization of advanced digital technologies, including AI, IoT, and big data analytics, to enhance real-time monitoring, prediction, and proactive response to climate-related hazards	Google's GraphCast for weather prediction [9]; ESA Copernicus IoT-enabled flood detection and heatwave alerts [10]; AI-based early bushfire detection systems [11]
Climate-driven disruption to cyber-physical infrastructure	Cascading disruptions triggered by climate-induced events impacting both digital infrastructure (communication networks, data centres) and physical critical infrastructure (energy grids, water supply systems)	2021 Texas power grid failure due to winter storms [12]
Cybersecurity vulnerabilities that amplify climate risks	Malicious cyber threats exploiting or exacerbating existing climate-related vulnerabilities, either by directly targeting critical climate-related technologies or during climate events to compound crises	2021 Colonial Pipeline ransomware attack disrupting fuel supplies [13]
Environmental and systemic risks introduced by cyber technologies	Systemic vulnerabilities arising from digital technologies themselves, including environmental impacts such as increased energy and water consumption from AI and IoT systems, as well as inequities arising from digital divides	Significant water and energy consumption by AI data centres exacerbating resource scarcity [14]; digital overloads such as the Watch Duty app becoming inaccessible during the 2025 Los Angeles wildfires [15]
Governance, policy, and cross-sector risk management	Integrated governance, regulatory frameworks, and cross-sector collaboration to effectively mitigate and manage complex interactions and cascading impacts across cyber, climate, and infrastructure sectors	SLACIP Act (2022) mandates for reporting cyber incidents [16] cross-sector cyber-climate simulation exercises [17]

4. **Environmental and systemic risks introduced by cyber technologies**: Energy- and water-intensive digital technologies exacerbate environmental strain and create inequities.
   (Explored in Sect. 5.5*)*
5. **Governance, policy, and cross-sector risk management:** Integrated frameworks are needed to mitigate cascading effects and manage interdependencies.
   (Explored in Sect. 5.6*)*

5.2 Nexus Type 1: Cyber-Enhanced Climate Forecasting and Mitigation

As climate patterns are changing, it is increasingly important for nations to adopt technological advancements that will help keep communities safe and functioning during adverse weather events. Not only are there more severe weather events, but land and water temperatures are rising, leading to an increase in bushfire activity and higher sea levels [18]. Protecting communities' health and safety before, during, and after natural hazard events hinges on ensuring resilience against infrastructure failure, a process that requires a firm grasp of economic costs, recovery timelines, and damage to critical infrastructure, particularly in regions recovering from past disasters or expanding their systems [19]. Cyber technology has been adopted in the climate context to support these functions.

With the increase in global connectivity and digitalization, the impact that cybersecurity has on climate change initiatives has become more pronounced. Specifically, the online infrastructure supporting climate change mitigation and sustainability efforts is vulnerable to cyber threats, as malicious actors seek to exploit system vulnerabilities, disrupt supply and energy production, manipulate climate data, and sabotage global environmental agreements.[3] Therefore, the protection of modern systems and networks is becoming an increasingly serious challenge. Selected emerging technologies can help address cybersecurity threats; however, they can also become a source of risk. By synthesizing diverse data sources and employing advanced analytical methods, these general applications illustrate how technology can transform our understanding and management of climate-related challenges. However, the true impact of these tools becomes even clearer when exploring the tangible, real-world examples of their deployment. From targeted early-warning systems to large-scale policy interventions, these specific case studies demonstrate the practical benefits and potential pitfalls of harnessing technology for climate resilience. We highlight four specific applications.

[3] https://www.captechu.edu/blog/impact-of-cybersecurity-on-climate-change.

5.2.1 Artificial Intelligence (AI)

In recent years, AI has been recognized as a powerful tool for technological progress, offering numerous possibilities for tackling climate-related challenges and improving climate monitoring and prediction [3, 20]. AI is considered a foundational technology with the potential to transform societies and economies worldwide. It has the ability to perform functions associated with the human brain, including perceiving, reasoning, learning, interacting, problem-solving, and exercising creativity, making it an incredible tool to apply in very complex situations, such as the environmental context. Key recent developments include foundation models, which are large-scale systems trained on very large volumes of diverse data that enable the application of knowledge learned in one context to a different context, making them more flexible and efficient than traditional task-specific models. Subfields of AI include computer vision, machine learning, and natural language processing [21].

One of the most noteworthy applications is the rapid processing of vast amounts of historical and real-time weather data. AI-driven models have the ability to detect subtle patterns and correlations that might otherwise go unnoticed, enabling more accurate and timely forecasts. Other general AI applications in the climate environment include:

- *Enhanced weather prediction*: AI algorithms can integrate information from satellites, ground-based sensors, and atmospheric data to forecast the development and trajectory of extreme weather events, such as hurricanes or droughts, with greater precision [22]. An example of this is Google's GraphCast for weather forecasting. This system leverages machine learning to analyze global weather patterns, generating highly accurate short-term forecasts faster and more cost-effectively than many traditional numerical models. This advancement aids in improving preparedness for sudden climate events [9].
- *Disaster risk management*: By analyzing topographical and ecological data in real time, AI models can identify regions at high risk for floods, landslides, or wildfires, facilitating proactive emergency preparedness [23]. An example of this is the application of AI-powered early detection of wildfires. Camera networks equipped with AI-based image recognition algorithms can detect smoke or unusual heat signatures in forested and rural areas. These systems can autonomously issue alerts to local authorities and first responders, thereby helping contain wildfires before they escalate [11].
- *Climate trend analysis*: Machine learning systems can sift through decades' worth of meteorological, hydrological, and geological data to uncover long-term trends, such as shifting rainfall patterns or expanding drought zones, to inform policymakers about future climate scenarios [23]. An example of this is the NASA-IBM geospatial AI foundation model. These organizations jointly released an open-source model trained on Harmonized Landsat and Sentinel-2 data. This foundation model allows researchers to study large-scale Earth observation datasets, extracting insights on land use changes, natural disaster impacts, and agricultural

patterns. Such analyses enrich climate science by offering new ways to monitor environmental shifts, including glacial retreat or deforestation [24].

- *Greenhouse gas tracking:* Satellite-based remote sensing data can be combined with ground-based measurements to monitor CO_2, methane, and other greenhouse gas emissions in near real-time. Policymakers can then identify high-emission sources and implement targeted interventions to reduce harmful outputs. An example of this is the Climate TRACE initiative that uses satellites and other remote sensing technologies to monitor greenhouse gas emissions from various sources worldwide [25].

Despite rapid progress in AI development, this technology is not well understood, and its capabilities are not yet fully appreciated. This can lead to unintended consequences. The specific risk associated with AI is the lack of explainability of how a specific conclusion is derived. Depending on the situation, explanations are not always necessary, but in some cases, explainability and, by extension, repeatability may be critical. Another perceived risk is perceived bias, which is introduced as a result of the datasets used to train the model and leads to skewed results. AI can also generate deepfakes – highly realistic but inauthentic audio and video – that can be used for nefarious purposes. Finally, as trust in AI grows, the risk of overlooking errors, mishaps, and unforeseen incidents also increases [21]. Particularly when users become overly dependent on predictions, this could have a significant economic impact if the predictions fail.

5.2.2 Big Data Analytics

Big data analytics focuses on discovering hidden trends and anomalies within large, complex datasets derived from satellites, sensors, and historical climate records. Although it employs techniques such as machine learning, data mining, and statistical modeling, big data analytics differs from AI in that it focuses on interpreting large datasets to gain insights. This can be used to guide more informed decisions in policy-making, resource management, and disaster response. Big data analytics is regarded as a significant economic disruptor, utilizing vast amounts of data that are being generated. Data from different sources and in different formats are being generated at an unprecedented scale [26]. Some applications of big data analytics in the climate environment include:

- *Climate trend discovery*: By integrating datasets that span decades or centuries, analysts can identify significant shifts in temperature patterns, precipitation levels, or greenhouse gas concentrations. This level of insight enables a clearer understanding of how specific regions are evolving under the pressures of climate change [27]. An example of this is the identification of a warming trend in the

Arctic region, which has led to the melting of sea ice. This trend has been documented through satellite observations, temperature records, and climate models.[4]

- *Real-time monitoring and alerts*: When streaming data from sensors and satellite feeds is combined with powerful analytics platforms, authorities can issue timely warnings for sudden changes, such as a rapid rise in river levels or unexpected temperature spikes, indicating imminent climate hazards. This capability significantly enhances proactive measures and preparedness in vulnerable communities [28]. An example of this is the Real-Time Climate Monitoring and Early Warning System (RT-CMEWS) that provides accurate climate predictions and timely early warnings for extreme weather events, achieving a prediction accuracy of 92.5%, with early warnings issued with an average lead time of 4 to 5 h [29].
- *Resource optimization*: Utilities and local governments utilize insights to tailor water distribution or energy usage more effectively, which is especially critical during droughts or heatwaves. By dynamically allocating resources where they are most needed, decision-makers can mitigate the detrimental effects of extreme weather and ensure that infrastructure continues to function efficiently [30]. An example of this is the use of satellite data and machine learning by the California Air Resources Board (CARB) to monitor and reduce methane emissions. By analyzing large datasets from satellites, CARB can identify methane hotspots and target specific areas for emission reduction efforts.[5]
- *Climate projections and scenario analysis*: By analyzing temperature, precipitation, wind, and sea-level changes under multiple emission scenarios [18], researchers can generate diverse models of possible future climate outcomes. These analyses take into account both technological and socio-economic factors, helping communities prepare for a wide range of potential scenarios. An example of this is the work done by the United Kingdom's Prudential Regulatory Authority. In 2021, they conducted the Climate Biennial Exploratory Scenario exercise to assess the financial risks posed by climate change, using big data scenario analysis to project different climate scenarios and potential impacts on the financial system over a 30-year horizon.[6]
- *Disaster response coordination*: Rapidly aggregating information from social media, satellite imagery, and on-the-ground sensors gives emergency managers a cohesive view of unfolding conditions. This integrated perspective enables emergency services to be directed more effectively, assess infrastructure damage more accurately, and guide relief efforts more efficiently. An example of this is the Pacific Disaster Center's DisasterAWARE platform that integrates data from various sources, including satellite imagery, social media, and sensor networks, to provide real-time situational awareness and decision support during disasters.[7]

---

[4] https://scitechdaily.com/earths-tipping-point-arctic-meltdown-signals-a-climate-catastrophe/.

[5] https://www.weforum.org/stories/2018/10/how-big-data-can-help-us-fight-climate-change-faster/.

[6] https://www.bankofengland.co.uk/stress-testing/2022/results-of-the-2021-climate-biennial-exploratory-scenario.

[7] https://www.disasteraware.com/.

Although big data analytics enable significant analytic capabilities that can improve operational drivers and support economic growth, the sheer volume and complexity of data associated with it introduce unique computational and statistical challenges [26]. Specific concerns include scalability and storage bottlenecks, as existing data warehouses are not geared towards storing this magnitude of data. Other concerns include noise accumulation, spurious correlation, incidental endogeneity, and measurement errors [31], as well as general security and privacy concerns related to data collection.

5.2.3 Internet of Things

IoT refers to the network of physical objects, such as devices, vehicles, and appliances, that are embedded with sensors. These objects connect to the internet and exchange data with other devices and systems. It underpins modern climate monitoring by linking vast networks of sensors and devices that continuously relay critical environmental data (e.g., weather patterns, infrastructure status, and resource availability) to centralized or distributed processing systems. It enables these 'smart' objects to collect and share information, making them more intelligent and responsive. This interconnectedness forms the backbone of many real-time climate prediction and response strategies [32].

By analyzing these real-time feeds, authorities and first responders can more accurately anticipate extreme events, such as hurricanes, floods, or wildfires, and coordinate preventive actions, including targeted evacuations or strategic power shut-offs [33]. These same technologies also improve communication with the public, delivering early warnings and guidance through various digital platforms, thereby enhancing community preparedness [34–36]. Some applications of IoT in the climate environment include:

- *Continuous environmental monitoring*: By measuring temperature, humidity, water levels, soil conditions, and atmospheric composition in real time, IoT-enabled sensors provide granular insights into local climate dynamics. Such data streams enable early anomaly detection, prompting faster emergency responses [32]. For example, the European Space Agency's Copernicus Programme utilizes IoT-enabled satellites and ground sensors to monitor environmental conditions and detect climate anomalies, such as extreme heatwaves and flooding, allowing for quicker disaster response [10].
- *Infrastructure integrity assessments*: IoT devices embedded in roads, bridges, pipelines, and power lines can track wear and tear or damage on a near-real-time basis. Identifying weak points early helps prevent cascading failures during extreme weather events such as hurricanes or floods, ultimately protecting communities and critical services. For example, the I-35W Saint Anthony Falls Bridge in Minneapolis is equipped with over 300 sensors to monitor strain, temperature, and movement. The data from these embedded systems allow engineers to

detect anomalies and schedule preventive maintenance before significant failures occur [37].
- *Smart resource management*: By integrating sensor data with predictive analytics, utility companies optimize electricity grids and water distribution networks, especially during high-stress periods such as heatwaves. Such data-driven decision-making balances demand and resource availability, reducing the risk of outages and shortages [38]. For example, in Australia, smart electricity meters are being rolled out across various states to enhance energy efficiency and grid reliability. The widespread deployment of smart meters allows real-time electricity usage monitoring, helping households and businesses optimize consumption and reduce energy costs [39].
- *Climate modeling enhancement*: Distributed IoT nodes feed localized sensor readings into larger climate models, improving accuracy and specificity for both current assessments and future projections [18]. These more precise models can inform policymakers and emergency planners about the likely impacts of climate change at the regional or even community level. An example of this is flood-prone regions relying on sensors placed along waterways to track changes in water levels and flow rates. Timely data on rising water levels allows authorities to issue evacuation notices and activate floodgates, potentially saving lives and minimizing property damage [40].

There is a big scope for adopting IoT much more widely. For example, the development of drones can help access remote areas that are difficult to cover with traditional cameras, and the deployment of 4G and 5G networks can offer higher bandwidth with lower latency and denser coverage. However, the adoption of such high-tech devices remains low due to concerns over security, privacy, and public safety. Overall, the most significant risk related to IoT is the security and privacy concerns associated with the collection of personal and identifiable data, as well as any commercially sensitive data [41].

5.2.4 Quantum Technologies

Quantum technology (QT) leverages quantum mechanic principles, including quantization, uncertainty principle, interference, entanglement, and decoherence, to produce useful devices and scientific advancements not possible with classical technologies. Two quantum algorithms have a direct and significant impact on cybersecurity—Shor's quantum factorization algorithm with polynomial time complexity involves decomposing an integer into a product of prime numbers, playing a crucial role in modern cryptography [42], and Grover's algorithm for searching unordered databases; its ability to accelerate search processes facilitates cryptanalysis [43]. By leveraging the principles of quantum mechanics, we can encode, process, and transmit information over distance. QT advances in processing and transmitting information have the potential to fundamentally reshape the foundations of data protection. However, such a technological revolution could also have a significant impact on climate change.

Currently, only a limited number of practical implementations directly based on quantum mechanics exist. Nevertheless, among them are solutions that enable an unprecedented level of confidentiality. Some theoretical applications of QT in the climate environment include:

- *Data security*: Quantum cryptography can provide a high level of data security by using cryptographic keys established through quantum states [44]. Currently, key distribution algorithms are vulnerable to network attacks, which can compromise data decryption if a key is compromised. However, QT allows for secure key establishment using quantum bits, protecting against eavesdropping. An example of this is the European Union's OpenQKD project. This project aims to establish a secure quantum communication infrastructure across Europe. One of its applications is to secure data transmission for environmental monitoring and climate research, ensuring that sensitive data related to climate models and environmental measurements are protected from cyber threats.[8]
- *Generating truly random numbers*: Random numbers are used in various fields, and their quality often directly affects the security of sensitive data. An example is the collaboration between the University of Geneva and the Swiss Federal Institute of Technology Lausanne. They implemented a Quantum Random Number Generator to secure data transmission for climate monitoring and environmental research to ensure that collected data is protected from potential cyber threats by using truly random numbers generated through quantum processes.[9]
- *Improved optimization*: Hybrid quantum-classical algorithms can be used to solve optimization problems. Quantum computers have the potential to trigger a transformation in the energy domain by reducing carbon footprints, supporting climate modeling and weather forecasting, and improving grid management [45]. In particular, quantum computing can be used to enable consumers to actively participate in managing electricity demand, with the goal of increasing renewable energy usage [46]. For example, quantum computers could be used to model climate and optimize agricultural processes, such as weather forecasting and crop management, leading to increased crop yields and food security. Quantum computers can also be used to monitor financial markets in real time, enabling traders to make informed investment decisions in a fraction of the time required by classical computers [47].

The practical application of quantum mechanics also has a downside—the risk of breaking currently used data encryption algorithms. Although the first quantum computers have already been constructed, their computational capabilities remain insufficient to pose a threat to currently used asymmetric encryption schemes. Scientists point to significant challenges related to decoherence, which currently prevents the efficient execution of complex quantum computations. The main difficulty

[8] https://cordis.europa.eu/project/id/857156.

[9] https://www.idquantique.com/the-importance-of-randomness-in-a-quantum-world/.

Table 5.2 Successful applications of AI, IoT, and big data in cyber-climate resilience

Technology	Use case	Outcome
AI	GraphCast weather prediction	Enhanced accuracy and speed of forecasts [9]
AI	Early detection of bushfires	Rapid identification and containment [11]
Big data	Climate TRACE greenhouse gas tracking	Real-time emissions monitoring for targeted policy [25]
Big data	CARB methane tracking (California)	Identification and reduction of emissions hotspots [30]
IoT	ESA Copernicus monitoring	Improved disaster response during floods/heatwaves [10]
IoT	Smart meter rollout (Australia)	Optimized energy use, improved grid reliability [39]

lies in performing operations on quantum states, which must be isolated from environmental disturbances. Maintaining quantum systems at extremely low temperatures enables quantum operations and helps preserve quantum states long enough to perform computations. This requires considerable electrical energy consumption. Therefore, the actual performance of quantum computers is questionable [48, 49]. Quantum cryptography also requires additional infrastructure to provide such a high level of protection. Dedicated channels must be established and used solely for key establishment purposes. With the current state of technology, no solutions exist that enable efficient infrastructure sharing or routing comparable to that of the Internet. In addition, if a quantum fast factoring algorithm enables efficient breaking of currently used asymmetric keys, the consequences would be extremely severe.

The operation of quantum computers has a significant carbon footprint and is not currently being leveraged to mitigate the environmental impact of traditional supercomputers. Moreover, we need to be careful of the Jevons' paradox, which means that a more efficient method, if not used with awareness of resources, can actually lead to more usage and cancel out any savings [50].

To consolidate the insights from the cyber-enhanced climate forecasting and mitigation nexus type, Table 5.2 provides a summary of representative case studies demonstrating how AI, big data, and IoT have been successfully applied to support cyber-climate resilience efforts across real-world settings.

Insights and future directions
Lessons learned

- Leveraging AI, big data, and IoT enhances hazard forecasting and resource efficiency, enabling faster, more precise decision-making in dynamic environments.
- Emerging technologies can introduce new risks, including algorithmic bias that can skew outcomes and expanded cybersecurity vulnerabilities that increase exposure to attacks.

Best practice

- Pair every operational AI model with an explainability audit and a human-override mechanism for critical life-safety decisions.

Key recommendations

- Establish explainability and transparency guidelines for AI-driven climate prediction systems to avoid reliance on opaque algorithms that may contain biases.
- Develop standards mandating cybersecurity and privacy assessments for IoT and big data analytics deployed in critical climate monitoring infrastructures.
- Incentivize the use of renewable energy sources and improved cooling technologies to offset the environmental impact of data-intensive technologies such as AI and IoT.

5.3 Nexus Type 2: Climate-Driven Disruption to Cyber-Physical Infrastructure

The interdependencies between cyber systems, climate resilience, and critical infrastructure form a complex network that significantly impacts the robustness and reliability of societal functions. Cyber infrastructure is the backbone of the cyber-physical-social-climate nexus, facilitating essential functions such as real-time communication, automation, data processing, and decision-making. These capabilities are vital for the effective management and resilience of physical and social infrastructures across various sectors, including energy grids, transportation networks, and emergency response systems. Moreover, cyber infrastructure supports advanced predictive analysis and scenario modeling, essential tools for risk mitigation in these critical sectors.

As critical infrastructure sectors increasingly adopt intelligent controls to automate essential services, the interdependencies between these sectors and the domains of the cyber-climate nexus become paramount. It not only supports but also ensures sustained interconnectivity among all infrastructure components, enabling physical systems to function optimally under diverse conditions [7]. In this sense, using technologies can support infrastructure adaptation. For example, it can lead to better flood preparation through water-sensitive urban design, improving the ability of roads to withstand flooding [18]. Moreover, the interconnected nature of these infrastructures means that disruptions can propagate beyond the initial points of impact, affecting broader networks and systems. For example, a cyberattack that disrupts water treatment can also impact industrial processes that depend on these water supplies, leading to economic repercussions and potential job losses. This cascade can extend even further, affecting supply chains and market stability on a regional scale. This section considers the various aspects of cyber, climate, and critical infrastructure interdependencies.

5.3.1 Cyber Resilience as Nexus Driver

Cyberinfrastructure is integral to the operation of modern critical infrastructures across key sectors such as energy, transportation, and emergency services. It facilitates essential functions such as real-time communication, automation, data processing, and decision-making, which are crucial for the efficient functioning and resilience of these systems. This interconnected nature of cyber systems means that vulnerabilities in one area can trigger cascading effects across others, potentially amplifying risks and leading to systemic failures. Considering the permeating application of cyber technologies in the climate context, cyber resilience is a big consideration for a successful cyber-physical-social nexus.

In the context of climate-related events, the National Institute of Standards and Technology (NIST) defines cyber resilience as the capacity to anticipate, withstand, recover from, and adapt to adverse conditions, stresses, attacks, or compromises on systems that utilize or are enabled by cyber resources. This definition emphasizes the need for systems to be not only robust against attacks but also adaptable to the dynamic conditions presented by climate change. Ensuring resilience in the face of infrastructure failure is vital for protecting communities during and after climate events [5]. For instance, during climate-related events such as hurricanes, floods, or wildfires, critical infrastructure systems such as energy grids, water supplies, and communication networks can become highly vulnerable. These systems rely heavily on cyber technologies for their operation and management. The ability to anticipate and prepare for these disruptions, withstand them without catastrophic failure, recover quickly, and adapt strategies in real time is crucial to maintaining functionality and ensuring service continuity during and after such events. Thus, cyber resilience in the context of climate-related events involves:

- *Anticipation*: Utilizing predictive analytics and threat intelligence to forecast and prepare for the impacts of severe weather events on cyber systems.
- *Withstanding*: Ensuring cyber infrastructure is robust enough to maintain critical operations despite the stresses caused by extreme weather, such as power outages or flooding.
- *Recovery*: Implementing rapid response measures to restore services and repair damage to cyber systems as swiftly as possible after a climate event.
- *Adaptation*: Learning from past incidents and evolving the systems to better handle future climatic disruptions, which may involve upgrading technology, refining emergency protocols, or rethinking system architecture to improve resilience.

The interconnectedness of cyber and physical systems can lead to cascading failures where an incident in one domain triggers disruptions across many others. Cyberattacks on infrastructure, such as power grids or water systems, not only halt those services but can also escalate into public health crises and economic shocks. In Australia, for example, electricity, gas, and water utilities generate the most cyber-incident reports [51]. Two recent extreme weather events illustrate how quickly such cascades unfold.

- *2023 ice-storm blackout, Canada.* In April 2023, freezing rain brought down thousands of trees and power lines, cutting electricity to around *1.1 million* Hydro-Québec customers (1.3 million affected overall). Hospitals in Montréal had to rely on diesel back-up generators, highlighting how a single weather event can quickly cascade from the physical infrastructure layer (overhead conductors) into disruption of the cyber layer that supports clinical care and other critical services.[10]
- *2021 winter-storm grid failure, US.* During the February 2021 Winter Storm Uri, frozen gas supply lines and iced wind turbines removed roughly *30 GW* of generation from the ERCOT system; at its peak, more than *4.5 million* households lost power. Rolling blackouts propagated outward, data centre clusters throttled load, and over 14 million Texans faced boil-water notices as electric pumps and telemetry failed, demonstrating how inadequate cyber-resilience planning magnifies climate-driven shocks across energy, communications, and water networks [12].

5.3.2 Cyber-Physical-Social System Interdependencies

Within the cyber-physical-social system, several types of interdependencies exist that can amplify risks and hazards beyond their initial ambit of impact. *Physical interdependencies* between systems refer to the dependency of one system on the output of another system, such as a data centre, which would be dependent on running water to cool its equipment. This also includes physical proximity between systems;

[10] https://news.hydroquebec.com/en/press-releases/1952/april-2023-ice-storm-hydro-quebec-takes-action-to-increase-system-resilience/?utm_source.

for example, a collapse of the water drainage system during a storm event can lead to traffic lane closures and subsequent traffic disruptions. *Virtual interdependencies* between systems might refer to a reliance on information flow. For example, a disruption to the mobile phone network might cause further disruptions because entities are unable to communicate actions and implement policies. *Social interdependencies* between systems include the cultural, political, and economic relationships between different stakeholders in a system. For example, aging transport systems could lead to private property damage during extreme events. These social interdependencies are often attitudinal and budgetary [7].

Climate changes accelerate the decentralised energy grid and increases exposure to cyber threats. We consider a number of real-world incidents to demonstrate the interdependent failures. Although many different combinations of cascading risks exist, the examples illustrate physical, climate, and cyber events and their impact on the cyber-physical-social nexus. These events underscore how climate impacts can exacerbate vulnerabilities in the cyber domain, in addition to the physical domains of critical infrastructure, emphasizing the importance of integrated resilience strategies that address the multifaceted nature of these challenges.

- *2001 train derailment, US*. This incident was most likely due to *physical infrastructure failure*. In 2001, a freight train derailed 11 of its 60 cars while passing through a tunnel in Baltimore. The train experienced an uncommanded emergency brake application (no human intervention), causing several tank cars carrying hazardous materials to rupture and alight. The fire in the tunnel triggered further *physical system* failure by triggering a water mains break due to the thermal expansion in the tunnel, causing flooding and power loss. Additionally, this led to a decline in water levels at nearby water treatment plants, affecting water services in the region. The fire further severed fiber-optic cables, resulting in disrupted internet services. This, in turn, directly impacted the *cyber systems* and connectivity in the region. *Socially*, the incident lasted five days, caused five minor injuries, and cost an estimated US$12 million in damages. The environmental impact of this event, resulting from the explosion of hazardous materials and the discharge of 14 million gallons of water from the ruptured water mains, was also significant. In this case, the cyber and climate elements occurred simultaneously but were not directly affected by each other [52, 53].
- *2011 tsunami, Japan*. This incident was a *climate-related event*. In 2011, a 9.0-magnitude earthquake triggered a 30-metre-high tsunami in Japan that washed up to 5 km inland. The *social impact* of this event was massive, resulting in the deaths of almost 16,000 people and significant environmental devastation. The event also damaged several nuclear power plants, leading to serious risks of radioactive contamination. The total estimated economic loss was about 16.9 trillion JPY (more than 155.7 billion USD). Many people were displaced and dispossessed. The deluge of water caused significant damage to *physical infrastructure*—128,530 houses were destroyed, 870 km of expressways and 939 water drainage system components were damaged, and the capacity of the fossil fuel-fired and geothermal power plants was reduced by 55%. In the aftermath of the

event, the debris generated by the event was a significant physical hurdle. An estimated impact of 1.3 trillion JPY (more than 12 billion USD) loss of lifetime infrastructure facilities (e.g., water supply, gas, electricity, communications, and broadcasting facilities) was recorded.[11] *Cyber systems* were also impacted. A survey of ICT divisions in 13 impacted municipalities revealed that ICT business continuity plans and data backup policies were not in place [54], which caused significant disruption to continuous virtual interdependencies.

- *2021 Colonial pipeline ransomware attack, US*. This incident was a *cyber-related event*. In 2021, the DarkSide group carried out a ransomware attack on Colonial Pipeline, resulting in significant disruptions worldwide as gas supplies became scarce, leading to price increases and shortages. Colonial Pipeline is America's largest refined products pipeline, delivering fuel to customers across 14 states. A data breach followed by a ransomware attack was the main attack vector, causing significant disruption to the company's *cyber systems*. 100 GB of data was stolen, and approximately $5 million was paid in ransom. *Physical systems* were vastly impacted, with 5500 miles of pipelines shut down in the US. Although the pipeline operation was restored within a week, the recovery time took longer than expected to stabilize to pre-attack fuel distribution levels. *Social systems* were affected globally. Gas prices rose an estimated average of 4 cents per gallon. The attack led to fuel shortages across the US, causing panic buying and long lines at gas stations. This cascaded throughout the economy, as transportation costs for goods and services, particularly those reliant on fuel-intensive industries such as logistics, construction, and aviation, also spiked. This had a cascading impact on international markets, leading to increased fuel prices and supply chain issues for industries reliant on petroleum products. The attack has significant international repercussions, primarily due to the interconnected nature of global supply chains and economies. As a result of this attack, the US Transportation Security Administration introduced new cybersecurity requirements for pipeline operators, including mandatory vulnerability assessments and improved incident response controls [13, 55].

Interdependency ties together various critical issues that span multiple domains, including geopolitical instability, technological advancements, and regulatory scrutiny. By fostering resilience, it becomes easier to withstand extreme weather events and reduce operational disruptions, whilst ensuring long-term sustainability and success. Targeted resilience strategies can enable communities to recover more effectively from adverse events.

[11] https://www.unep.org/topics/disasters-and-conflicts/country-presence/japan/great-east-japan-earthquake-and-tsunami.

Insights and future directions
Lessons learned

- Extreme weather can trigger cascading failures – disrupting power lines, hospitals, water systems, and telecoms within hours – highlighting the tight interdependence of critical infrastructure.
- Sectoral silos obscure critical infrastructure interdependencies, making it difficult to anticipate how disruptions in one system, such as energy, can ripple into others, such as healthcare, water, or communications.
- Lack of shared visibility across sectors delays coordinated response, as agencies and operators may not have real-time insight into cascading impacts or shared vulnerabilities.

Best practice

- Conduct cross-sector tabletop exercises simulating concurrent cyber and climate disruptions, and publicly share the lessons learned to strengthen collective preparedness and transparency.

Key recommendations

- Require interdependency mapping and scenario exercises for all critical infrastructure sectors to anticipate cascading failures caused by cyber-climate disruptions.
- Integrate multi-hazard cyber-climate contingency planning into standard risk assessment frameworks for emergency management and infrastructure development.
- Implement mandatory cyber resilience checks as part of infrastructure investment and maintenance cycles, especially after significant climate events.

5.4 Nexus Type 3: Cybersecurity Vulnerabilities that Amplify Climate Risks

Despite its benefits, cyber infrastructure is highly vulnerable to a range of threats, including human errors and malicious attacks. Common cybersecurity risks such as weak security practices, inadequate encryption, and poor access controls expose systems to significant threats, including ransomware, phishing, and Distributed Denial of Service (DDoS) attacks. These vulnerabilities are exacerbated by outdated firewalls, improperly applied patches, and the growing interconnectivity of systems, which increases the risk of cascading failures. Disruptions in communication networks can cripple entire sectors that rely on real-time data, amplifying risks across the cyber-physical-social-climate nexus [56].

Identifying specific cybersecurity challenges and implementing remediation for these will enhance critical infrastructure planning, policymaking, and targeted interventions. Considering the deepened integration of digital tools into power grids, water treatment plants, and transportation networks, a range of existing and new threats needs to be considered. We consider three main categories of cybersecurity challenges. By understanding and addressing these categories, organizations can better protect themselves against the evolving landscape of cyber threats.

5.4.1 Deliberate Cyber Attacks

Cyber attacks remain a significant concern for any digitized system, as adversaries are constantly looking for new ways to attack systems and disrupt services. Cyber attacks, such as hacking, phishing, and ransomware, are deliberate attempts by malicious actors to gain unauthorized access to systems, steal sensitive information, or disrupt operations. These attacks can have severe consequences, including financial loss, reputational damage, and compromised data integrity. Essential systems rely on advanced control mechanisms that are susceptible to cyber intrusions—a single compromised IoT sensor or edge device can distort critical predictive models in moments when swift, accurate decisions are vital. Moreover, malware attacks or DDoS campaigns can cripple emergency responses, underscoring the importance of embedding robust cybersecurity measures within climate-focused applications [57].

The variety of cyber attack vectors is vast. A select set of attack types is detailed here, along with examples of how they would be realized in a climate context.

- *Network attacks*: These attacks exploit the vulnerabilities in the networks and communications within Distributed Energy Resources (DERs)/Consumer Energy Resources (CERs). Key examples include insecure communications, network infiltration through co-networked smart devices, and Denial of Service (DoS) attacks. In terms of connectivity, modern-day digitized devices are expected to have prolonged connectivity with external parties for regulatory and user access purposes. For example, the default network connectivity of most CER devices is typically enabled through users' home networks. Interactions with third parties open opportunities for infiltration, as home networks are notorious for poor cyber hygiene [58].
- *Device attacks*: These attacks exploit the security flaws in the physical hardware and embedded software of DERs/CERs. Key examples include app permission attacks, sensor spoofing, port tampering, hardware trojans, and supply chain attacks by embedding malicious code or backdoors through manufacturers. There have been numerous reported attacks on DERs/CERs. For example, the Dutch Government Inspectorate for Digital Infrastructure (RDI) reported that none of the nine

examined solar inverters met the requirements for cybersecurity, as they can be hacked, remotely disabled, or exploited for DDoS attacks.[12] The US Cyber and Infrastructure Security Agency (CISA) issued two warnings relevant to inverter products from manufacturer Enphase, which contain security bugs and hardcoded credentials that could allow an attacker to gain root access to the affected products.[13] For example, EV charging stations merge payment, IT, and OT functions, creating a broad attack surface. Most public chargers still run *OCPP 1.6J* over WebSockets, without TLS and mutual-certificate checks that an on-path attacker can replay the handshake, inject rogue firmware, or falsify `MeterValues` [59]. SO 15118 introduced X.509-based "Plug & Charge" authentication, but early implementations allowed session hijacking under certain conditions, particularly in trusted environments where TLS was optional. This flaw stemmed from insufficient session resumption protections and was recognized as a security vulnerability. Later updates, such as ISO 15118-20, addressed these issues by requiring stricter enforcement of TLS and improved session validation mechanisms [60]. Physical tampering amplifies risk, with unencrypted flash allowing intruders to clone contract certificates and launch "charge-for-free" abuse [61].

- *Service attacks:* These attacks exploit the weaknesses in the services that manage and interact with critical infrastructure. Key examples include weak credentials (e.g., hard-coded/weak passwords), weak authentication, and cloud/aggregator attacks that target the remote control parties and false data injection. Attackers can overwhelm servers by launching a DDoS attack on technologies used to manage tasks during or before natural disasters, such as early warning networks, preventing critical alerts from being disseminated before a wildfire or other disaster spreads. This could result in the loss of lives and property due to insufficient time for emergency response and evacuation. This threat was realized in 2016 when the Mirai botnet targeted Dyn, a major DNS provider. This attack disrupted internet services across the United States, affecting various critical systems, including disaster management and emergency response services.[14]
- *System compromise and data manipulation*: Attackers compromise IT infrastructure, such as early warning systems, and manipulate data with malicious intent or personal and financial gain. Examples of how this can play out include sending false hurricane alerts or delaying legitimate flood warnings by manipulating communication protocols or injecting false sensor data. This could result in public confusion, mistrust, improper evacuation responses, and inadequate disaster preparedness.[15] In another example, Fire Rescue Victoria in Australia was the victim of a cyber attack in 2022, causing a disruption to its entire IT system, including its computer dispatch system. This necessitated them to revert to non-automated

---

[12] https://nltimes.nl/2023/05/30/many-solar-panel-inverters-netherlands-vulnerable-hackers-sabotage.

[13] https://www.cisa.gov/news-events/ics-advisories/icsa-23-171-01.

[14] https://coverlink.com/case-study/mirai-ddos-attack-on-dyn/.

[15] https://observer.com/2017/02/noaa-fake-global-warming-data-paris-agreement-climate-change/.

systems and to alert firefighters of emergencies by mobile phone or radio[16] during Victoria's peak wildfire season.

- *Advanced Persistent Threats (APT):* APT groups are, in many cases, state-sponsored and emerging as a significant threat to the cybersecurity of critical infrastructure. There have been numerous reports of APT-led cyberattacks that have successfully infiltrated telecommunications networks, as well as governmental and institutional systems in the US and Europe. These attacks have enabled the attackers to control some of the largest botnets, which they have used to orchestrate DDoS attacks [62]. This is particularly concerning in the energy sector, where the abundance of DERs/CERs can facilitate APT attacks on energy infrastructure, resulting in significant and cascading risks across multiple sectors.

5.4.2 Vulnerability Exploits and Unintended Consequences

Cyber infrastructures can no longer be considered independent entities but are embedded within most other infrastructures due to their pervasiveness [7]. Technology advances enable efficient data exchange and system automation, facilitating communication and decision-making for physical systems and directly affecting overall infrastructure performance. Advanced technologies, such as Supervisory Control and Data Acquisition (SCADA) systems and Industrial Control Systems (ICS), are used across critical sectors to monitor, control, and optimise physical systems, reducing human intervention and increasing operational efficiency. Societal changes imposed by urbanization and infrastructure development increase the exposure of assets and people. Unfortunately, the increasing interdependencies between infrastructures and their services facilitate the cascading of disaster impacts across sectoral and geographical boundaries, supply chains, and infrastructure networks [63]. Because of these vulnerabilities, cyber systems are exploitable for launching further attacks on critical infrastructure that can cripple operations, disrupt service delivery and cause long-term damage to infrastructure.

Of particular interest is identifying vulnerabilities early on that might amplify the impact of any natural disaster or extreme weather event, thereby helping to improve the overall resilience of critical infrastructure. This will not only enhance the security and resilience of critical infrastructure and its stakeholders but also enable systems to adapt to market changes, respond to natural hazards or market dynamics, and address long-term challenges [16]. A select set of vulnerabilities is detailed here, along with examples of how they would be realized in a climate context.

- *Configuration issues:* These issues arise when system or application settings are not properly set, leading to security gaps that can be exploited by attackers. An example of this is the 2016 power outage in South Australia. The outage was

[16] https://www.abc.net.au/news/2022-12-16/fire-rescue-victoria-confirms-cyber-attack/101781530.

triggered by a severe storm, but the software configuration of the state's power grid management system failed to properly respond to the event, resulting in a statewide blackout.[17]

- *Software vulnerabilities:* These vulnerabilities refer to flaws or bugs within software that can be exploited by malicious actors to gain unauthorized access or cause damage. An example of this is the extensive incorporation of solar rooftops and household battery systems into energy grids. Notwithstanding the benefits of decentralized energy grids, there is also a downside of embedding potential vulnerabilities and unknown weaknesses into a critical energy infrastructure that may be exploitable by malicious actors. Moreover, the supply chains of DERs/CERs are highly concentrated in a few countries, which amplifies the risks of supply chain attacks by embedding stealthy backdoors into devices to trigger zero-day vulnerabilities.[18]
- *Emergent vulnerabilities:* These vulnerabilities arise from complex interactions within systems that were not anticipated or considered during design and implementation. They can result from the combination of multiple components, processes, or environmental factors that together create a security risk that was not previously identified. A recent realization of this was demonstrated during the January 2025 fires in Los Angeles County, United States. With fires spreading rapidly in the area, residents turned to digital tools such as Watch Duty[19] to track fires and manage emergency communication. However, with increased app use during a crisis, Watch Duty was downloaded 1.4 million times in the 48 h before 10 January 2025, putting incredible strain on the servers and leading to inaccessibility.[20]
- *Human vulnerabilities:* Human error remains one of the most significant cybersecurity challenges. Mistakes made by individuals, such as weak passwords, falling for phishing scams, or misconfiguring security settings, can lead to security breaches and data loss. Addressing human error requires continuous education, training, and the implementation of robust security policies and practices. An example of this is the use of social engineering attacks during natural disasters. CISA advised[21] users to stay vigilant against malicious cyber activity following natural disasters, as attackers exploit these events to target victims and concerned individuals using social engineering. For example, during Hurricane Katrina, cybercriminals launched phishing attacks and fraudulent donation scams, taking advantage of chaos to deceive individuals and organizations.[22]

---

[17] https://www.abc.net.au/news/2019-08-07/sa-blackout-explained-as-regulator-launches-legal-action/11390760.

[18] https://www.certa.ai/resources/concentration-risk.

[19] https://app.watchduty.org/.

[20] https://www.pymnts.com/technology/2025/tracking-fires-cyber-fraud-digital-tools-face-scrutiny-los-angeles-wildfires-crisis-response/.

[21] https://www.cisa.gov/news-events/alerts/2023/08/31/cisa-warns-hurricane-related-scams.

[22] https://www.everbridge.com/blog/double-trouble-when-climate-change-and-cyber-crime-collide/.

Insights and future directions
Lessons learned

- Cyberattacks like ransomware or APTs during natural disasters can have amplified, cross-sector impacts, as overstretched defenders and interdependent systems accelerate the spread and severity of disruptions.

Best practice

- Adopt secure-by-design procurement standards, including mandatory third-party penetration testing and enforced signed firmware updates.

Key recommendations

- Implement strict cybersecurity baselines for DER/CER technologies to prevent vulnerabilities that could threaten grid stability during climate-related disruptions.
- Introduce cyber incident response and recovery plans tailored specifically to climate-related emergencies, and tested via regular simulation exercises.
- Provide targeted cybersecurity training to critical infrastructure operators, emphasizing social engineering tactics that exploit climate emergencies.

5.5 Nexus Type 4: Environmental and Systemic Risks Introduced by Cyber Technologies

As digital transformation accelerates, the intersection of cyber risks and climate threats presents a new paradigm of systemic vulnerabilities. Disruptions from extreme weather events, such as wildfires, floods, and hurricanes, are increasingly impairing cyber-physical systems. Meanwhile, cyberattacks targeting energy grids, emergency response systems, and climate monitoring networks exacerbate environmental crises. The intersection of cybersecurity and climate resilience exposes significant biases and information gaps that must be addressed to fortify infrastructures against both cyber and climate risks effectively. This section highlights the vulnerabilities and biases within the technological space and measures employed in climate resilience efforts, underscoring the importance of a comprehensive perspective on systemic problems.

5.5.1 Systemic Vulnerabilities and Environmental Costs

Integrating advanced technologies such as AI, machine learning, and IoT into climate resilience efforts has transformed how societies monitor, predict, and respond to climate phenomena. These technologies enable real-time data analysis, predictive modelling, and automated response mechanisms that can significantly enhance the effectiveness of disaster preparedness and emergency response strategies. For example, machine learning algorithms can analyse vast amounts of meteorological data to predict weather patterns and potential climate disasters more accurately. This allows for more effective evacuation plans, resource allocation, and risk mitigation strategies tailored to anticipated climate events.

However, deploying these technologies comes with substantial environmental costs that are often overlooked in the rush to implement digital solutions. Data centres, which house the infrastructure necessary for big data analytics and machine learning operations, consume large amounts of electricity and require extensive cooling systems to prevent overheating. These energy requirements contribute significantly to the carbon footprint of digital technologies—ironically, the tools used to combat climate change can exacerbate it. For instance, the International Energy Agency (IEA) estimates that electricity demand from data centres, cryptocurrencies, and AI could reach about 1000 Terawatt Hours (TWh) by 2026, compared to 460 TWh in 2022 [64]. The report also states that electricity consumption from data centres, AI, and the cryptocurrency sector could double by 2026. This surge in energy use contributes significantly to greenhouse gas emissions, primarily when powered by fossil fuels. Similarly, data centres use water during construction and, once operational, to cool electrical components. According to one estimate, it is projected to consume six times more water than Denmark, a country of 6 million people [65]. This massive water usage, primarily for cooling data centers, exacerbates water scarcity issues in many regions.

While digital solutions provide enhanced capabilities for managing and responding to climate threats, their heavy reliance on energy-intensive operations can contribute to environmental degradation. This paradox highlights the need for more sustainable approaches to deploying digital technologies in climate resilience efforts. Solutions such as improving energy efficiency in data centres, utilising renewable energy sources, and developing new, less energy-intensive computing technologies are critical to reducing the environmental impact of digital climate solutions.

An example of such infrastructure strain and environmental trade-offs is present in AI-centric data centres. These data centres are now essential for climate modelling, hazard forecasting, and real-time emergency analytics, yet their environmental footprint still matters. Robust bottom-up studies estimate global data centre electricity use at approximately 205 TWh in 2018 (about 1% of world demand) and indicate that it could remain near this level for the next several years if current efficiency trends continue, even as workloads double again [66]. While conventional air-cooled data centres normally run with Power Usage Effectiveness (PUE) figures in the 2.0–2.5 range, immersion-cooled installations have already demonstrated PUE as low as 1.02, and more than 40% of the reported systems achieve PUE below 1.1 [67]. However, concentrated GPU clusters in large-scale data centers are intensifying electricity and water demands, raising concerns about infrastructure and sustainability in water-stressed regions such as Arizona [14].

In addition, climate-driven heatwaves threaten service continuity by overheated servers, causing them to throttle or shut down. This stalls AI systems used for coordinating evacuation and resource triage, disrupting critical operations, and increasing carbon emissions as infrastructure strains to stay online. Practical defences include energy- and carbon-aware scheduling that shifts intensive workloads to cooler, low-carbon periods, along with enhanced cooling methods such as airflow management and liquid cooling [68]. Consequently, mandatory disclosure of site-level energy and water metrics would align these sustainability levers with broader cyber-climate-resilience goals.

5.5.2 Technological Disparities and the Digital Divide

The digital divide also significantly impacts climate resilience by creating disparities in access to technology across different regions. According to the Federal Emergency Management Agency's 2022 National Preparedness Report, lack of access to technology significantly increases vulnerability to disasters and slows recovery processes [69]. This divide is not just a matter of connectivity but extends to the availability of advanced technologies for monitoring and adapting to climate change.

In regions with scarce technological resources, there is a marked limitation in the ability to forecast, monitor, and respond to climate phenomena effectively. For instance, without access to real-time weather data and predictive analytics, these regions cannot adequately prepare for impending climate events, resulting in greater vulnerability to disasters and slower recovery processes. In one such incident, during the devastating 2022 US floods, communities without access to real-time digital alerts found themselves unprepared, leading to tragic losses that could have been mitigated with timely information.[23] Entire communities faced economic setbacks, highlighting the critical role of digital access in disaster preparedness and response.

Underdeveloped regions often face several challenges due to the lack of technological resources [70]. There is a reduced capacity for early warning systems, which are vital for timely evacuations and preparations ahead of severe weather events. In addition, insufficient technological infrastructure can hinder effective data collection and analysis, which is critical for understanding and responding to climate trends over the long term. These regions may lack the skilled workforce needed to manage and operate advanced technological systems, further impeding their climate resilience capabilities.

Comprehensive strategies are crucial to bridging the digital divide and promoting equitable access to technology. Investments in infrastructural development provide

[23] https://www.technologyreview.com/2023/06/20/1074244/climate-vulnerability-digital-divide/.

not only physical frameworks, such as broadband and data centres, but also foster local capabilities to use these technologies effectively. Parallel efforts in capacity building are essential, focusing on both the operational use of technologies and the analytical skills needed to interpret data for climate resilience [71].

5.5.3 Barriers to Cross-Sector Collaboration

Effective cyber-climate resilience requires robust collaboration across multiple sectors. However, several barriers hinder seamless cooperation. Key obstacles include regulatory fragmentation, misaligned stakeholder priorities, data silos, and inconsistent standards.

Regulatory fragmentation remains a significant challenge, as distinct compliance frameworks for environmental management, infrastructure systems, and digital services often result in conflicting mandates or duplicative efforts. For example, energy utilities operate within governance structures for electricity and telecommunications that are managed separately from environmental regulations, complicating coordinated responses during crises [8].

Misaligned priorities between stakeholders can further hinder collaboration. Private-sector operators typically prioritize service continuity and cost minimization, while government emergency services and environmental agencies may focus on risk reduction and long-term resilience. This divergence can delay coordinated decision-making, particularly under the pressure of compounded cyber-climate threats. Public and private actors often operate with conflicting incentives, leading to coordination breakdowns, reluctance to share costs, and a tendency to prioritize individual goals over collective resilience [72].

Moreover, *data silos and confidentiality concerns* impede timely information sharing. Critical information often remains restricted within organizational boundaries, limiting situational awareness during emergencies and reducing the effectiveness of collective response efforts. This issue is exacerbated by the lack of standardized risk communication frameworks, which hampers cross-sector dialogue. Overcoming these barriers requires establishing integrated governance structures, shared data platforms, and harmonized regulatory standards that foster aligned incentives and transparent collaboration across sectors.

5.5.4 Addressing Gaps by Enhancing Collaborative Efforts

The lack of integrated data between cybersecurity threat intelligence and climate data also creates a critical gap in enhancing system-wide robustness. This disconnection hinders effective predictive modelling and real-time response to the intertwined challenges of cyber threats and climate impacts [70]. A focused approach is necessary to enhance collaborative efforts and integrate data across domains, thereby bridging the information gap. This necessitates the creation of systems for the seamless integration of this data, enabling stakeholders to conduct comprehensive, informed risk assessments that simultaneously consider both cyber threats and climate impacts.

Additionally, enhanced collaborative governance frameworks are essential in fostering an environment where cybersecurity, climate science, and policy-making experts can come together to develop holistic strategies.[24] Such collaboration not only helps in aligning strategies across different domains but also in sharing crucial data and insights that can lead to better preparedness and response mechanisms against the complex challenges at the intersection of cybersecurity and climate change. Moreover, there is a pressing need for the public and private sectors to work together more closely [73]. This partnership could involve co-developing technological solutions, sharing best practices, and co-investing in research to understand and mitigate the combined risks of cyber and climate threats. Encouraging private sector involvement through policies and incentives can enhance these collaborative efforts, leading to more robust and resilient infrastructure capable of withstanding the evolving threats in both cyber and climate arenas. Together, these strategies ensure that gaps in information do not turn into vulnerabilities and that the interconnected systems crucial for our societal functioning are protected against the increasing overlap of cyber and climate risks.

Insights and future directions
Lessons learned

- AI data centers' high energy and cooling water demands can negate climate benefits unless offset by renewable energy use and efficiency improvements.

Best practice

- Create regional cyber-climate data trusts to securely pool threat intelligence, climate projections, and critical infrastructure metadata for improved risk coordination and resilience.

Key recommendations

- Mandate environmental footprint disclosures (energy, water) for data centers and related technologies to inform sustainable cyber-climate investment decisions.
- Launch targeted digital equity initiatives to close the technology gap in climate-vulnerable communities, ensuring equitable access to critical resilience infrastructure.

[24] https://www.fortinet.com/blog/industry-trends/cybersecurity-and-climate-change.

- Foster cross-sector data-sharing platforms that integrate cybersecurity threat intelligence with climate risk analytics for holistic system-wide monitoring.

5.6 Nexus Type 5: Governance and Interdependency Risk Management

The increasing convergence of cyber vulnerabilities and climate risks necessitates a comprehensive governance approach that integrates resilience-building measures across critical infrastructure, policy frameworks, and global security strategies. Climate change is no longer a distant environmental concern; it is reshaping geopolitical stability, operational security, and economic structures worldwide. Rising temperatures, extreme weather events, and shifting environmental patterns are amplifying systemic risks that threaten infrastructure integrity, national security, and socio-economic stability [17]. This section outlines key policy and governance recommendations to fortify cyber-climate resilience, enhance adaptive risk management, and ensure the long-term stability of critical infrastructure.

To effectively navigate resource constraints and maximize impact, we categorize our recommendations into three priority tiers: **High-leverage fixes (Tier 1)**, which offer substantial cross-sectoral impacts and should be mandated at a policy or regulatory level; **Quick wins (Tier 2)**, representing low-cost, rapidly implementable actions that can swiftly enhance resilience; and **Foundational enablers (Tier 3)**, which, while requiring sustained investment and capacity-building, lay the groundwork for long-term cyber-climate resilience. Each recommendation in this section is tagged accordingly to guide policy prioritization and implementation strategy.

5.6.1 Integrated Cyber-Climate Risk Assessments (Tier 1)

Global governance frameworks must transcend traditional sectoral boundaries to acknowledge the complex interplay between climate hazards, cyber threats, and critical infrastructure resilience. Both the Australian *National Climate Resilience and Adaptation Strategy (2021–2025)* and the *Critical Infrastructure Resilience Strategy (2023)* underscore the necessity for proactive policymaking, regulatory advancements, and cross-sector collaboration to enhance resilience in an increasingly uncertain world [74, 75]. Similarly, the *Department of Defense Climate Risk Analysis (DCRA)* highlights climate change as a critical national security concern, necessitating its integration into risk assessments, contingency planning, and military readiness strategies. The *DCRA* emphasizes that climate-driven geopolitical instability, resource scarcity, and supply chain disruptions compound security risks, necessitating an integrated approach to risk assessment [17].

To effectively mitigate cascading failures, policymakers should adopt multi-hazard risk assessment models that integrate cyber and climate risk variables. AI-driven threat intelligence and predictive analytics should be leveraged to simulate cyber-climate threat interactions, enhancing preparedness across infrastructure networks. Additionally, governance frameworks should mandate scenario-based planning that incorporates future climate projections and evolving cyber threats, ensuring dynamic adaptation to emerging risks [76]. Regulatory bodies should require all critical infrastructure operators to conduct comprehensive cyber-climate risk assessments and report their findings to national resilience agencies. This proactive approach will strengthen national security by reducing systemic vulnerabilities and enabling swift response mechanisms in the face of compound crises. Future risk governance must prioritize continuous monitoring, cross-sector collaboration, and data-driven decision-making to sustain resilience in an evolving cyber-climate landscape [16].

5.6.2 Cyber-Resilient Critical Infrastructure Development (Tier 1)

Ensuring infrastructure resilience against cyberattacks and climate-related disruptions requires robust investment in redundancy, decentralization, and adaptive security architectures. The *DCRA* identifies critical vulnerabilities in energy, water, and telecommunications networks, emphasizing the need for resilient infrastructure capable of maintaining functionality under stress conditions.

To enhance resilience, governments and industry leaders should invest in edge computing and distributed systems that enable uninterrupted data processing during cyber incidents or environmental disasters. AI-driven self-healing energy grids should be deployed to autonomously reroute power supplies in response to disruptions, while zero-trust security architectures should be implemented to prevent unauthorized access to critical infrastructure systems. Moreover, the proliferation of IoT-based environmental monitoring systems and AI-driven climate models necessitates stringent cybersecurity standards to prevent cyber intrusions that could compromise climate-sensitive technologies. Policymakers should enforce cybersecurity mandates for climate-related digital infrastructure, ensuring data integrity, resilience, and security against cyber-physical threats [77, 78].

5.6.3 International Collaboration and Policy Harmonization (Tier 1)

Given the transnational nature of cyber and climate threats, international cooperation is critical to fostering resilience across borders. Policies should promote participation in international research collaborations and knowledge-sharing platforms, and the development of global cybersecurity standards aligned with climate objectives to ensure consistent protection across borders. The *DCRA* highlights that geopolitical adversaries may exploit climate-induced instability to gain influence, necessitating coordinated global action to safeguard critical infrastructure and promote regional stability [17].

At a global level, forums such as the G20 and United Nations (UN) Climate Change Conferences can embed cybersecurity into broader climate and economic initiatives [79, 80]. Agreements such as the EU's NIS 2 Directive and the Council of Europe's Budapest Convention on Cybercrime serve as templates for integrating threat intelligence and harmonizing incident response protocols [81, 82]. Aligning renewable energy standards with the Paris Agreement and the UN Sustainable Development Goals offers a path to secure supply chains while reducing environmental impact [83].

Equally vital are cross-border data-sharing mechanisms that enable the real-time exchange of vulnerability and threat information, allowing nations to preempt cyberattacks before they exacerbate climate-related crises. Addressing capacity gaps, particularly in developing regions most vulnerable to climate impacts, requires sustained training, technology transfer, and public-private partnerships that bolster cybersecurity infrastructure and human capital. When these strategies converge within a cohesive, rules-based framework, nations can more effectively mitigate the cascading risks posed by an interconnected world, ensuring that cyber-enabled disruptions do not undermine the drive toward a low-carbon future.

5.6.4 Implementing a Cyber-Climate Resilience Framework (Tier 1)

To transition from conceptual recommendations to practical implementation, policymakers and industry stakeholders should adopt a structured cyber-climate resilience framework encompassing the following actions:

1. **Mandatory cybersecurity audits**: Require routine cybersecurity audits for critical infrastructure, particularly for climate-sensitive systems such as power grids, water management facilities, and emergency response networks. Audits should comply with standards such as ISO/IEC 27001 [84] and the NIST Cybersecurity Framework [85].
2. **Joint cyber-climate scenario exercises**: Regularly conduct cross-sector simulation exercises involving combined climate-related events and cyber incidents.

These exercises, modeled after NATO's crisis management approaches [86], can help stakeholders understand interdependencies, test response protocols, and identify critical vulnerabilities in real-time.

3. **Transparency guidelines**: Develop clear guidelines mandating public and private entities to disclose cybersecurity incidents and climate-related disruptions. These can be modeled after existing disclosure and privacy frameworks such as the General Data Protection Regulation (GDPR) in the EU [87], and Australia's Security of Critical Infrastructure (SOCI) Act [88].
4. **Shared risk dashboards**: Implement shared digital platforms aggregating real-time data on climate risks and cybersecurity threats. Securely pooling data on infrastructure vulnerabilities, climate projections, and cyber threat intelligence can enable coordinated, timely, and risk-informed responses.
5. **Incentivize research and development**: Utilize international platforms such as the NATO Science for Peace and Security (SPS) Programme [86] to foster public-private partnerships addressing cyber-climate resilience. SPS-supported projects can accelerate innovations in resilient infrastructure, cybersecurity, and climate risk analytics.

Implementing these actions will create a robust foundation for addressing interconnected cyber and climate threats, ensuring more resilient, secure, and adaptable infrastructure systems.

5.6.5 Secure Data Sharing and Governance Mechanisms (Tier 2)

Effective governance of cyber-climate risks requires seamless data-sharing frameworks that enable real-time threat intelligence exchange across cybersecurity agencies, climate research institutions, and infrastructure operators. A standardized national data trust framework can be established to facilitate secure information sharing, enabling governments and private entities to collaborate in mitigating emerging threats. The *SLACIP Act (2022)* introduced mandatory reporting mechanisms for cyber incidents affecting critical infrastructure, but its scope should be expanded to include climate-related cyber risks. Integrating cyber-climate risk intelligence into regulatory frameworks will ensure timely threat detection and coordinated response strategies. Additionally, strengthening international cooperation in cyber-resilient climate modelling will promote interoperability and global information-sharing capabilities [16]. Policy frameworks must be flexible and forward-looking to address evolving cyber-climate risks. This involves incorporating scenario-based risk assessments that account for future climate projections and emerging cyber threats, and establishing mechanisms for rapid policy updates in response to new technological developments or threat landscapes.

5.6.6 *Enhanced Cyber-Climate Literacy and Workforce Preparedness (Tier 3)*

A resilient cyber-climate security framework requires a well-trained workforce equipped to navigate the complexities of evolving cyber and environmental threats. The *DCRA* underscores the importance of upskilling defense, intelligence, and infrastructure personnel to manage multi-domain security risks, advocating for interdisciplinary training programs that integrate cybersecurity, climate science, and disaster resilience [17]. Governments should develop specialized cybersecurity curricula focused on protecting renewable energy grids, smart environmental monitoring systems, and climate-sensitive infrastructure. Additionally, introducing climate risk modules into cybersecurity certifications will ensure that professionals understand the interdependencies between digital security and environmental resilience [76].

Public awareness initiatives should also be launched to educate businesses, policymakers, and local communities on the risks of cyber-enabled climate disruptions. Enhancing cyber-climate literacy at all societal levels will foster proactive resilience measures and strengthen national preparedness against systemic threats. Collaboration with Indigenous communities should also be prioritized, incorporating traditional ecological knowledge into holistic climate adaptation strategies [75].

Enhancing cyber-climate literacy across sectors is crucial. This can be achieved by developing specialized training programs for cybersecurity professionals focused on climate technologies and sustainable energy systems, and creating awareness campaigns to educate policymakers and the public about the interconnections between cyber and climate risks. CISA and FEMA are currently funding the upskilling of critical infrastructure operators in cybersecurity[25] and released a cyber incidents planning guidance for emergency managers.[26]

5.6.7 *Long-Term Resilience and Future Outlook (Tier 3)*

Cybersecurity in combination with climate risks has been formally included as an emerging risk in government audit and risk management guidance [89]. As the frequency and intensity of climate disasters increase and overwhelm disaster response capacity and resources, improving our understanding of climate risks will be critical for developing national priorities for climate adaptation action [8].

Interdependencies enable cascading effects to span across the infrastructure system; that is, a disruption in one critical infrastructure sector spreads across the network, disrupting other sectors and their vital societal functions [63]. Interdependencies enable climate risks to be systemic, that is, highly interconnected, transbound-

---

[25] https://www.thefastmode.com/expert-opinion/33496-cryptography-at-the-heart-of-cybersecurity-state-and-local-governments-lead-the-charge.

[26] https://www.fema.gov/press-release/20231107/fema-and-cisa-release-first-ever-cyber-incidents-planning-guidance-emergency.

ary, or cross-sectoral in their scope of consequences, leading to complex dynamics and unknown tipping points with ripple effects on systems outside the system of origin [90]. Understanding a system's boundaries, interconnections, and interdependencies is key to understanding the root causes of systemic risk. This will enable the design and implementation of effective risk measures and interventions that may lead to a more resilient state [8].

By identifying in advance what the impact of the initial weather event would be on the extended system if the impacts were to cascade across the network of interconnected systems, the overall resilience can be strengthened to enable the system to resist, absorb, accommodate, adapt to, transform and recover from the effects of a hazard in a timely and efficient manner [75]. This has the overall benefit of enhancing communication, planning, and targeted interventions. This is especially pertinent for regions rebuilding after a natural disaster and those planning infrastructure expansions [91].

5.7 Conclusion

Cybersecurity threats affect all critical infrastructure sectors, particularly during extreme weather events, resulting in power outages, delays, structural damage, and public panic. This chapter adopts a systems-level perspective that integrates insights from diverse fields, including cybersecurity, engineering, data science, and environmental policy. Rather than treating cyber threats and climate hazards as separate issues, we examine their overlapping and potentially compounding effects. This holistic approach helps reveal hidden vulnerabilities and leverages shared opportunities for strengthening infrastructure resilience. While discussions of climate change adaptation and cybersecurity threats exist independently in the literature, there is a limited focus on the direct interplay between them, particularly on how cyber incidents might exacerbate the impact of climate hazards or vice versa. By uniting these threads and providing real-world examples, we offer an integrated discussion for understanding and mitigating cascading disruptions across critical infrastructures.

In this chapter, we explored the intricate and evolving relationship between cyber technology and climate resilience, emphasizing the cascading impacts that cyber vulnerabilities can have on climate-related disasters. As digital technologies become more embedded in critical infrastructure, the risk of cyber incidents amplifying climate hazards grows significantly. From cyberattacks on emergency response systems to vulnerabilities in decentralized energy grids, the intersection of cyber and climate risks presents complex challenges that demand proactive mitigation strategies. We highlighted the role of emerging technologies, such as AI, quantum computing, and IoT, in enhancing climate resilience, while also acknowledging the potential cybersecurity risks they introduce. Additionally, we identified key biases and information gaps that must be addressed to ensure equitable and effective cybersecurity measures across all sectors and regions.

Moving forward, integrating cybersecurity into climate adaptation policies will be crucial to safeguarding critical infrastructure and ensuring long-term resilience. Governments, industry leaders, and researchers must collaborate to develop cyber-secure, climate-resilient infrastructures that can withstand both environmental and digital disruptions. Standardized cybersecurity frameworks, cross-sector risk assessments, and adaptive governance strategies will be essential in navigating this complex landscape. By prioritizing cybersecurity within climate resilience efforts, we can enhance societal preparedness, minimize cascading disruptions, and create a more secure and sustainable future in the face of escalating climate challenges. This chapter examines a highly relevant topic for the general populace, as it demonstrates how digital systems, which millions rely on daily, can both mitigate and exacerbate the effects of climate disasters, ultimately shaping public safety, economic stability, and social well-being.

Acknowledgements The research supporting this chapter was funded by CSIRO's Critical Infrastructure Protection and Resilience (CIPR) mission, which aims to strengthen resilience and foster informed dialogue around the critical infrastructure that underpins economic, environmental, and societal systems.

References

1. International Telecommunication Union (2020) Digital component in climate-related patents. https://www.itu.int/dms_pub/itu-d/oth/07/31/D073100000A0009PDFE.pdf
2. Ahire DB, Gond DVJ, Ahire NL (2022) IoT based real-time monitoring of meteorological data: a review. In: Proceedings of the 3rd international conference on contents, computing and communication (ICCCC-2022)
3. Stackpole B (2025) Climate change and machine learning—the good, bad, and unknown. https://mitsloan.mit.edu/ideas-made-to-matter/climate-change-and-machine-learning-good-bad-and-unknown
4. Radel FW (2025) California wildfires: scammers may take advantage. https://www.usatoday.com/story/money/2025/01/11/california-wildfires-cyber-scams/77612316007/
5. Liu W, Shan M, Zhang S, Zhao X, Zhai Z (2022) Resilience in infrastructure systems: a comprehensive review. Buildings 12(6):759
6. EpiSensor (2025) The future of environmental monitoring: IoT and big data analytics. https://episensor.com/knowledge-base/the-future-of-environmental-monitoring-iot-and-big-data-analytics/
7. Mohebbi S, Zhang Q, Wells EC, Zhao T, Nguyen H, Li M et al (2020) Cyber-physical-social interdependencies and organizational resilience: a review of water, transportation, and cyber infrastructure systems and processes. Sustain Cities Soc 62:102327
8. Romanach L, Boulaire F, Fleming A, Capon T, Bluhm S, Lin BB (2024) Australia's national climate risk assessment: identifying climate risk interdependencies within the infrastructure and built environment system for effective climate adaptation. Clim Risk Manag 46:100670
9. Lam R, Sanchez-Gonzalez A, Willson M, Wirnsberger P, Fortunato M, Alet F, Ravuri S, Ewalds T, Eaton-Rosen Z, Hu W et al (2023) Learning skillful medium-range global weather forecasting. Science 382(6677):1416–1421
10. Bereta K, Caumont H, Daniels U, Goor E, Koubarakis M, Pantazi DA, Stamoulis G, Ubels S, Venus V, Wahyudi F (2019) The Copernicus app lab project: easy access to Copernicus data. In: EDBT, pp 501–511

11. Giannakidou S, Radoglou-Grammatikis P, Lagkas T, Argyriou V, Goudos S, Markakis EK, Sarigiannidis P (2024) Leveraging the power of internet of things and artificial intelligence in forest fire prevention, detection, and restoration: a comprehensive survey. Internet of Things 26:101171
12. Busby JW, Baker K, Bazilian MD, Gilbert AQ, Grubert E, Rai V, Rhodes JD, Shidore S, Smith CA, Webber ME (2021) Cascading risks: understanding the 2021 winter blackout in Texas. Energy Res Soc Sci 77:102106
13. Mittal M (2024) Colonial pipeline cyberattack drives urgent reforms in cybersecurity and critical infrastructure resilience. Int J Oil Gas Coal Eng 12(5)
14. Lei N, Lu J, Shehabi A, Masanet E (2025) The water use of data center workloads: a review and assessment of key determinants. Resour Conserv Recycl 219:108310
15. PYMNTS (2025) Tracking fires to cyber fraud: Digital tools face scrutiny in LA's crisis response. https://www.pymnts.com/technology/2025/tracking-fires-cyber-fraud-digital-tools-face-scrutiny-los-angeles-wildfires-crisis-response/
16. Cyber and Infrastructure Security Centre (CISC) (2023) Critical infrastructure resilience strategy. Tech Rep. https://www.cisc.gov.au/resources-subsite/Documents/critical-infrastructure-resilience-strategy-2023.pdf
17. U.S. Department of Defense (2021) Department of defense climate risk analysis. Tech. Rep. https://media.defense.gov/2021/Oct/21/2002877353/-1/-1/0/DOD-CLIMATE-RISK-ANALYSIS-FINAL.PDF
18. Infrastructure Victoria (2024) Weathering the storm: adapting Victoria's infrastructure to climate change. Tech. Rep. Infrastructure Victoria
19. Alkhaleel BA (2024) Machine learning applications in the resilience of interdependent critical infrastructure systems—a systematic literature review. Int J CritAl Infrastruct Prot 44:100646
20. Rolnick D, Donti PL, Kaack LH, Kochanski K, Lacoste A, Sankaran K, Ross AS, Milojevic-Dupont N, Jaques N, Waldman-Brown A, Luccioni AS, Maharaj T, Sherwin ED, Mukkavilli SK, Kording KP, Gomes CP, Ng AY, Hassabis D, Platt J, Creutzig F, Chayes J, Bengio Y (2022) Tackling climate change with machine learning. ACM Comput Surv 55
21. Stanford Emerging Technology Review (2025) Artificial intelligence. https://setr.stanford.edu/technology/artificial-intelligence/2025
22. Dewitte S, Cornelis JP, Müller R, Munteanu A (2021) Artificial intelligence revolutionises weather forecast, climate monitoring and decadal prediction. Remote Sens 13(16):3209
23. Albahri A, Khaleel YL, Habeeb MA, Ismael RD, Hameed QA, Deveci M, Homod RZ, Albahri O, Alamoodi A, Alzubaidi L (2024) A systematic review of trustworthy artificial intelligence applications in natural disasters. Comput Electr Eng 118:109409
24. Li W, Lee H, Wang S, Hsu CY, Arundel ST (2023) Assessment of a new GeoAI foundation model for flood inundation mapping. In: Proceedings of the 6th ACM SIGSPATIAL international workshop on AI for geographic knowledge discovery, pp 102–109
25. Climate TRACE (2025) Climate trace. Accessed 17 June 2025. https://climatetrace.org/
26. Cui B, Mei H, Ooi BC (2014) Big data: the driver for innovation in databases. National Sci Rev 1:27–30
27. Faghmous JH, Kumar V (2014) A big data guide to understanding climate change: the case for theory-guided data science. Big data 2(3):155–163
28. Yu M, Yang C, Li Y (2018) Big data in natural disaster management: a review. Geosciences 8(5):165
29. Kolhe PN, Deshmukh RK (2024) Development of a real-time climate monitoring and early warning system: integrating IoT and machine learning for enhanced predictive accuracy. https://www.jetir.org/papers/JETIR2401760.pdf
30. Bassoo V, Ramnarain-Seetohul V, Hurbungs V, Fowdur T, Beeharry Y (2018) Big data analytics for smart cities. In: Internet of things and big data analytics toward next-generation intelligence, pp 359–379
31. Fan J, Han F, Liu H (2014) Challenges of big data analysis. Natl Sci Rev 1:293–314
32. Mrabet M, Sliti M (2024) Climate change mitigation and adaptation through optical-wireless communication networks: applications, challenges, and opportunities. Front Clim 6:1480190

33. Strahan K, Whittaker J, Handmer J (2018) Self-evacuation archetypes in Australian bushfire. Int J Disaster Risk Reduct 27:307–316
34. Haggag M, Siam AS, El-Dakhakhni W, Coulibaly P, Hassini E (2021) A deep learning model for predicting climate-induced disasters. Nat Hazards 107:1009–1034
35. Zanella A, Bui N, Castellani A, Vangelista L, Zorzi M (2014) Internet of things for smart cities. IEEE Internet Things J 1(1):22–32
36. Intergovernmental Panel on Climate Change (IPCC) (2023) Climate change 2021—the physical science basis: working group I contribution to the sixth assessment report of the intergovernmental panel on climate change. Cambridge University Press
37. Brown RJ, McCoy R, Shield CK, Linderman LE, Hedegaard BD et al (2020) Ten-year review of monitoring system on i-35w saint Anthony falls bridge
38. Garland J, Baker K, Livneh B (2024) The climate-energy nexus: a critical review of power grid components, extreme weather, and adaptation measures. Infrastruct Sustain Environ Res
39. Chandrashekeran S (2020) From responsibilization to responsiveness through metrics: smart meter deployment in Australia. Geoforum 116:110–118
40. Zakaria MI, Jabbar WA, Sulaiman N (2023) Development of a smart sensing unit for LoRaWAN-based IoT flood monitoring and warning system in catchment areas. Internet Things Cyber-Phys Syst 3:249–261
41. Abiodun OI, Abiodun EO, Alawida M, Alkhawaldeh RS, Arshad H (2021) A review on the security of the Internet of Things: challenges and solutions. Wirel Pers Commun 119:2603–2637
42. Shor PW (1997) Polynomial-time algorithms for prime factorization and discrete logarithms on a quantum computer. SIAM J Comput 26(5):1484–1509
43. Jaques S, Naehrig M, Roetteler M, Virdia F (2019) Implementing Grover oracles for quantum key search on AES and LowMC. IACR Cryptology ePrint Archive
44. Mehic M, Niemiec M, Rass S, Ma J, Peev M, Aguado A, Martin V, Schauer S, Poppe A, Pacher C, Voznak M (2020) Quantum key distribution: a networking perspective. ACM Comput Surv 53
45. Ricciardi Celsi M, Ricciardi Celsi L (2024) Quantum computing as a game changer on the path towards a net-zero economy: a review of the main challenges in the energy domain. Energies 17(5)
46. Bucher D, Nüßlein J, O'Meara C, Angelov I, Wimmer B, Ghosh K, Cortiana G, Linnhoff-Popien C (2024) Incentivizing demand-side response through discount scheduling using hybrid quantum optimization. IEEE Trans Quantum Eng 5:1–15
47. Mafu M, Senekane M (2023) Quantum technology for development framework as a tool for science diplomacy. Front Res Metr Anal
48. Chen S (2023) Are quantum computers really energy efficient? Nat Comput Sci 3:457–460
49. Jaschke D, Montangero S (2023) Is quantum computing green? An estimate for an energy-efficiency quantum advantage. Quantum Sci Technol 8(2)
50. Gutiérrez-Finol GM, Ullah A, González-Béjar M, Gaita-Ariño A (2025) A call for frugal modelling: two case studies involving molecular spin dynamics. Green Chem
51. Cyber and Infrastructure Security Centre (CISC) (2024) 2023–2024 cyber threat trends for critical infrastructure. Tech Rep. https://www.cyber.gov.au/sites/default/files/2024-11/2023-24-cyber-threat-trends-for-critical-infrastructure.pdf
52. Kane BR, Webber S, Tucker KH, Wallace S, Chang J, McCarthy D, Murphy D, Egel D, Wingfield T (2024) Threats to critical infrastructure: a survey. https://www.rand.org/content/dam/rand/pubs/research_reports/RRA2300/RRA2397-2/RAND_RRA2397-2.pdf
53. National Transportation Safety Board (2004) Railroad accident report: derailment of CSX transportation coal train and fire in Tunnel Baltimore, Maryland July 18, 2001. https://www.ntsb.gov/investigations/AccidentReports/Reports/RAR0401.pdf
54. Sakurai M, Kokuryo J (2012) Municipal government ICT in 3.11 crisis: lessons from the great East Japan earthquake and tsunami crisis
55. Homeland Security Digital Library (2021) 2021 Colonial Pipeline ransomware attack. https://www.hsdl.org/c/timeline/2021-colonial-pipeline-ransomware-attack/

56. Cyber and Infrastructure Security Centre, Department of Home Affairs, Australian Government (2024) Critical infrastructure annual risk review: second edition, report, November 2024. Accessed Feb 07 2025. https://www.cisc.gov.au/resources-subsite/Documents/critical-infrastructure-annual-risk-review-2024.pdf
57. Dobson S, Hutchison D, Mauthe A, Schaeffer-Filho A, Smith P, Sterbenz JP (2019) Self-organization and resilience for networked systems: design principles and open research issues. Proceedings of the IEEE 107(4):819–834
58. Whitty MT, Moustafa N, Grobler M (2024) Cybersecurity when working from home during COVID-19: considering the human factors. J Cybersecur 10:tyae001
59. Priyasta D, Septiawan R et al (2023) Enabling EV roaming through cascading websockets in ocpp 1.6. Journal Européen des Systèmes Automatisé 56(3):437
60. Bao K, Valev H, Wagner M, Schmeck H (2018) A threat analysis of the vehicle-to-grid charging protocol iso 15118. Comput Sci Res Dev 33(1):3–12
61. Johnson J, Anderson B, Wright B, Quiroz J, Berg T, Graves R, Daley J, Phan K, Kunz M, Pratt R et al (2022) Cybersecurity for electric vehicle charging infrastructure. Tech Rep, Sandia National Lab.(SNL-NM), Albuquerque, NM (United States)
62. Jeyashankar A (2025) Comprehensive list of APT threat groups, motives, and attack methods. https://www.socinvestigation.com/comprehensive-list-of-apt-threat-groups-motives-and-attack-methods/
63. Barquet K, Englund M, Inga K, André K, Segnestam L (2024) Conceptualising multiple hazards and cascading effects on critical infrastructures. Disasters 48(1):e12591
64. IEA (2024) Electricity 2024: executive summary. Accessed Feb 13 2025. https://www.iea.org/reports/electricity-2024/executive-summary
65. Li P, Yang J, Islam MA, Ren S (2023) Making AI less" thirsty": uncovering and addressing the secret water footprint of ai models. arXiv preprint arXiv:2304.03271
66. Masanet E, Shehabi A, Lei N, Smith S, Koomey J (2020) Recalibrating global data center energy-use estimates. Science 367(6481):984–986
67. Kong R, Zhang H, Tang M, Zou H, Tian C, Ding T (2024) Enhancing data center cooling efficiency and ability: a comprehensive review of direct liquid cooling technologies. Energy 308:132846
68. Krishnaram PCM (2025) Green AI: optimizing energy efficiency of workloads for sustainable data centers. In: 2025 IEEE green technologies conference (GreenTech). IEEE, pp 148–152
69. National Preparedness Assessment Division (2022) FEMA 2022 national preparedness report. Tech Rep. Accessed Feb 13 2025. https://www.fema.gov/sites/default/files/documents/fema_2022-npr.pdf
70. Manteaw BO, Amoah A-B, Ayittah B, Enu KB (2022) Climate-informed decision-making in data-poor environments: managing climate risk through citizen science networks. Front Clim 4:835768
71. Granit IM (2020) Digital divide: effects on distribution of wealth and resources and climate change. Undergrad J Polit Policy Soc 3(1):7–38
72. Givens AD, Busch NE (2013) Realizing the promise of public-private partnerships in U.S. critical infrastructure protection. Int J Crit Infrastruct Prot 6(1):39–50
73. Bai Y, Lu C, Dong X, Li Y (2024) Role of collaborative governance in unlocking private investment in sustainable projects. Hum Soc Sci Commun 11(1):1–9
74. Department of Agriculture (2021) Water and the environment. Australian Government, National Climate Resilience and Adaptation Strategy, pp 2021–2025. https://www.awe.gov.au/science-research/climate-change/adaptation/strategy
75. Department of Home Affairs (2023) National disaster risk reduction framework 2023. Tech Rep, Department of Home Affairs. https://www.homeaffairs.gov.au/emergency/files/national-disaster-risk-reduction-framework.pdf#:~:text=The%20National%20Disaster%20Risk%20Reduction%20Framework%20is%20a,levels%20of%20government%2C%20business%20and%20the%20community%20sector
76. Australian Government Research Council (2023) Critical infrastructure and climate resilience: adapting to evolving threats. https://www.infrastructure.gov.au/climate-resilience

77. Australian Cyber Security Centre (2023) 2023 Cyber security strategy. https://www.cyber.gov.au/
78. Ezurike BO, Nmadu D, Ajah SA, Uma UU, Nwawelu UN, Ogah OE, Osuji CC, Emeh JC, Okoronkwo CA (2024) Recent advances on the challenges of resilient automation systems of the power grid: critical review. Smart Grids Sustain Energy 9(2):42
79. Smith G (2011) G7 to g8 to g20: evolution in global governance
80. COP CC (2021) United nations climate change conference
81. Singh C (2023) The European approach to cybersecurity in 2023: a review of the changes brought in by the Network and Information Security 2 (NIS2) Directive 2022/2555. Int Co Commer Law Rev 5:251–261
82. Campina A, Rodrigues C (2022) Cybercrime and the council of Europe Budapest convention: prevention, criminalization, and international cooperation. In: The book of full papers-7th international zeugma conference on scientific researches, vol 1, pp 112–123, IKSAD
83. Gomez-Echeverri L (2018) Climate and development: enhancing impact through stronger linkages in the implementation of the Paris agreement and the sustainable development goals (SDGS). Philos Trans R Soc A: Math, Phys Eng Sci 376(2119):20160444
84. ISO: Global standards for trusted goods and services (2022) Iso/IEC 27001:2022 information security, cybersecurity and privacy protection—information security management systems—requirements. https://www.iso.org/standard/27001
85. National Institute of Standards and Technology (NIST) (2018) Framework for improving critical infrastructure cybersecurity. Version 1.1. https://doi.org/10.6028/NIST.CSWP.04162018
86. NATO (2023) NATO science for peace and security programme. https://www.nato.int/cps/en/natohq/78209.htm
87. European Union (2016) General data protection regulation (GDPR). https://gdpr-info.eu
88. Australian Government (2018) Security of critical infrastructure act 2018 (Australia). https://www.legislation.gov.au/Details/C2022C00389
89. New South Wales Treasury (2020) Internal audit and risk management policy for the General Government Sector (TPP20-08). Tech Rep, New South Wales Treasury. https://arp.nsw.gov.au/assets/ars/attachments/tpp20-08_internal-audit-and-risk-management-policy_rev1-2.pdf
90. Helbing D (2013) Globally networked risks and how to respond. Nature 497(7447):51–59
91. Schweikert A, Nield L, Otto E, Deinerty M (2019) Resilience and critical power system infrastructure: lessons learned from natural disasters and future research needs. https://documents1.worldbank.org/curated/en/245911560800329681/pdf/Resilience-and-Critical-Power-System-Infrastructure-Lessons-Learned-from-Natural-Disasters-and-Future-Research-Needs.pdf

Part III
Technical Resiliency and Risk Mitigation Alternatives

Chapter 6
Mapping Critical Infrastructure Interdependencies for Building Resilience to Climate and Other Risks

Mariana Dias Baptista, Lygia Romanach, M. A. P. Chamikara, Geoff Lee, Susannah Tweddle, Sarah Kaur, and Marthie Grobler

Abstract Critical infrastructure (CI) resilience is essential in our interconnected world. With many systems inextricably linked to one another, it is essential to understand the interdependencies between them and how a disruption in one system can cascade to disrupt another. Particularly in a changing climate, physical and transition risks can amplify other risks, such as cyber events. To strengthen CI against climate and other risks, national strategies should ensure that all systems and sectors work together. Recent climate crises have underscored the importance of reliable access to energy and communication as a fundamental necessity, particularly during floods or wildfires, as well as the use of digital technologies to overcome certain barriers, such as during the COVID-19 pandemic. This chapter considers CI interdependencies between systems for building resilience to climate and other risks with a focus on energy, telecommunications and data storage or processing, as these three CI sectors function as enabling infrastructure. Understanding how interconnected risks to these sectors impact other CI sectors can improve predictions and forecasts for better strategies to build resilience. To this end, this chapter presents the findings of a desktop review that maps CI interdependencies within three key sectors through a global lens, with Australia as the primary use case. While mapping CI interdependencies is a challenge worldwide, Australia provides valuable context for analysis due to its unique geographic, economic, and climatic conditions. By examining how sectors interact and influence one another within Australia, we can identify vulnerabilities,

M. Dias Baptista (✉)
CSIRO's Environment, Canberra, Australia
e-mail: mariana.diasbaptista@csiro.au

L. Romanach
CSIRO's Environment, Brisbane, Australia

M. A. P. Chamikara · G. Lee · M. Grobler
CSIRO's Data61, Melbourne, VIC, Australia
e-mail: marthie.grobler@data61.csiro.au

S. Tweddle
CSIRO's Energy, Perth, Australia

S. Kaur
Portable, Melbourne, Australia

M. Huber et al. (eds.), *Critical Infrastructure Systems and Extreme Threat Nexus: Preparing for Resilience and Security*, NATO Science for Peace and Security Series C: Environmental Security, https://doi.org/10.1007/978-3-032-13170-6_6

knowledge gaps, and data limitations that are relevant to national resilience efforts and broader global discussions on climate adaptation and infrastructure interdependence. The insights gained from this case study contribute to a more comprehensive understanding of CI risks and adaptive strategies applicable across different regions. With CI sectors facing similar challenges across the globe, lessons from this chapter would be applicable to other countries and regions.

Keywords Interdependencies · Critical infrastructure · National-level resilience · Resilience strategy · Energy · Communications · Data storage or processing

6.1 Introduction

Worldwide, climate-related hazards such as wildfires, cyclones, droughts, floods, heatwaves, and sea-level rise have been causing extensive damage to critical infrastructure (CI), disrupting access to essential services and vital societal functions [1]. Simultaneously, the increasing interdependencies between physical and digital assets have elevated the risk of cyber-attacks, leading to compounded and more significant impacts on CI functioning [2]. These changes in the global security scenario have encouraged countries to focus on the protection and resilience of CI, adopting an all-hazards approach to responding to current and future risks [3].

While it is not possible to make infrastructure 100% resistant to disruption, it is imperative to improve the resilience of networks, especially the key infrastructures that serve at-risk communities [4]. Recent studies aim to improve CI resilience by examining how individual elements and systems can withstand and recover from disasters. In this context, resilience has been defined as a system's ability to prepare and plan for, absorb, recover from, and then adapt to adverse events [5] or a change in conditions [6]. Resilience can also be regarded as a system's capacity to absorb stress or shock and return to a state of stability [7]. In this sense, CI resilience can be defined as the capacity of CI to withstand disruption, operate effectively in crisis, and deal with and adapt to shocks and stresses, including the flexibility to adapt to present and future conditions.

To make CI more resilient, it is important to focus not just on individual assets, but also on the environment they operate in. Knowing a system's boundaries, interconnections, and interdependencies is crucial for creating effective resilience measures [8] because disruptions in CI can have widespread, cross-sectoral effects beyond the original system [9, 10]. Therefore, it is important to consider the broader network that supports each asset to identify the interdependencies that exist among assets both within their own sector and beyond [2]. Consequently, it is important for CI resilience strategies to include cross-sector collaborations to address the interconnectedness of the systems and sectors [11].

While cross-sector collaboration has been happening for disaster recovery, there are concerns that most CI sectors are still operating in silos when developing adaptation and/or mitigation strategies to improve CI protection and resilience [12]. Worldwide, there is no shortage of examples of how CI sectors rely upon one another. This is particularly evident in the energy, telecommunications, and data storage or processing sectors. For example, power outages caused by climate events such as wildfires tend to have cascading impacts on communication, data storage or processing, and several other CI sectors such as transport, health, and financial services [13]. Similarly, during the COVID-19 pandemic, access to energy, telecommunications, and data storage or processing was crucial to provide online access to essential services, such as health and education. To minimize disruption to all sectors, maintaining resilience through fast asset recovery and service continuity within these three sectors should be a top priority of a CI ecosystem [14].

To contribute to this literature, this chapter presents a desktop review of CI sectors' interdependencies for three core CI sectors: energy, communications, and data storage or processing. By mapping CI interdependencies, this chapter aims to provide CI decision-makers with cross-sector information to help improve the design and implementation of CI response strategies to both climate and cyber risks. Identifying how CI sectors impact one another is critical to strengthening cross-sector collaboration, improving data sharing between CI sectors and addressing current CI resilience gaps.

6.2 Methodology

A team of researchers with expertise in CI resilience, especially within the energy and data sectors, conducted a desktop review of the key known CI dependencies, with Australia as the primary use case. Australia provides valuable context for analysis due to its unique geographic, economic, and climatic conditions. To explore the level of understanding within both the public and private sectors, the review was primarily focused on searching for publicly available information, including both grey and academic literature. It therefore included online reports and working papers from government agencies, industry, and other non-government organizations. Information from news bulletins was used to provide examples of how climate and cyber threats have and are currently impacting the functioning of CI in Australia. Academic literature was not the primary focus of the review, but was used to support current knowledge and initiatives within the government and industry agencies.

For each sector, the review included a summary of interdependencies as well as examples and response strategies, when available. Developing a high-level, preliminary qualitative mapping of dependencies is critical to improving information sharing and decision-making across sectors. It can, for example, be used to model the cascading impacts of CI disruption across sectors, leading to better risk quantification and management of CI sectors nationally. It also provides an important basis for engaging stakeholders across CI sectors to identify any missing dependencies, to explore the criticality of the identified dependencies, and to identify where stronger cross-sector collaboration is needed to develop effective national-level response strategies.

6.3 Case Study: Australian CI Sectors

As part of the Critical Five, Australia shares a definition and understanding of the importance of CI with Canada, New Zealand, the United Kingdom, and the United States. This definition states that *"Critical infrastructure, also referred to as nationally significant infrastructure, can be broadly defined as the systems, assets, facilities and networks that provide essential services and are necessary for the national security, economic security, prosperity, and health and safety of their respective nations"* [3]. Furthermore, Australia's Security of Critical Infrastructure Act 2018 (SOCI Act) defines CI as the physical facilities, supply chains, information technologies, and communication networks, which, if destroyed, degraded, or rendered unavailable for an extended period, would significantly impact the social or economic wellbeing of the nation, or affect Australia's ability to conduct national defense and ensure national security [15].

The way countries separate their CI sectors may vary. For example, the United States outlines sixteen CI sectors [16] while Australia outlines eleven. Table 6.1 describes each of Australia's eleven SOCI Act sectors considered in this chapter's desktop review. In Australia, CI is exposed to a highly variable climate across its vast continental dimensions. Wildfires and tropical cyclones are of high concern for infrastructure and the built environment, with other climate hazards such as coastal and estuarine flooding, coastal erosion, drought, extratropical storms, extreme heat, and flooding also expected to be of high concern by 2090. Australia's first National Climate Risk Assessment has identified 56 significant climate risks affecting various systems under the current climate and projected changes for 2050 and 2090 [17]. On top of climate-related hazards, Australian CI sectors and organizations are increasingly subjected to cyber risks because they provide critical services, hold sensitive data, and are subject to sector interdependencies [18]. As operational technology systems are increasingly interconnected, systems' vulnerability also increases, and consequently, exposure to cyber risks such as malware, cyberespionage, and cyberattacks. It is therefore essential for CI organizations to comprehend and map their networks effectively, implement a robust event logging system, and establish a comprehensive asset registry to enhance their operational resilience and security [18].

Before outlining the interdependencies identified for energy, communications, and data storage or processing from the desktop review, we define CI interdependency.

6.3.1 CI Interdependency

CI dependency is defined as a *"linkage or connection between two infrastructures, by which the state of one infrastructure influences or is reliant upon the state of the other"* [19]. When this relationship is bidirectional, that is, when the state of each infrastructure influences the state of the other [20], such as for continued service provision [21], it is characterized as an interdependency. Understanding an interde-

Table 6.1 Australia's SOCI act sectors

Sector	Description
Communications	Provides essential transmission systems and services to the nation through the telecommunications, broadcast, international submarine cable, and postal sub-sectors
Data storage or processing	Enables end-users to store or back up data and is provided on a commercial basis
Defence industry	Provides essential services to the Australian Defence Force and Department of Defence capabilities
Energy	Sustains society and the economy through the generation, transmission, and distribution of electricity and the production and delivery of natural gas and liquid fuels
Financial services and markets	Provides the infrastructure necessary for the exchange of goods, services, and financial assets
Food and grocery	Sustains society by ensuring people have access to food. It encompasses food for consumption inside and outside of the home
Health care and medical	Includes a multifaceted network of public and private providers, settings, participants, and supporting mechanisms that maintain the health and well-being of the Australian community
Higher education and research	Provides essential services to many Australians' education needs, and vital research to advance a range of fields
Space technology	Provides input, advice, support, and guidance on current, emerging, and future issues and trends relating to the operation, integration, and use of space-based systems, technologies, and information by Australian CI
Transport	Sustains society and the economy through the movement of people and essential goods; Includes aviation and maritime assets
Water and sewage	Oversees the provision of safe drinking water as well as the appropriate collection and treatment of wastewater and stormwater

pendency, therefore, requires analysis of the one-way dependencies that comprise it [22]. Interdependency between actors and infrastructures can occur due to physical, geographical, informational, and/or legislative relationships [12].

Interdependencies can intensify risks, create new threats, increase vulnerabilities, and lead to cascading failures [23]. For example, the occurrence of extreme events, such as heavy rainfall, can lead to inundations and permanent damage or temporarily disrupt the energy systems, such as electricity and/or liquid fuel supply. Interdependencies between the energy sector and the transport sector can then affect supply chains, indirectly impacting the food and health sectors. Disruption to food distribution can have public health consequences, with cascading effects ultimately affecting the entire economy.

Therefore, to map CI interdependencies, this chapter examines both the unidirectional links (dependencies) and bidirectional links (interdependencies) that form the foundation for considering the effects of CI interdependencies on vulnerabilities, resilience, and consequences when conducting risk assessments [23]. Failing to identify interdependence and shared responsibilities can transfer risk from one sector to another, leading to cascading impacts and hindering efforts to improve the resilience of the broader system.

6.3.2 The Energy Sector

The energy sector sustains society and the economy through the generation, transmission, and distribution of electricity and the production and delivery of natural gas and liquid fuels. Asset classes for energy include electricity, the energy market operator, gas, and liquid fuels [24]. Energy customers are comprised of industrial, commercial, and residential customers [4].

This desktop review identified that dependencies for the energy sector were extensively reported and quite distinct depending on the sub-sector affected. Therefore, for this sector, the information collected allowed for the presentation of both intra-sector and inter-sector dependencies. Australia's energy transition is also likely to change the scale and nature of certain interdependencies, as the electricity system becomes more decentralized and increasingly electrified and digitalized. These shifts will alter the system's overall vulnerability profile. For example, while digital technologies enhance efficiency and control, they also open new avenues for cyber threats.

Table 6.2 outlines the intra-sector dependencies identified during the desktop review. The review highlighted interactions where the electricity sector both impacts and is impacted by the liquid fuel sector. It also noted that the gas sector affects both the electricity and liquid fuel sectors. These inter-sector dependencies are visually represented in Fig. 6.1.

Table 6.2 Summary of energy intra-sector dependencies

Dependency	Summary and/or examples
Electricity → Liquid fuels	Disruption to power can disrupt refineries and terminals' operations as well as the distribution and sale of liquid fuels. In January 2024, two regions in Western Australia suffered through days of power outages due to a significant storm event and, as a result, experienced a fuel shortage [25].
Liquid fuels → Electricity	Disruption to the liquid fuels supply impacts the ability to operate diesel generators during an emergency when electricity power from the grid is disrupted. In 2006, the isolated Pilbara community of Cotton Creek received emergency fuel supplies after floods cut off the area. Fire and emergency services arranged a delivery from the Capricorn Roadhouse to ensure the community had enough fuel for generators until the next scheduled delivery [26].
Gas → Electricity	Gas shortages impact the electricity supply as gas is an important baseload fuel to manage renewable energy intermittency. Australia's east coast gas system faces shortages in winter due to higher demand. The market has low redundancy, with heavy reliance on a few key sources. In June 2024, an outage at the Longford gas plant during a cold snap caused a spike in demand. Simultaneously, calm weather reduced wind farm output, increasing gas use for electricity. The Australian Energy Market Operator (AEMO) had to intervene to manage the situation [27].
Gas → Liquid fuels	If gas-fired electricity generators go down, some can be run on diesel to maintain the power supply. However, this will cause a surge in demand for diesel that may have flow-on effects for other sectors.

Note The dependency should be interpreted as from 'sector A' to 'sector B'. For example, Electricity → Liquid fuels should be interpreted as the Liquid fuels sector being dependent on the Electricity sector

Table 6.3 outlines the inter-sector dependencies where other CI sectors impact the energy sector. The review identified eight CI sectors as impacting the electricity sector, including communications, data storage or processing, financial services and markets, food and grocery, higher education and research, space technology, transport, and water. Six CI sectors were identified as impacting the gas sector: communications, data storage or processing, financial services and markets, higher education and research, transport and water, and seven CI sectors as impacting the liquid fuels

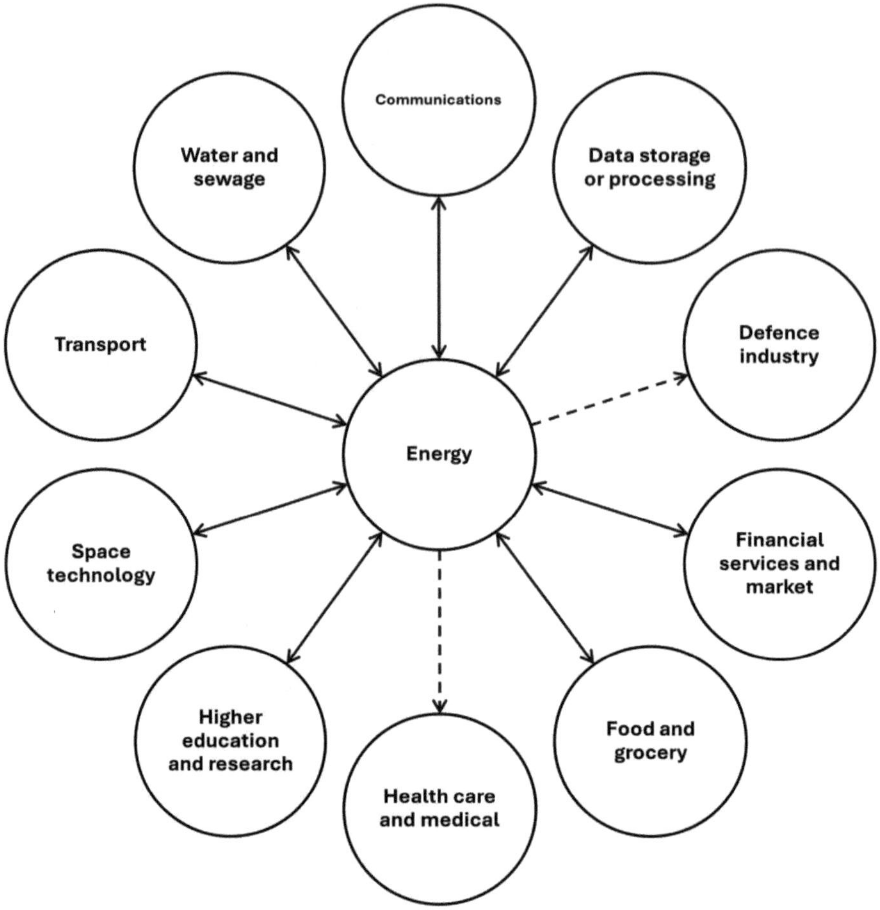

Fig. 6.1 Energy sector dependencies (dashed arrows) and interdependencies (solid arrows)

sector: communications, financial services and markets, food and grocery, higher education and research, space technology, transport, and water.

Table 6.4 outlines the inter-sector dependencies where the energy sector impacts other CI sectors. The review identified that the electricity and liquid fuels sectors impact all eleven CI sectors, and that the gas sector impacts four CI sectors: defense, financial services and markets, healthcare and medical, and water.

6.3.3 The Communication Sector

The communications sector includes telecommunications, broadcasting, international submarine cables, and postal services. It covers assets like broadcasting,

Table 6.3 Summary of energy inter-sector dependencies (impact on sector, e.g., communications → electricity)

Dependency	Summary and/or examples
Communications → Electricity; Gas; Liquid fuels	All three energy system sub-sectors rely on the telecommunication system for operation [28]. For example, **Electricity:** The electricity system relies on the telecommunication system to access data that informs its operation. During an emergency, large volumes of calls, messages, and internet usage can overload the network and stop critical calls from getting through. The extra load on the network can also exhaust backup power supplies [28, 29]. During electricity outages, if one network goes down and temporary mobile roaming is activated, it can overload the surviving network system and make the network unusable for all users due to access and/or core congestion, resulting in worse outcomes for all customers (e.g., slow data speeds; battery drain of consumer devices; inability to contact emergency services). Temporary mobile roaming is also likely to require additional power reserves, potentially accelerating the depletion of power reserves (e.g., battery back-ups). This could lessen the duration of the network(s) that remain operational during a disaster. Distributed energy resources (DERs) such as rooftop solar require communication networks that provide near real-time monitoring, coordination, and control for grid operations and stability [30]. **Liquid fuels:** Lack of mobile coverage may impact the capacity of service stations to sell fuel (due to EFTPOS outages). Climate disasters such as cyclones, floods, and bushfires tend to impact liquid fuels distribution and sales due to disruptions to road, power, and telecommunications [31].
Data storage or processing → Electricity; Gas; Liquid fuels	All energy system sub-sectors rely on the data storage or processing sector to store or process data that informs their operation. For example, the energy sector increasingly deploys smart grids and real-time energy trading platforms that rely on robust data flows and analytics. In addition, disruptions to climate monitoring and forecasting services impact electricity demand and supply forecasts. Weather forecasting is critical to estimate peak energy demands (e.g., due to heat or cold waves) as well as forecast and integrate renewables and mitigate overloads in the National Electricity Market (NEM) [32]. The increasing penetration of variable renewable generation in the grid will place even greater importance on weather forecasting to ensure the reliability of the electricity supply [33].
Financial services and markets → Electricity; Gas; Liquid fuels	Disruptions to financial services and markets impact the capacity of the energy sector to invest in the development and deployment of new technology, as well as to invest in human capability needed for the sector to operate.

(continued)

Table 6.3 (continued)

Dependency	Summary and/or examples
Food and grocery → Electricity; Liquid fuels	Reduction in food crops (e.g., sugarcane, wheat, and sorghum) yields destined for bioenergy and/or biofuel (e.g., ethanol) production can impact the electricity and the liquid fuels sectors. Biomass such as firewood and bagasse (i.e., sugar cane pulp) constitutes about one third of all renewable energy consumption in Australia, while biofuels constitute about 1% [34]. The impact might increase if the use of biomass and biofuels becomes more prevalent.
Higher education and research → Electricity; Gas; Liquid fuels	Disruptions to research and higher education impact the capacity of the energy sector to develop new technology and human capability. The Australian Renewable Energy Agency funding cuts in 2013 and 2016 had an impact on the support of critical research and early-stage development in new renewable energy and storage technologies [35, 36].
Space technology → Electricity; Liquid fuels	Disruption to space technology impacts satellites and grid systems [15] as well as clean energy production forecasts and renewable energy planning. Disruption to space technology impacts transport planning as satellites are key for road mapping and maintenance, which helps reduce fuel consumption.
Transport → Electricity; Gas; Liquid fuels	Extreme weather events can block transport routes, impacting the access of response crews to damaged energy infrastructure that needs repairing. In terms of liquid fuels, the distribution can also be disrupted due to damage to petrol stations and depots, leading to delayed replenishment of supplies and loss of access to fuel depots and service stations. Other threats might include failures due to fires at refineries, trucking fleet issues, contractual disputes, and/or malicious attacks on infrastructure, either physical or cyber. Coastal infrastructure like ports and terminals may face increased risks from storm surges or coastal inundations, affecting the liquid fuels supply chain. Australia's fuel imports are geographically focused, with Australia exposed to potential fuel supply disruption in Asia and the Middle East. The most extreme liquid fuel security threat to Australia occurs if it is unable to receive any imports of crude oil and refined products for an extended period [37]. Australia would be completely cut off from the world by a global conflict or multiple major disasters. If fuel imports stopped, other imports like food and medicines would also stop, and exports would halt. In this situation, Australia would rely on local stocks and oil production, and fuel would need to be rationed [38]. The country has minimum stock levels of transport fuels to ensure domestic reserves [39].

(continued)

Table 6.3 (continued)

Dependency	Summary and/or examples
Water → Electricity; Gas; Liquid fuels	Energy producers and oil refineries depend heavily on the water sector for production, processing, and cooling. Shifts in water systems impact electricity generation, especially from hydro- and coal-fired power stations, which depend on abundant water [40]. Lower rainfall reduces water availability, impacting electricity supply. In Australia, hydropower only accounts for 5–7% of electricity supply [41]. Countries that rely heavily on hydroelectric power, such as New Zealand, will be more impacted by reductions in rainfall. In Australia, there's growing awareness of the water challenges in developing a green hydrogen market, given the country's dry climate. Onshore oil and gas development activities use water primarily in the creation of water-based fluids for drilling and hydraulic fracturing [42]. Competition over water could lead to higher costs or be subject to greater regulatory restrictions [43, 44]. Most of the water for oil and gas extraction in the Pilbara region comes from groundwater, which is replenished by cyclonic rainfall between November and March. Less rainfall will increase competition for water among oil and gas, mining, manufacturing, farming, and urban areas [44]. Refineries may experience additional maintenance and operational, and capital expenses to manage decreased water availability [45].

Note Communications → Electricity can be interpreted as the Electricity sector being dependent on the Communications sector

Domain Name Systems (DNS), and telecommunications [50]. The sector enables the transmission, reception, and delivery of information or data through services such as network, data, voice, mobile, cloud, and wireless services. Key assets include physical infrastructure, data centers, networks, and operational port systems [20]. The telecommunications sector in Australia is predominantly privately owned and operated. However, access to radio-frequency spectrum is regulated and auctioned by the Australian Communications and Media Authority (ACMA), and the National Broadband Network (NBN), a government-owned entity, provides wholesale broadband services and fiber infrastructure. The sector relies on various actors, including infrastructure owners, service providers, equipment manufacturers, and cybersecurity providers. While Australia's mobile network, supported by 4G and 5G towers, is primarily provided by three organizations (Telstra, Optus, and TPG Telecom), consumer network access is provided through more than 150 retail service providers. Mobile coverage reaches 99% of the Australian population and one-quarter of Australia's land area, although rural and remote areas experience challenges with reliability and coverage of internet, mobile, and fixed phone services. Passive infrastructure,

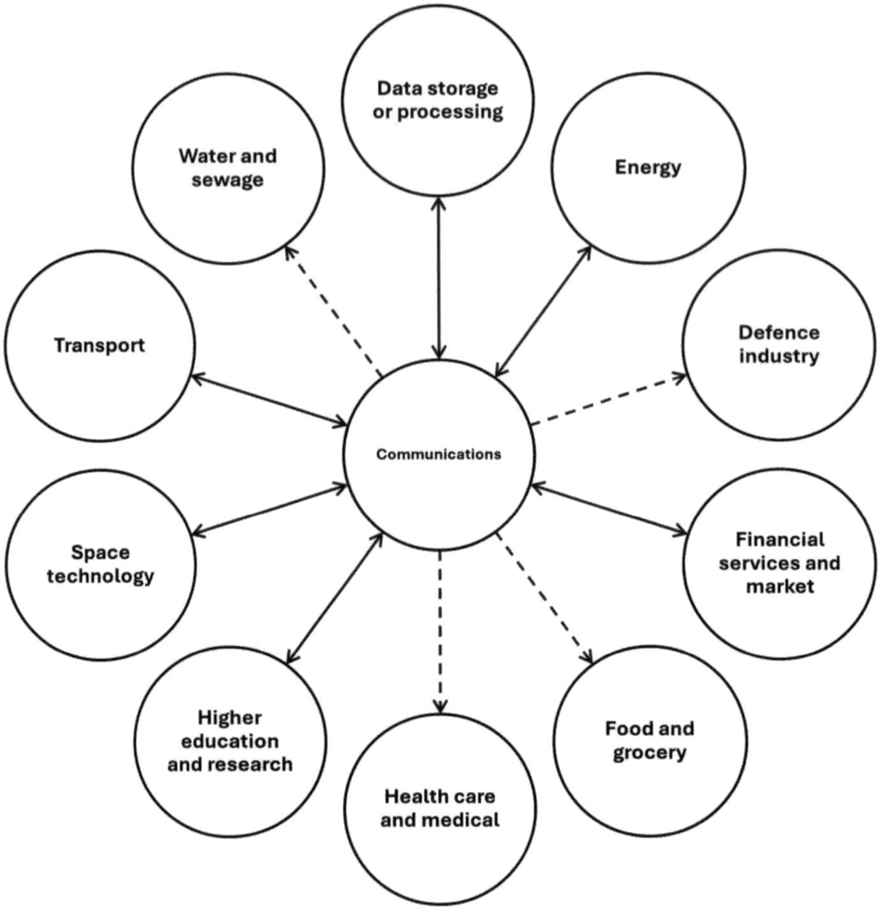

Fig. 6.2 Communications sector dependencies (dashed arrows) and interdependencies (solid arrows)

such as towers, represents a significant cost in investing in additional coverage or increased capacity [4].

Inter-sector dependencies for the communications sector identified in the literature review are illustrated in Fig. 6.2. In terms of inter-sector dependencies where other CI sectors impact the communications sector, Table 6.5 outlines six CI sectors as impacting the communications sector: data storage or processing, energy, financial services and markets, higher education and research, space technology, and transport. In terms of inter-sector dependencies where the communication sector impacts other CI sectors, Table 6.6 outlines the communications sector impacting all eleven CI sectors.

Table 6.4 Summary of energy inter-sector dependencies (impact from sector, e.g., electricity → communications)

Dependency	Summary and/or examples
Electricity → Communications	Electricity disruption impacts the telecommunications sector due to the power dependency of the network infrastructure, including fixed and mobile towers, as well as data centres. Loss of power might impact mobile phone coverage, affecting individuals and business operations, and can collapse the public communication network, which in turn can impact the electricity network [29, 46]. In the 2019–2020 wildfires, most outages (89%) were caused by power failures.
Liquid fuels → Communications	Disruptions in the supply of liquid fuels can affect transportation of supplies, critical staff, and vehicles used for repair and maintenance of communications infrastructure [47].
Electricity → Data storage or processing	Power loss halts data processing and can corrupt data. Data centres require electricity to operate. E.g., the 2016 South Australia blackout disrupted telecommunications and data operations. A 2023 lightning strike in Sydney led to data centre outages and triggered emergency shutdowns due to temperature spikes [48].
Liquid fuels → Data storage or processing	Disruption of liquid fuels can affect logistics and transport of maintenance staff and supplies for data infrastructure.
Electricity; Gas; Liquid fuels → Defence	Cyberattacks and energy disruptions are tools of geopolitical strategy that affect military infrastructure, operations, and logistics. Defence vehicles and transport networks depend on fuel and power [47].
Gas → Financial services and markets	Gas shortages can affect energy-intensive industries (e.g., bricks, cement, paper, steel), disrupting the broader economic system [17].
Electricity; Liquid fuels → Financial services and markets	Power or fuel shortages can hinder business continuity and delivery of services [47].
Electricity → Food and groceries	Power disruptions affect food manufacturing, storage, distribution, and retail.
Liquid fuels → Food and Grocery	Fuel shortages disrupt agricultural production and the entire food supply chain (farm → factory → retail).
Electricity → Health	Outages can impact household health: refrigeration, cooking, lighting, heating, and life-support. Hospitals often have backups, but remote customers may not. There are several cases involving fines for energy providers who cut power to life-support customers without notice [49].

(continued)

Table 6.4 (continued)

Dependency	Summary and/or examples
Gas → Health	Gas shortages can affect basic needs like hot water, cooking, and heating, compromising well-being.
Liquid fuels → Health	Fuel disruption can delay emergency services and transport of medical supplies.
Liquid fuels → Higher education and research	Fuel disruption can affect staff and student travel, delivery of goods, and operation of education facilities.
Electricity → Higher education and research	Outages can disrupt heating/cooling and access to internet needed for education delivery.
Electricity; Liquid fuels → Space technology	Power and fuel shortages can impact the operation of satellites and other space systems.
Electricity → Transport	Disruption affects fuel pumping systems and transport infrastructure (e.g., traffic signals, ticketing). This cascades into disruptions across other sectors.
Liquid fuels → Transport	Fuel shortages impact freight and passenger travel. Aviation fuel shortages at one airport can ripple across the country [47].
Electricity → Water	Power is essential to operate water treatment and distribution systems. Load shedding or blackouts can interrupt water services.
Gas → Water	Gas is used in water treatment processes; disruptions may reduce efficiency or stop operations.
Liquid fuels → Water	Fuel is required for the delivery of water treatment materials and maintenance equipment.
Liquid fuels → Transport	**Road transport:** The supply chains of most commodities are dependent on fuel supply [47]. **Air transport:** Disruptions in supply for jet fuel can have nationwide implications, as most flights are interstate, and shortages at one airport can impact the system as a whole.
Electricity → Water	Disruption to the electricity supply can impact the capacity to maintain water treatment and distribution, impacting access to safe water for consumption and wastewater treatment. Fluctuations in electricity supply (e.g., load shedding) can also disrupt water operations.
Gas → Water	Gas is supplied to operational processes, so gas disruption can lead to decreased efficiency or disruption to water operations.
Liquid fuels → Water	Disruptions in the supply of liquid fuels can impact the domestic transportation of supplies needed for water treatment.

Table 6.5 Summary of communications inter-sector dependencies (impact on sector, e.g., data storage or processing → communications)

Dependency	Summary and/or examples
Data storage or processing → Communications	The communications sector relies on data analysis, storage, and processing to optimize network performance, route traffic efficiently, and maintain service quality. Big data insights help telecommunication providers manage bandwidth usage, detect faults, and predict maintenance needs. Data centers and analytics platforms also assist in lawful intercept requirements and prevent service disruptions. Disruptions to data processing and cloud storage impact telecommunications as they are integral to traditional telecommunication infrastructure [20]. Subsea cables, which handle over 95% of global digital data transfer, are vulnerable to storms, quakes, and sabotage [51].
Energy → Communications	Please refer to Table 6.3
Financial services and markets → Communications	The communications sector depends on labor, consumers, supply chains, and investment [20]. **Labor:** Skill shortages in software, programming, and cybersecurity can affect infrastructure maintenance. Sovereign capability and offshore security risks are driving demand for local expertise. **Business and consumers:** Many businesses and individuals rely exclusively on digital payments, which depend on stable telecommunications networks. **Supply chain:** Reliance on single-source global technology suppliers creates vulnerability. Semiconductor shortages during COVID-19 highlighted risks. **Investment:** Foreign direct investment from the USA, UK, and Singapore shapes network expansion. Geopolitics or policy shifts may reduce funding. The 2021 Regional Telecommunications Review highlighted poor rural access and the need for mobile and broadband investment.
Higher education → Communications	Disruptions to research and education delay innovation and reduce the availability of skilled graduates in the communications sector. COVID-19 caused project delays and graduate shortages across technical fields.

(continued)

Table 6.5 (continued)

Dependency	Summary and/or examples
Space technology → Communications	Satellite infrastructure supports communications and disaster coordination [52]. Remote and rural regions depend on satellites [53]. Space technology disruptions also affect wireless internet [15].
Transport → Communications	**Road transport:** Blocked roads hinder repair crews' access to damaged infrastructure and prevent delivery of generators and batteries for temporary telecommunication power. **Supply chains:** Australia's communication sector depends on imported technology. Global supply chain disruptions, especially during COVID-19, led to higher costs and limited equipment access [20, 54].

6.3.4 *The Data Storage or Processing Sector*

The data storage or processing sector enables end-users to store or back up data and is provided on a commercial basis. The data can relate to individuals, businesses, or governments. The sector is fundamental to Australia's digital ecosystem, which underpins the nation's economy, public services, and CI. It encompasses a range of facilities and services, including large-scale data centers, cloud computing platforms, managed hosting, and distributed 'edge' infrastructure that brings processing capabilities closer to end-users [4]. These components support the secure storage, management, and analysis of data essential for key functions such as financial transactions, healthcare operations, emergency management coordination, utility provision, logistics optimization, and other vital services. The sector also ensures data security to manage risks relating to loss of data [62]. Predominantly privately owned and operated, the sector functions within a regulatory framework established by national legislation and standards (e.g., Hosting Certification Framework) to protect data [63]. Major global cloud providers, such as Amazon Web Services, Microsoft Azure, and Google Cloud, have set up data centers in or near metropolitan areas to meet data sovereignty requirements and minimize latency for local users.

The desktop review identified critical inter-sector interdependencies between the data storage or processing sector and other sectors in the context of Australia's CI. As depicted in Fig. 6.3 and Tables 6.7 and 6.8, the interdependencies between data storage or processing and other sectors are significant, as they have the potential to either mitigate or exacerbate disruptions within the sectors. Hence, these connections play a pivotal role in ensuring sector resilience and service continuity. The review highlights that the data storage or processing sector relies heavily on all other CI sectors. Key dependencies include energy (e.g., electricity for powering infrastructure), communications (e.g., for data transmission), and financial services (e.g., for opera-

Table 6.6 Summary of communications inter-sector dependencies (impact from sector, e.g., communications → data storage or processing)

Dependency	Summary and/or examples
Communications → Data storage or processing	The communications sector is the backbone of data transmission across Australia, facilitating internet services, emergency communications, and daily connectivity. Disruptions can impede operations across other critical sectors. Australia has implemented measures to safeguard this infrastructure through the SOCI Act 2018 [54]. Broadcasting services' reliance on IP networks and IT increases vulnerability to cyber-attacks. In March 2021, the Nine Network cyber-attack disrupted operations for over 24 hours and cost over AUD$1 million.
Communications → Energy	Please refer to Table 6.2
Communications → Defence	Disruption to secure lines of communication can impact national security. For example, resilient satellite communications capability is critical to support Defence needs.
Communications → Financial services and markets	Significant mobile outages can impact contact center operations at major banks, prevent customer service access, disrupt SMS-based two-factor authentication, and affect EFTPOS terminals [55].
Communications → Food and grocery	Farmers rely on ICT to improve yield, cut costs, and enhance market access. Disruptions to communication services can impair agricultural production, logistics, and food supply chains. Events like the 2019–20 wildfires and floods demonstrated this vulnerability [56–58]
Communications → Health	Emergency services depend on telecommunication access to restore service and coordinate response. Gaps between public safety networks and mobile operators reduce response effectiveness. The 2023 Optus outage hindered access to patient data and medical communications. Telehealth services also require stable internet [4, 14, 59].
Communications → Higher education and research	Disruptions to telecommunications impact education services, particularly online education in regional areas [14].

(continued)

Table 6.6 (continued)

Dependency	Summary and/or examples
Communications → Space technology	Ground stations rely on telecommunication networks for satellite control and data transmission. Ground station or terrestrial network failure can lead to spacecraft failure. Satellite constellations like Starlink also rely on terrestrial internet backhaul and encryption keys supplied via telecommunication networks.
Communications → Transport	Traffic management systems require telecommunications to operate. The 2023 Optus outage disrupted transit services, delayed logistics, and stranded commuters [60].
Communications → Water	Telecommunications are vital for the highly digitalized water sector. They enable automation, remote monitoring, and analysis of water quality, flow, and infrastructure status. Disruptions can hinder both operations and safety [61].

tional transactions) (see Table 6.7). At the same time, the data storage or processing sector itself is integral to the functioning of all other CI sectors, providing essential support to communications, health, financial services and markets, and energy (see Table 6.8). Consequently, these interdependencies highlight the importance of robust cybersecurity measures and risk management strategies to protect CI across interconnected sectors.

Mapping interdependencies is subject to the understanding of what constitutes a dependency [12]. The dependency of other CI sectors on the data sector is quite clear. Pervasive digitalization has made electricity grids, water utilities, transport networks, health systems, and supply chains reliant on cloud-based enterprise-resource planning, records management, remote operation, and backup environments. However, the extent to which the data sector depends on other CI sectors tends to be more nuanced. Considering the CI dependency definition, *"linkage or connection between two infrastructures, by which the state of one infrastructure influences or is reliant upon the state of the other"* [19], dependencies can be defined quite broadly. That is, a dependency may exist when one sector influences another, for example, by driving innovation and growth in another sector. With that definition in mind, severe disruption in the health or food sectors, for instance, would deprive data-service providers of mission-critical streams, such as epidemiological data, food-supply telemetry, and real-time emergency-management inputs, thereby compromising national service continuity. Consequently, both policy design and technical hardening of the data sector must account for the dynamic requirements, privacy constraints, and security obligations that originate in other sectors such as health and food to ensure resilient, end-to-end protection of the data sector.

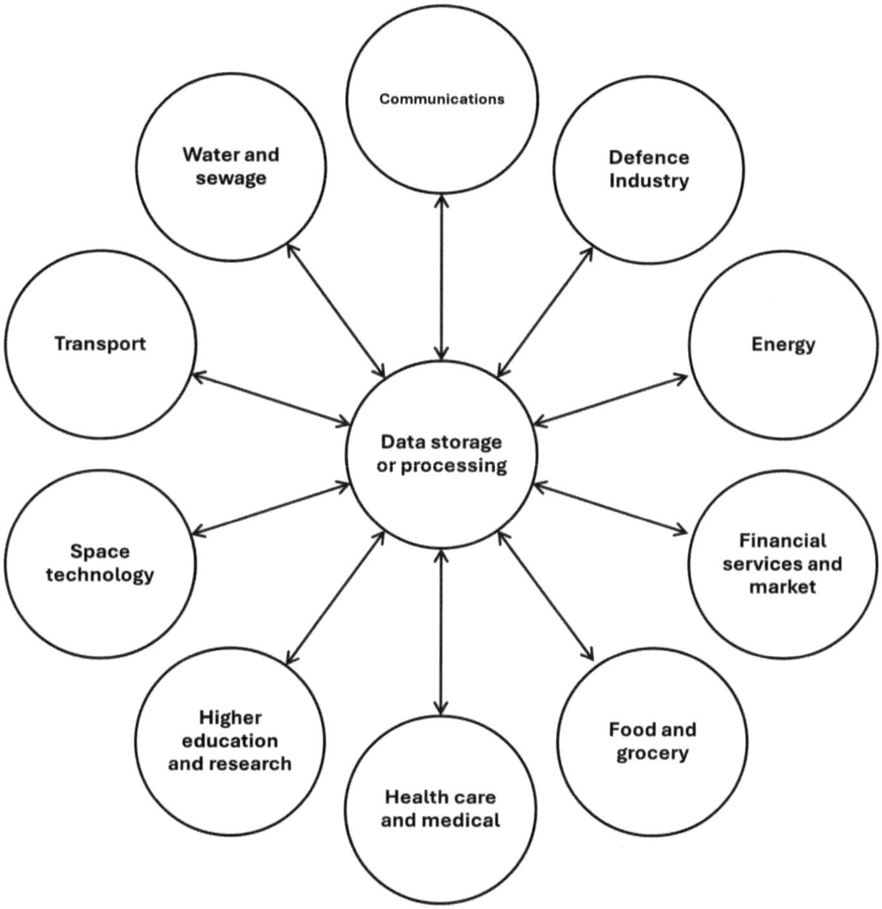

Fig. 6.3 Data storage or processing sector interdependencies (solid arrows)

6.4 Discussion

As identified through the desktop review and summarized in Table 6.9, the energy, communications, and data storage or processing sectors impact one another bidirectionally. As the backbone of the digital economy, these three sectors are highly dependent on each other, and support the operation of all eight other CI sectors included in the desktop review. It is therefore not surprising that a recent review of policy documents found an increased awareness of the importance of protecting critical systems, with a focus on the energy, communication, and data storage and processing sectors in over 100 countries worldwide [78].

Table 6.7 Summary of data storage or processing inter-sector dependencies (impact on sector, e.g., communications → data storage or processing)

Dependency	Summary and/or examples
Communications → Data storage or processing	Please refer to Table 6.5
Defence → Data storage or processing	The Australian Defence Force and Department of Defence maintain high-security networks that indirectly benefit the data sector through collaboration and resilience standards. Cyber intelligence from the Australian Signals Directorate identifies threats to data centers. Strict defence procurement standards also improve data-centre hardware and software resilience [64].
Energy → Data storage or processing	Please refer to Table 6.4
Financial services and markets → Data storage or processing	Financial sector disruption limits data-sector investment and capability. Banks require low-latency, secure systems, pushing innovation. Regulation (e.g., APRA Standard 234) mandates information-security controls, often driving private cloud upgrades [65].
Food and grocery → Data storage or processing	Online grocery growth increases demand for scalable data-processing and cloud solutions, prompting infrastructure innovation [66].
Health → Data storage or processing	The data sector must comply with strict privacy rules under the My Health Records Act 2012 and Privacy Act 1988. These frameworks require investments in secure infrastructure [63].
Higher education and research → Data storage or processing	Research and education disruptions reduce the sector's ability to develop technology and a skilled workforce.
Space technology → Data storage or processing	Satellite systems generate Earth observation and positioning data that must be processed and stored efficiently. Space assets feed data centres with large volumes of real-time imagery and sensor data [67].
Transport → Data storage or processing	**Road transport:** Access for repair crews and staff may be delayed by road blockages. **Supply chains:** The data sector depends on global imports. Geopolitical or policy changes increase costs and delay access to hardware [20]. COVID-19 exposed vulnerabilities in both physical and digital supply chains.
Water → Data storage or processing	Water is critical for cooling and facility operations. Water-based chillers or evaporative cooling systems are common in data centres. Droughts or water restrictions can disrupt operations or growth plans [68].

Table 6.8 Summary of data storage or processing inter-sector dependencies (impact from sector, e.g., data storage or processing → communications)

Dependency	Summary and/or examples
Data storage or processing → Communications	Please refer to Table 6.4
Data storage or processing → Defence	Modern defense operations rely on secure data storage, processing, and analytics for intelligence, surveillance, and mission planning. Big data enhances threat detection, logistics, and simulations. Shared situational awareness improves strategy [63].
Data storage or processing → Energy	Please refer to Table 6.3
Data storage or processing → Financial services and markets	Finance depends on real-time data for trading, risk management, and compliance. AI detects fraud and market manipulation [69]. A lightning strike near a Sydney data center disrupted cloud customers, including the Bank of Queensland [48].
Data storage or processing → Food and grocery	**Precision agriculture:** Satellite imagery and Internet of Things (IoT) sensors (e.g., soil moisture) support decisions on planting and fertilization [70]. **Retail and distribution:** Online ordering and logistics use real-time analytics for inventory, pricing, and forecasting [71].
Data storage or processing → Health	Healthcare uses big data for diagnostics and research [72]. Telehealth, wearables, and electronic health records require secure data systems. In 2022, the Medibank breach affected 9.7 million customers, leaking personal records [73].
Data storage or processing → Higher education and research	Universities use High Performance Computing and large-scale storage for genomics, climate models, and international collaboration [74].
Data storage or processing → Space technology	Satellite data feeds into data centers for science and emergency management [67, 75].
Data storage or processing → Transport	IoT sensors, smart traffic systems, and route planning rely on robust data infrastructure [76]. A lightning event in Sydney (2023) affected Jetstar customers accessing bookings [48].
Data storage or processing → Water	Smart meters and SCADA systems use data analytics to detect leaks and anticipate pump failures. AI has the ability to improve water quality and reliability [77].

Table 6.9 Summary of inter-sector dependencies identified

Australian CI Sectors	Communications	Data storage or processing	Energy
Communications			
Data storage or processing			
Energy			
Defence industry			
Financial services and markets			
Food and grocery			
Health care and medical			
Higher education and research			
Space technology			
Transport			
Water and sewage			

Note: Red (darker shading) means dependency is bidirectional; orange (lighter shading) means dependency is unidirectional from communications, data storage or processing, or energy to other CI sectors

6.4.1 CI Sectors and Interdependencies Across the World

Different nations may define their CI sectors differently. As presented earlier, Australia outlines 11 CI sectors while the United States outlines sixteen. Despite such differences, the CI interdependencies identified for the Australian eleven CI sectors would likely be replicated in other countries, with some contextual differences. For example, when Hurricane Maria struck Puerto Rico in 2017, dependencies between the energy, communications and water sectors meant that when 95% of the island lost power, it led to cascading impacts on the communication and water sectors, leaving 95% of the population without mobile phone service and half the population without tap water [79]. Similarly, in June 2021, an unprecedented heatwave in the Pacific Northwest of the United States left tens of thousands of customers without energy, with cascading effects on the health sector, which resulted in at least 600 excess deaths, and over 3500 emergency department visits for heat-related illnesses [80]. Stone et al. [81] found the estimated rate of heat-related mortality more than doubled when power outages occurred during heatwave conditions in the US.

Across the world, the likelihood of compounding events, or the occurrence of more than one event at a time, has been increasing due to climate change, exacerbating the impacts of extreme events with cascading infrastructure failure. For example, the impact of the heatwave compounded by abnormal drought conditions in the US Western region has been affecting the regional capacity of electrical generation and distribution systems to handle surges in electricity demand [82]. Indeed, high-end scenarios have predicted that approximately 20% of land areas of the planet will be experiencing two compound drought and heatwave events of around 25 days per year by the end of the 21st century [83].

As evidenced by the examples presented in this chapter, the criticality of CI interdependencies is easily observed and experienced during and after climate disasters. For example, during the Australian 2019–2020 wildfires, direct fire damage to telecommunication infrastructure accounted for only 1% of outage incidents, with cascading effects from power outages responsible for 88% of telecommunications disruption [46]. Across the world, the systematic mapping of CI dependencies is essential for mitigating vulnerabilities and enhancing operational continuity. When the focus is on single risks and/or sectors, the extent and nature of the adaptation challenge tends to be underestimated [84]. Identifying CI interdependencies is a foundational step in developing targeted resilience strategies that address both direct and cascading risks. While mapping cascading and compounding risks across sectors is more complex than mapping single risks within sectors, it is critical to improve the information available for decision-making [85] and allow countries to both recalculate what is in their national interest and to define what investments should be prioritized to improve CI resilience [84].

6.4.2 *Whole-of-System Approach*

A whole-of-system approach is needed to develop resilience strategies that recognize the knock-on effects from climate and cyber hazards across sectors and borders. Such an approach starts by recognizing that CI interdependency is heavily influenced by the design, location, and hardening of infrastructure, as well as by how society uses CI across a range of everyday activities and not just during extreme climate events [86]. Importantly, the dependencies also show how CI resilience demands more than the assessment of the direct physical impacts of extreme climate on infrastructure assets, to include the interconnectedness with human and climate systems [29].

A whole-of-system approach to resilience also includes developing both reactive and preventive risk management strategies that go beyond sectors to improve CI resilience at the national level. For instance, power outages remain one of the most significant threats to both telecommunications and data centers, with disruptions capable of triggering widespread service failures in financial transactions, cloud computing, and government operations. Therefore, investing in backup power for communication and data storage infrastructure is critical to minimizing the cascading impacts of electricity disruption. Similarly, identifying regional locations that will benefit from standalone renewable energy generation and storage, such as microgrids, might help minimize the impact of electricity disruption as well as its cascading effects to other CI infrastructure, especially when such regions are also key providers of other essential services such as healthcare, essential supplies, and financial systems.

Likewise, understanding the reliance of data centers on telecommunications infrastructure, such as undersea cables, mobile networks, and satellite communications, allows for pre-emptive diversification of network providers and the implementation of software-defined networking solutions to enable dynamic traffic rerouting during service disruptions. A key benefit of interdependency mapping is the ability to anticipate and mitigate the risks posed by single points of failure. By mapping CI interdependencies, policymakers and industry leaders can develop sector-wide risk frameworks, conduct joint scenario planning, and establish interoperable incident response protocols to reduce the potential for such cascading failures [22].

6.4.3 Cross-Sector Collaborations

Mapping the interdependencies of energy, communications, and data storage or processing with all other CI sectors allows commonalities to be identified and the exploration of where resilience strategies are redundant and where gaps exist. For example, power and communications outages, as well as blocked transport roads, tend to impact all sectors similarly. In terms of response strategies, back-up power supplies and geographical diversity are strategies deployed by all three sectors to minimize disruption following extreme weather events. However, the effectiveness of these strategies depends on effective collaboration between organizations and sectors to, for example, find alternative routes and/or divert services from affected areas. Detailing the interdependencies for all sectors can therefore help identify the common challenges across sectors to develop meaningful collaborations. A whole-of-system approach requires a coordinated approach for cross-sector and geographical collaboration as these risks often do not fall exactly within individual systems and/or jurisdictions [84]. Having a shared global perspective on CI could support better international cooperation, boost information sharing, and lead to best practices for CI protection and resilience worldwide [78].

Interdependency mapping also strengthens cross-sectoral collaboration by clarifying shared responsibilities for resilience. The criticality of energy, communications, and data infrastructure to sectors such as healthcare, national security, and emergency response means that disruptions can have life-threatening consequences, exacerbating the urgency for coordinated resilience planning. For example, in 2022, 18 communities across Northern New South Wales (Australia) had no telecommunications access through mobile, internet, or landline due to flooding. Flood warnings, calls for help, and contact between family and friends were virtually impossible [87]. Telecommunications companies said the scale of the outages was due to power loss, flood damage to land-based infrastructure like fiber-optic cables, and sites like mobile base stations being inaccessible, illustrating the need for coordinating efforts between the telecommunication, energy, and transport sectors.

6.4.4 Public–Private Collaborations

Resilience frameworks and policies should encourage public-private participation from all sectors. For instance, Australia's 2023 CI Resilience Strategy provides a framework for how industry, state and territory governments, and the Australian Government can work together to mature the security and resilience of CI, and to anticipate, prevent, prepare for, respond to, and recover from all-hazards [63]. Another initiative is the Critical Infrastructure Program for Modelling and Analysis (CIPMA) [2], which aims to enhance CI protection by examining system relationships and dependencies to demonstrate how a failure in one sector can greatly affect the operations of CI in other sectors [88]. Similar initiatives have been created worldwide to incentivize the private sector participation, such as the National Critical Information Infrastructure Protection Center (NCIIPC) in India [89], and the President's National Infrastructure Advisory Council in the US [85].

CI resilience often depends on non-traditional assets that may not be captured in formal asset registers, particularly relevant when climate-related disasters disrupt primary infrastructure. During floods in Australia, the banking sector demonstrated this through mobile banking solutions, with organizations learning to coordinate their deployment of alternative service delivery mechanisms to maximize coverage while maintaining backup capacity. While this evidence comes from the banking sector, the principle of alternative asset activation applies to other sectors. The role of public-private collaboration and asset sharing, loaning, or procurement during climate-related hazard events such as the Queensland floods in 2010–2011 saw retailers from the food and grocery sector providing modified logistical support to supply essential items to flood-affected areas, and of the loan of a helicopter from another private sector to facilitate air rescue [90].

6.4.5 Data Integration and Sharing

Fragmented responsibilities around risks pose challenges for data quality and sharing, which are also critical to developing effective resilience strategies. To improve data gathering and sharing, two initiatives have been implemented in Australia. The Trusted Information Sharing Network (TISN) brings together many different stakeholders from across the CI community and is the primary way for industry and all levels of government to work together to enhance the security and resilience of CI. Through the secured network of the TISN, members of the CI community collaborate to strengthen the resilience of their organizations and industry sectors in the face of all hazards [91]. Additionally, using existing networks, such as Information Sharing and Analysis Centers (ISACs), to coordinate data sharing and quality can enhance the effectiveness of cross-sector information sharing and offer protection from potential legal liabilities associated with such exchanges [85].

The role of informal networks in data sharing should also not be underestimated. While formal frameworks and regulations provide essential structure, the effectiveness of informal networks in supporting CI resilience becomes particularly crucial during climate-related disasters. Informal peer networks enable rapid information sharing and response coordination that often outpaces official channels [90, 92], suggesting that regular, informal communication channels between sectors create resilience through trust-based collaboration that complements formal regulatory frameworks. This becomes vital when climate impacts create cascading effects across interconnected sectors, allowing organizations to share real-time intelligence about emerging threats and coordinate responses more effectively. In addition, resilience in the wider environment of the communities affected by climate hazard events can include the non-digital communications of neighbors [93] and community-embedded emergency management professionals and volunteers, especially in regional or rural areas [94]. Looking beyond formal CI sectors and at the broader context and environment where climate resilience is enacted reveals that for data collection, data sharing, as well as communications, people who are not part of the CI sectors can be activated to play a part in managing and mitigating risks.

6.5 Conclusion and Recommendations

By mapping interdependencies between the energy, communication, and data storage or processing sectors, this chapter presented real-world examples that illustrate the urgent need for collaboration and data sharing across sectors and borders to improve CI protection and resilience. It is therefore recommended that public and private institutions work together both nationally and internationally to develop the structures needed to foster cross-sector collaborations. To better prepare, mitigate, and adapt to climate and/or cyber risks, it is also recommended to invest in data tools that can integrate public and private data from across CI sectors.

Integrating data across sectors is critical to consider the compounding and cascading risks that impact the whole system. It is also key to better understand current gaps and to identify where further collaborations are needed. Investing in access to robust data and knowledge sharing across systems is critical for comprehensive CI resilience policy development and decision-making. By working together, a 'whole-of-system' approach to resilience should become the new business-as-usual [84], creating a more robust and resilient infrastructure capable of withstanding future challenges.

Acknowledgements The research supporting this chapter was funded by CSIRO's Critical Infrastructure Protection and Resilience (CIPR) mission, which aims to strengthen resilience and foster informed dialogue around the critical infrastructure that underpins economic, environmental, and societal systems.

References

1. Intergovernmental Panel on Climate Change (IPCC) (2023) Climate change 2022 — Impacts, adaptation and vulnerability. Cambridge University Press
2. Infrastructure Australia (2021) A pathway to infrastructure resilience advisory paper 1: opportunities for systemic change. Tech Rep
3. Public Safety Canada (2014) Critical 5: forging a common understanding for critical infrastructure shared narrative. Tech Rep
4. Infrastructure Australia (2021) Reforms to meet Australia's future infrastructure needs. Tech Rep
5. Tiwari S, Schelly C, Ou G, Sahraei-Ardakani M, Chen J, Jafarishiadeh F (2022) Conceptualizing resilience: an energy services approach. Energy Res Soc Sci 94:102878
6. U.S. Department of Homeland Security (2010) DHS risk lexicon. Tech Rep
7. Ewa WO, Ugwu OO, Okafor FO (2024) Resilient-vulnerability analysis of critical infrastructure, key resources, assets, and facilities. Innov Infrastruct Solut 9(4)
8. Hochrainer-Stigler S, Trogrlić Šakić R, Reiter K, Ward PJ, de Ruiter MC, Duncan MJ, Torresan S, Ciurean R, Mysiak J, Stuparu D, Gottardo S (2023) Toward a framework for systemic multi-hazard and multi-risk assessment and management. Science 26(5):106736
9. Renn O (2021) New challenges for risk analysis: systemic risks. J Risk Res 24(1):127–133
10. Renn O, Laubichler M, Lucas K, Kröger W, Schanze J, Scholz RW, Schweizer P-J (2022) Systemic risks from different perspectives. Risk Anal 42(9):1902–1920
11. Romanach L, Boulaire F, Fleming A, Capon T, Bluhm S, Lin BB (2024) Australia's national474 climate risk assessment: identifying climate risk interdependencies within the infrastructure and built environment system for effective climate adaptation. Clim Risk Manag 46:100670
12. Sonesson TR, Johansson J, Cedergen A (2021) Governance and interdependencies of critical infrastructures: exploring mechanisms for cross-sector resilience. Saf Sci 142
13. CBC News (2013) Quebec forest fire causes widespread blackouts for a 2nd day. CBC News. https://www.cbc.ca/news/canada/montreal/quebec-forest-fire-causes-widespread-blackouts-for-a-2nd-day-1.1334290
14. Marshall A, Wilson C-A, Dale A (2023) New pathways to crisis resilience: solutions for improved digital connectivity and capability in rural Australia. Media Int Aust 189(1):24–42
15. Cyber and Infrastructure Security Centre (2023) Risk assessment advisory for critical infrastructure space technology sector. Tech Rep. Department of Home Affairs. Australian Government. https://www.cisc.gov.au/resources-subsite/Documents/raa-space-technology.pdf
16. Cyber and Infrastructure Security Centre (n.d) Risk assessment advisory for critical infrastructure: Energy sector. https://www.cisc.gov.au/resources-subsite/Documents/raa-energy.pdf
17. Department of Climate Change, Energy, the Environment and Water (2024) National climate risk assessment: first pass assessment report. Tech Rep. Department of Climate Change, Energy, the Environment and Water. Commonwealth of Australia
18. Australian Signals Directorate (2024) Annual cyber threat report 2023–24. Tech rep. Australian Signals Directorate. www.cyber.gov.au
19. Rinaldi SM, Peerenboom JP, Kelly TK (2001) Identifying, understanding, and analyzing critical infrastructure interdependencies. IEEE Control Syst Mag 21(6):11–25
20. Curtis H, Harpley C (2023) Telecommunications risk and resilience profile: Pestle and gap analysis. http://creative-commons.org/liecenses/by-sa/4.0/
21. Quin C, Bryant B (2021) NSW critical infrastructure resilience strategy guide. Tech Rep
22. Lewis LP, Petit F (2019) Critical infrastructure interdependency analysis: operationalising resilience strategies
23. Petit F, Verner D (2016) Critical infrastructure interdependencies assessment. Tech Rep
24. Cyber and Infrastructure Security Centre (2023) Energy. https://www.cisc.gov.au/information-for-your-industry/energy
25. Smith EJB, Edmondson C, Collins A, Harradine N, Bertoglio G (2024) No fuel in Kalgoorlie, Coolgardie, as tens of thousands of customers remain without power across WA's Goldfields, Wheatbelt—ABC News. https://www.abc.net.au/news/2024-01-18/kalgoorlie-power-lightning-western-power-phone-service-wa/103363060

26. ABC News (2006) Fuel replenished in flood affected community. ABC News. https://www.abc.net.au/news/2006-03-20/fuel-replenished-in-flood-affected-community/822370
27. Mercer D (2024) Gas shortage threatens businesses, jobs and households can expect further bill pain. https://www.abc.net.au/news/2024-06-29/gas-crisis-in-australia-s-east-threatens-industry/104016398
28. Royal Commission into National Natural Disaster Arrangements (2020) Royal Commission into National Natural Disaster Arrangements—Report. Tech Rep
29. Perera ATD, Hong T (2023) Vulnerability and resilience of urban energy ecosystems to extreme climate events: a systematic review and perspectives. Renew Sustain Energy Rev 173:113038
30. U.S Department of Energy (2023) Communications in the electric grid: an evolving interdependent ecosystem between the grid and communications utilities. Tech Rep
31. Council New Zealand (2023) National vulnerability assessment part C: infrastructure sectors and hazards assessment. Tech Rep
32. Clean Energy Regulator (2025) Data services beta platform now live. Accessed 08. https://cer.gov.au/news-and-media/news/2024/december/data-services-beta-platform-now-live
33. AEMO (2023) Forecasting increasingly critical to harnessing wind and solar for power systems. Australian Energy Market Operator. https://aemo.com.au/newsroom/news-updates/forecasting-increasingly-critical-to-harnessing-wind-and-solar-for-power-systems
34. Department of Climate Change, Energy, the Environment and Water, "Renewables" (2024). Australian Energy Statistics. https://www.energy.gov.au/energy-data/australian-energy-statistics/renewables
35. Parkinson G (2013) Funding cuts to ARENA confirmed, $435m gone. https://reneweconomy.com.au/funding-cuts-to-arena-confirmed-435m-gone-32670/
36. Vorrath S, Parkinson G (2016) Coalition, Labor agree to slash $500m from ARENA budget. https://reneweconomy.com.au/coalition-labor-agree-to-slash-500m-from-arena-budget-83469/
37. Department of Environment and Energy (2019) Liquid fuel security review: interim report. Tech Rep. Department of Environment and Energy. Australian Government
38. Department of the Environment and Energy (2019) Liquid fuel security review: interim report. Tech Rep. Department of the Environment and Energy. Australian Government
39. Department of Climate Change, Energy, the Environment and Water (2022) Australia's fuel reserves boosted to strengthen resilience and supply. Australian Government
40. Steffen W, Vertessy R, Dean A, Hughes L, Bambrick H, Gergis J, Rice M (2018) Deluge and drought: Australia's water security in a changing climate. Tech Rep
41. Australian Renewable Energy Agency (2025) Hydropower / Pumped hydro energy storage. https://arena.gov.au/renewable-energy/pumped-hydro-energy-storage/
42. Varianou Mikellidou C, Shakou LM, Boustras G, Dimopoulos C (2018) Energy critical infrastructures at risk from climate change: a state of the art review. Saf Sci 110:110–120
43. Downes SM, Steel A, Favaro E, Wood M (2021) Disruption and damages: climate-related risks to the Australian oil and gas sector. The APPEA J 61(2):395–399
44. Smith M (2013) Assessing climate change risks and opportunities for investors: oil and gas sector. Tech Rep
45. International Energy Agency (2022) Climate resilience for energy security. Tech Rep
46. Australian Communications and Media Authority (2020) "Impacts of the 2019–20 bushfires on the telecommunications network, report. Australian Communications and Media Authority, Report for the Minister for Communications, Cyber Safety and the Arts
47. Wisetjindawat W, Burke MI, Fujita M (2017) Analysing impacts of natural disasters on logistics activities: flood risks and petroleum fuels in Queensland, Australia. Transp Res Procedia 25:1198–1212
48. Abbott B (2023) Impact of recent Australian data centre outages on businesses. TechRepublic. https://www.techrepublic.com/article/data-centre-outages-impact-australia/
49. Boisvert E, Mullins S (2022) Energyaustralia fined $12 m for failing to register people on life-saving medical equipment. Accessed 08 Sept 2025

50. Cyber and Infrastructure Security Centre (2023) Communications. https://www.cisc.gov.au/resources-subsite/Documents/raa-communications.pdf
51. Curtis H, Harpley C, Weaver J, Hawkins Z, Jackson J (2024) Australian telecommunications sector resilience profile. http://creative-commons.org/liecenses/by-sa/4.0/
52. Australian Space Agency (2021) Advancing space communications technologies and services roadmap 2021–2030. Tech Rep
53. CSIRO (2018) Space: a roadmap for unlocking future growth opportunities for Australia 2018. Tech Rep
54. Cyber and Infrastructure Security Centre (2022) Regulation impact statement: a risk management program framework for critical infrastructure assets. Tech Rep. https://oia.pmc.gov.au/sites/default/files/posts/2025/03/2022%20RIS%20-%20RMP%20for%20Critical%20Infrastructure%20Assets.pdf
55. Environment and Communications References Committee (2024) Optus network outage. Tech Rep. The Senate, Commonwealth of Australia
56. Salim R, Mamun SAK, Hassan K (2016) Role of communication technologies in broadacre agriculture in Australia: an empirical analysis using panel data. Aust J Agric Resour Econ 60(2):243–264
57. Parliament of Australia (2023) Australian food story: feeding the nation and beyond. Tech Rep
58. Marshall A, Wilson C-A, Dale A (2023) Telecommunications and natural disasters in rural Australia: the role of digital capability in building disaster resilience. J Rural Stud 100:102996
59. ABC News (2023) More than 10 million customers were affected by the Optus outage. Here's how it played out. ABC News. https://www.abc.net.au/news/2023-11-09/how-the-optus-outage-played-out/103079768
60. Environment and Communications References Committee (2023) Inquiry into the telecommunications legislation amendment (enhancing consumer safeguards and other measures) Bill 2023. https://www.aph.gov.au/Parliamentary_Business/Committees/Senate/Environment_and_Communications/Safeguards. Accessed 08 Sept 2025
61. Latimer C (2023) How 5G could steer the future of water management. Australian Water Association. https://www.awa.asn.au/resources/latest-news/how-5g-could-steer-the-future-of-water-management
62. Cyber and Infrastructure Security Centre (2023) Data storage and processing. https://www.cisc.gov.au/information-for-your-industry/data-storage-and-processing
63. Department of Home Affairs (2023) Critical infrastructure resilience strategy. Tech Rep. Australian Government
64. Rajaretnam T (2020) A review of data governance regulation, practices and cyber security strategies for businesses: an Australian perspective. Int J Technol Manag Inf Syst 1–17
65. Australian Prudential Regulation Authority (2025) Operational risk management. https://www.apra.gov.au/operational-risk-management#:~:text=July%202022,standard%20to%20strengthen%20operational%20resilience. Accessed: 08 Sept 2025
66. Raman S, Patwa N, Niranjan I, Ranjan U, Moorthy K, Mehta A (2018) Impact of big data on supply chain management. Int J Logist Res Appl 21(6):579–596
67. Dhu T, Dunn B, Lewis B, Lymburner L, Mueller N, Telfer E, Lewis A, McIntyre A, Minchin S, Phillips C (2017) Digital earth Australia-unlocking new value from earth observation data. Big Earth Data 1(1–2):64–74
68. Soares IV, Yarime M, Klemun M (2024) Balancing the trade-off between data center development and its environmental impacts: a comparative analysis of data center policymaking in Singapore, Netherlands, Ireland, Germany, USA, and the UK. Environ Sci Policy 157:103769
69. Australian Payments Plus (2025) Real-time, data-rich payments. https://www.auspayplus.com.au/solutions/nppa. Accessed: 08 Sept 2025
70. Bramley R, O'Shaughnessy A (2013) Precision agriculture in Australia: present status and recent developments. Engenharia Agrícola 33(3):575–588
71. Maganja D, Ceddia A (2020) Changes in household food grocery shopping patterns in melbourne, Australia during Covid-19 restrictions. Aust N Z J Public Health 100088

72. Umair M, Choi O (2021) Impact of Covid-19 on IoT adoption in healthcare, smart homes, smart buildings, smart cities, transportation and industrial IoT. Sensors 21(11):3838
73. Office of the Australian Information Commissioner (2025) Medibank data breach: alleged timeline." https://www.oaic.gov.au/_data/assets/pdf_file/0037/228979/Medibank-data-breach-alleged-timeline-infographic.pdf. Accessed 08 Sept 2025
74. AARNet (2025) Solutions made for research and education. https://www.aarnet.edu.au/. Accessed 08 Sept 2025
75. Australian Space Agency (2025) Australia's emergency services rely on space. https://www.space.gov.au/Australian-emergency-services-rely-on-space
76. Dia H, Geurs K (2023) Creating our future transport and mobility workforce: understanding the workforce implications of transport digitalisation and automation in Australia, research report, Swinburne University of Technology. Prepared for the Department of Infrastructure, Transport, Regional Development and Communications. Accessed 08 Sept 2025
77. Koech R, Choi O (2021) Smart water metering: adoption, regulatory and social considerations. Australas J Water Resour 25(2):173–182
78. Weber V, Riera MP, Laumann E (2023) Mapping the world's critical infrastructure sectors. Tech Rep, DGAP
79. Baker GH, Volandt S (2018) Cascading consequences: electrical grid critical infrastructure vulnerability. Domestic Preparedness. https://domesticpreparedness.com/articles/cascading-consequences-electrical-grid-critical-infrastructure-vulnerability
80. Popovich N, Choi-Schagrin W (2021) Hidden toll of the Northwest heat wave: hundreds of extra deaths. https://www.nytimes.com/interactive/2021/08/11/climate/deaths-pacific-northwest-heat-wave.html
81. Stone B, Gronlund CJ, Mallen E, Hondula D, O'Neill MS, Rajput M, Grijalva S, Lanza K, Harlan S, Larsen L, Augenbroe G, Krayenhoff ES, Broadbent A, Georgescu M (2023) How blackouts during heat waves amplify mortality and morbidity risk. Environ Sci Technol 57(22):8245–8255
82. Patel S (2021) Rolling blackouts triggered as historic heatwave grips Pacific Northwest. https://www.powermag.com/rolling-blackouts-triggered-as-historic-heatwave-grips-pacific-northwest
83. Tripathy KP, Mukherjee S, Mishra AK, Mann ME, Park Williams A (2023) Climate change will accelerate the high-end risk of compound drought and heatwave events. In: Proceedings of the National Academy of Sciences of the United States of America, vol. 120, no. 28, p. e2219825120
84. Townend R, Aylett C, Benzie M (2023) Cascading climate risks: strategic recommendations for European resilience. Tech Rep
85. NIAC (2023) Cross-sector collaboration to protect critical infrastructure: barriers and recommendations for improvement. Tech Rep
86. National Resilience Taskforce and Commonwealth of Australia (2018) Profiling Australia's vulnerability: the interconnected causes and cascading effects of systemic disaster risk. Tech Rep
87. ABC News (2022) Five key issues raised in 2022 NSW flood inquiry. https://www.abc.net.au/news/2022-08-17/key-takeaways-from-2022-nsw-flood-inquiry/101342094
88. Geoscience Australia (2009) Critical infrastructure protection modelling and analysis program (CIPMA). https://ecat.ga.gov.au/geonetwork/srv/api/records/a05f7892-ef2d-7506-e044-00144fdd4fa6
89. Sadonian L (2025) NCIIPC explained: safeguarding India's critical infrastructure. https://www.upguard.com/blog/nciipc-explained
90. Bajracharya B, Hastings P (2015) Public–private partnerships in emergency and disaster management: examples from the Queensland floods 2010–11. Aust J Emerg Manag 30(4)
91. Cyber and Infrastructure Security Centre (2024) Trusted information sharing network. https://www.cisc.gov.au/how-we-support-industry/partnership-and-collaboration/trusted-information-sharing-network

92. Heffernan T, Shearing C, Sanderson D, De Sisto M (2024) Building cross-sector recovery collaborations after Australian bushfires: the importance of embracing and linking diverse capitals and capacities. Environ Hazards
93. Boon H (2014) Investigation rural community communication for flood and bushfire preparedness. Aust J Emerg Manag 29(4)
94. Blaustein J, Miccelli M, Hendy R, Burns KH (2023) Resilience policing and disaster management during Australia's Black Summer bushfire crisis. Int J Disaster Risk Reduct 95:103848

Chapter 7
Autonomous Uncrewed Systems for Offshore Wind Farm and Undersea Infrastructure Maintenance: Pathways to Climate and Cyber-Physical Resilience

Ville Lehtola, Johannes Singer, Bart De Schutter, Joris Sijs, Jussi Simola, and Jakub Montewka

Abstract Maintaining offshore infrastructure—such as wind farms, undersea communication cables, and energy pipelines—is vital to securing energy supply, connectivity, and critical logistics networks amidst growing climate instability. As climate change accelerates, with intensified extreme weather events and sea-level rise, the resilience of these infrastructures becomes a central concern. This chapter positions Autonomous Uncrewed Systems (AUS) as a cyber-physical-social enabler capable of transforming how we inspect, maintain, and protect offshore assets in this challenging context. By embedding AUS within the CPSICC (Cyber-Physical-Social Infrastructure Climate Change) nexus framework, we examine how these systems simultaneously enhance physical infrastructure resilience, strengthen cybersecurity, and reduce human exposure to hazardous, repetitive, and environmentally challenging tasks. We specifically address challenges faced by AUS in offshore deployments, including rough seas, sensor reliability, communication loss, and potential interference from hostile actors. The discussion includes AUS-based inspection and maintenance strategies, which address both physical threats (e.g., climate and adverse weather) and cyber threats (e.g., malicious use of AI and cybersecurity breaches). In particular, the AI must be able to take over in case communications fail, which

V. Lehtola
Department of Earth Observation Science, University of Twente, Enschede, The Netherlands

V. Lehtola (✉) · J. Simola
Faculty of Information Technology, University of Jyväskylä, Jyväskylä, Finland
e-mail: v.v.lehtola@utwente.nl

J. Singer
Fugro Innovation and Technology, Fugro, Nootdorp, The Netherlands

B. De Schutter
Delft Center for Systems and Control, Delft University of Technology, Delft, The Netherlands

J. Sijs
Department of Cognitive Robotics, Delft University of Technology, Delft, The Netherlands

J. Montewka
Institute of Naval Architecture, Gdańsk University of Technology, Gdańsk, Poland

M. Huber et al. (eds.), *Critical Infrastructure Systems and Extreme Threat Nexus: Preparing for Resilience and Security*, NATO Science for Peace and Security Series C: Environmental Security, https://doi.org/10.1007/978-3-032-13170-6_7

raises multiple requirements related to robotic perception and cognition, as well as self-sufficient mission control. This leads to multiple research gaps. Concretely, our primary case study is the North Sea, where international navigation and offshore infrastructure are present in abundance. Moreover, we examine regulatory challenges posed by evolving cross-border frameworks and AUS resilience strategies to counter risks concerning navigation safety and threats from hostile actors in international waters. By examining AUS as a cyber-physical-social asset under climate stressors, this chapter directly supports the CPSICC initiative's goal to develop resilient, adaptive, and intelligent systems that secure critical infrastructure at the intersection of technology, society, and environment.

Keywords Autonomous uncrewed systems · Offshore wind farms · Undersea infrastructure · Maritime perception and control · Cyber-physical systems · Climate resilience

7.1 Introduction

Consider the following scenario. An Uncrewed Surface Vessel (USV) is steaming in the North Sea, controlled remotely from a land office via a satellite link. Its mission is to perform maintenance and safety inspections of critical offshored infrastructure such as wind farms and undersea pipelines, see Fig. 7.1. The satellite link exchanges video feed and joystick commands between the USV and the command center. Suddenly, the communication is lost, for example, because the satellite link is down. An autonomous mode is to be triggered on the USV, to continue operations. Note that any simple "halt" algorithm is out of the question, because oftentimes just stopping is not an option. For instance, its route may intersect shipping lanes such as the Traffic Separation Scheme (TSS) routes, since the vessel is based in a harbor. It must navigate safely and predictably as a part of the traffic, ensuring it neither provokes or inconveniences crews aboard other vessels. To this end, the USV must have knowledge of its own state and capabilities and be able to identify how it should carry out the mission by itself.

Autopilot is a commonly used functionality in the airline industry. There, the autopilot is activated by a human to deal with routine situations, i.e., when the plane is supposed to be flown straight for an extended period. The need for autonomy for USVs arises in opposite situations: not by humans but by external factors, and not to handle routine tasks but to manage the mission execution itself. For instance, while a remotely piloted vessel relies on continuous operator input, an autonomous USV must not only maintain safe navigation but also achieve mission-specific objectives under uncertain conditions, such as adverse weather or complex maritime traffic patterns. Hence, the USV autopilot must vigilantly observe when to take over the mission control to guarantee operational safety.

Contrary to common belief, long distance (over the horizon) remote operations inherently require autonomy capabilities (Autonomy Assisted Remote Con-

Fig. 7.1 Uncrewed surface vessels (USVs) inspecting a wind farm in calm weather. An uncrewed underwater vehicle (UUV) is launched for subsurface inspections. Fugro, a multinational company, has multiple of these Blue Essence systems in operation

trol, AARC). Even if the communication link remains active, round-trip delays and limited bandwidth slow the operator's response time. This renders most interaction tasks challenging or impossible without autonomy in an environment that contains sea waves and currents, and potential other external factors. These limitations in communication pose critical safety risks especially when the USV is close to a structure or must interact with it, when it is performing a docking maneuver or when it is avoiding an unexpected floating obstacle. In such time-sensitive scenarios, local autonomy is analogous to a human's reflex arc: the system must perceive, interpret, and react instantly, without waiting for remote input.

The necessity for autonomy becomes even more evident in multi-vehicle missions. Relying on separate human operators for each vehicle results in slow coordination and prone to communication bottlenecks. A prime example is launching or retrieving an Uncrewed Underwater Vehicle (UUV) from a USV: rapid, local decisions are crucial for safe docking in dynamic sea conditions.

We consider the North Sea as the operational environment, for a two-fold reasoning. There-in lies a vast demand for USV operations, but also it is one of the most challenging maritime environments in the world, as will be explained next.

7.1.1 Challenges Posed by Climate and Environment

The North Sea is home to some of the largest offshore wind farms in the world, contributing significantly to Europe's renewable energy supply. These installations are strategically located to harness the region's strong and consistent winds. Examples include the Dogger Bank Wind Farm (UK), the Horns Rev Wind Farm (Denmark), and the Hollandse Kust Zuid (Netherlands). The infrastructure includes turbines, substations, and transmission cables, often extending hundreds of kilometers to connect with mainland grids.

The strong, consistent winds near the wind farms, although beneficial to electricity production, pose challenges to maintenance operations—including those involving USVs. Main challenges include strong winds and other harsh weather conditions such as high waves, and sudden storms, which are common in the North Sea, but also strong tidal currents. These environmental factors can impair sensor functionality, degrade communication links, and more. Some of the USVs are capable of launching an Uncrewed Underwater Vehicle (UUV) for underwater inspections. Docking the UUV amongst the waves also requires careful planning. Furthermore, climate change is expected to increase the frequency and intensity of extreme weather events, which likely to exacerbate the challenges of USV operations. Therefore, operating safely near the wind farms requires careful (autonomous) navigation and more so if the pattern of currents and waves is complex, or altered by the infrastructure.

The North Sea is one of the world's busiest maritime regions, with high-density traffic routes, including major shipping lanes and traffic separation schemes (TSS). USVs must navigate safely among large commercial vessels, fishing boats, and recreational craft, often in poor visibility or adverse weather conditions. The need for real-time traffic assessment and compliance with international maritime regulations, such as COLREGs [1], further complicates autonomous decision-making.

The call for USV inspection capabilities is not limited to wind farms, especially if an USV can launch an UUV for underwater inspections. A dense network of submarine communication cables lies beneath the North Sea, forming a backbone for internet and telecommunications across Europe. Also, numerous oil and natural gas pipelines traverse the seabed, connecting offshore extraction facilities to onshore processing plants. These pipelines are crucial for Europe's energy supply.

The North Sea's critical infrastructure is a potential target for both accidental damage and intentional sabotage. USVs operating in this environment could contribute to their protection. A fleet of USVs could act to detect and keep track of accidental interactions, such as collisions and anchor dragging, which are to be avoided at all costs. This role necessitates sophisticated sensor suites for threat detection and collaboration with human operators or other autonomous systems in real time. This adds another layer of complexity, as situational awareness is shared with different parties. This layer is best viewed from the perspective of cybersecurity.

Overall, the North Sea represents a uniquely demanding operational theater for USVs. Navigating its intricate mix of critical infrastructure, environmental hazards, and high traffic density requires advanced autonomy, precise situational awareness, and a resilient design capable of adapting to one of the most challenging maritime environments in the world.

---

[1] International Maritime Organization (IMO), Convention on the International Regulations for Preventing Collisions at Sea (COLREG)

7.1.2 Safety and e-Navigation for USVs

Modern maritime vessels continue to face longstanding navigational threats, such as collisions and groundings. These risks are expected to persist in future sea transport systems, including those that incorporate USVs. In situations where a conventional ship faces challenges beyond the officer of the watch's (OOW) capabilities, the captain is summoned for assistance to ensure safety. But, when an unclear situation arises regarding an USV, what action should be taken? What are the safety guidelines?

In the context of maritime safety and maritime transportation, several classification schemes for levels of autonomy have been proposed by authorities and classification societies. For example, the International Maritime Organization (IMO) adopted the following four-level classification [1].

1. *Ship with automated processes and decision support.* Seafarers are on board to operate and control shipboard systems and functions. Some operations may be automated and at times be unsupervised but with seafarers on board ready to take control.
2. *Remotely controlled ship with seafarers on board.* The ship is controlled and operated from another location. Seafarers are available on board to take control and to operate the shipboard systems and functions.
3. *Remotely controlled ship without seafarers on board:* The ship is controlled and operated from another location. There are no seafarers on board.
4. *Fully autonomous ship:* The operating system of the ship is able to make decisions and determine actions by itself.

Therefore, these USVs are at autonomy level three unless they incorporate an autopilot system, in which case they move to level four. This represents a substantial leap in capability, which is why classification societies offer more detailed categorizations [1], such as the one proposed by Lloyd's Register of Shipping in Table 7.1.

Regardless of the granularity of the classification system, one may notice a common thing, there are three broad types of operational modes (OM) for USVs, as follows: Manual Control; Supervisory Control; and Autonomous Control. These has been widely discussed in the scientific literature [1, 2]. However, it is not clear under what conditions and in what situations the USV shall operate in a given OM. To this end, classification society DNV proposes four operational conditions for an autonomous ship: normal (NC), abnormal (AC), minimum risk (MRC), and last resort (LRC) [3]. Normal conditions are those defined in the concept of operation for a given autonomous ship. Once the ship goes beyond NC it begins to operate in abnormal conditions. From there it can go back to NC, once the situation improves, remain in abnormal, though acceptable, conditions or move towards MRC if the conditions deteriorate and the safety of ship may be compromised. The MRC is a state what minimizes the risk when operating in abnormal conditions. However, if conditions continue to deteriorate, the ship can rely to last resorts to remain safe.

To ensure the USV remains safe under anticipated conditions, the most suitable operational mode should be selected. According to military standards MIL-STD-882D, safety is defined as *"freedom from those conditions that can cause death,*

Table 7.1 Autonomy levels (AL) for maritime operations

Level	Description
AL-0	*No autonomous function*—decision-making is performed manually, i.e., a human controls all actions at the ship level
AL-1	*On-ship decision support*—All actions at the ship level are taken by a human operator, but a decision-support tool can present options or otherwise influence the actions chosen (e.g., DP capability plots (Dynamic Positioning), route planning)
AL-2	*On and off-ship decision support*—All actions at the ship level are taken by a human operator on board the vessel, but decision-support tools (on or off the ship) can present options or otherwise influence the actions chosen (e.g., DP capability plots, OEM recommendations (Original Equipment Manufacturer), weather routing)
AL-3	*'Active' human in the loop*—Decisions and actions at the ship level are performed autonomously with human supervision. High-impact decisions are implemented to allow human operators the opportunity to intercede and override them. Data may be provided by systems on or off the ship
AL-4	*Human on the loop: Operator/supervisory*—Decisions and actions are performed autonomously with human supervision. High-impact decisions are implemented to allow human operators the opportunity to intercede and override them
AL-5	*Fully autonomous (rarely supervised)*—Decisions are made and actioned by the system, i.e., the impact is at the total ship level. Human supervision is minimal or only occasional
AL-6	*Fully autonomous (unsupervised)*—Decisions are made and actioned entirely by the system, with no active human supervision

injury, occupational illness, damage or loss of equipment or property, or environmental harm" [4]. Since achieving absolute freedom from these conditions is unfeasible, safety is effectively "*freedom from conditions of unacceptable mishap risk.*" Here, *mishap* may be understood as any unexpected or unintended event that may disrupt a ship's operations or compromise safety. Consequently, the *mishap risk* serves as a safety metric that assesses the danger level associated with system design and operations. Hazards within the system design and operations cause potential mishaps. The mishap risk can be utilized to gauge and track the ship's operational safety to aid in selecting the most appropriate operational mode under given or expected conditions. For this purpose, risk acceptance criteria must be established to define negligible, acceptable, and unacceptable risk zones. All decisions regarding the operational mode should occur within the negligible and acceptable risk zones, with the unacceptable risk zone serving as a last resort. However, a science-based approach for defining and applying these risk zones is not currently available.

The goal is to establish a risk metric that provides thorough coverage in defining the risk zones while remaining easy to interpret. To this end, there are multiple concepts and definitions of risk, from which the most suitable can be selected for the given purpose [5]. In addition, there exists a wide range of scientific approaches to the risk analysis process, its description, the elements it shall encompass, and the interpretation of the results obtained. As USVs will operate within existing sea transportation systems involving numerous stakeholders, whose values may be at risk in the event of a mishap, at least two parameters—namely the likelihood and consequence of a relevant mishap arising from a predefined set of hazards—shall be accounted for, together with associated uncertainties. However, the presence of a human in the loop, and thus the selected operational mode (OM), can affect both the likelihood and the consequences of a mishap [6]. The direction and magnitude of this effect are currently unknown, contributing significantly to overall uncertainty. Therefore, properly accounting for this effect is essential, as empirical knowledge of transportation systems modified by autonomous ships is limited and relevant operational data are likely to be scarce. Further research in this area is needed.

When it comes to the definition of relevant hazards and the evaluation and judgment of likelihood or consequences, there are methods that can be employed, as recently elaborated in [7]. However, a key role in decision making is played by *epistemic uncertainty*, i.e., the things that could (in principle) be known but are not known. This uncertainty shall be the main driver for the selection of the proper scientific methods and tools for risk analysis and assessment knowing the risk analysis is meant primarily to support decision making process.

Awareness of mishap risk can support the determination of an appropriate OM for an autonomous ship under given conditions, as the chosen OM in turn affects the risk profile. Such risk awareness is expected to be an essential element for operators and competent authorities overseeing maritime traffic in designated areas [8].

7.2 Situational Awareness: Exteroception and Interoception

In deciding on whether to go on autopilot or not, what becomes crucial is the robot's own situational awareness. Situational awareness for autonomous vessels encompasses two primary dimensions: *exteroception*, which pertains to the perception of the external environment such as position and bearing of other ships, and *interoception*, which refers to the internal state of the USV, such as its position, velocity, and system status. Achieving robust situational awareness involves integrating data from

Table 7.2 Consolidated sensing requirements, reproduced from [9]

Target	Detection range	Main reference
Ship >50m with Class-A AIS	8NM	[10], IMO RADAR range
Ship12-50m with Class-A AIS	5NM	COLREGs Rule 22
Boat < 12m with Class-B AIS	5NM	COLREGs Rule 22
Boat <12m without AIS	2NM	COLREGs Rule 22
Kayak-size object	1NM	[10]
Fixed radar mark	8NM	IMO RADAR range
Ice buoy	3NM	IMO RADAR range
Port/marina buoy	1NM	IMO RADAR range

a variety of onboard sensors, such as radar, LiDAR, cameras, and GNSS, with computational models that interpret this information in the context of the vessel's mission objectives and operational constraints [9]. This holistic understanding enables the USV to perform robust navigation in dynamic and uncertain environments [11–13].

Multiple target types must be detectable at sea, see Table 7.2, to guarantee safe navigation. Sea captains tend to verify their observations on the situation from more than one independent sources, using direct visual observations as well, if available. Hence, further requirements for a multi-sensor system could be, for example, formulated as follows: (1) Targets shall be detected with at least two separate sensors (e.g. RADAR and camera); (2) targets shall also be classified to a probability of at least 80%; and (3) Computing shall be done in real time. In short, all detections must be reliable and available within the required time frame.

To operate effectively in autonomous mode, a USV must possess comprehensive situational awareness, see Fig. 7.2. Exteroception, involving sensors such as cameras, LiDAR, and radar, provides critical information about the surrounding environment, including obstacles, traffic patterns, and weather conditions. Advanced sensor fusion techniques enable the integration of this data into a coherent environmental model. Meanwhile, interoception involves monitoring the USV's internal state, such as engine health, fuel levels, and navigational system accuracy. Together, these dimensions form the basis for predictive and adaptive decision-making in complex maritime environments.

7.2.1 Perception Tasks for Exteroception

One of the most important abilities in maritime exteroception is the ability to see far and reliably. Seeing far refers to detecting small targets that could be near the horizon, while seeing reliably involves covering the entire surroundings of the USV without leaving blind spots (see Table 7.2). This dual challenge necessitates the processing of multiple high-resolution images captured in all directions, which presents a new

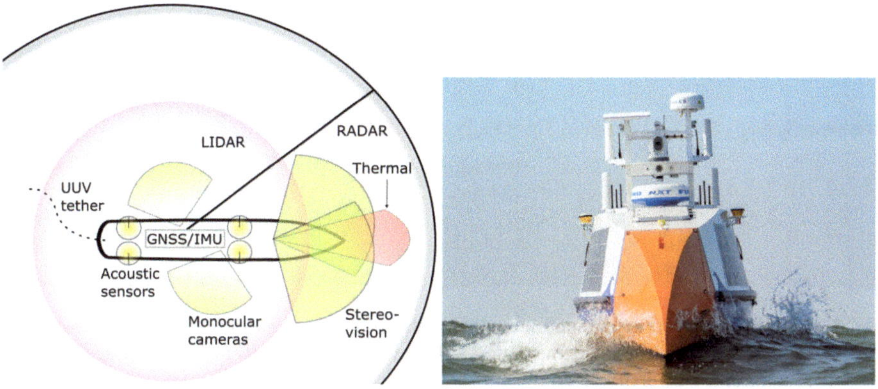

Fig. 7.2 Left: Illustration of a multi-sensor layout for an USV, from the top, i.e., bird eye view (BEV). The dotted path indicates a tether to the UUV. Right: Image of DEMCON Unmanned Systems V5750 Class and Type Approved USV "VO:X Barentsz" (Van Oord) featuring redundant situational awareness (360 o RGB and thermal cameras, PTZ camera, Radar, LiDAR, weather station, dual GNSS) and multiple communication links (LTE, Starlink, SATCOM, 2.4/5 GHz mesh).

problem in perception [15]. Most prior work has focused on smaller imagery—typically around 224 × 224 or 384 × 384 pixels [16] ($\approx$ 0.15 megapixels)—to reduce GPU memory usage and computing costs. However, for USV applications, images of 8 megapixels or more are preferable, posing a significant challenge for processing and detection tasks.

Overall, perception techniques have evolved significantly in a short time, transitioning from traditional machine learning and convolutional neural networks to vision transformers [17] and, more recently, vision Mamba [18]. The latter two methods, transformers and Mamba, can be understood as state-space models [19]. Furthermore, the emergence of foundation models [20] has transformed AI perception workflows, making them the starting point for most tasks, while follow-up tasks are considered downstream operations. However, these advances bring a second challenge that some models such as transformers and foundation models may be slow, whereas models for edge computing onboard the USV need to operate in real time.

Exteroception tasks can be broadly categorized into data segmentation, object detection, object identification, and object tracking. **Segmentation** is a fundamental task that can be performed by foundation models or faster task-specific models to facilitate follow up tasks [14], e.g. in directing the attention to the part of the image that shows water, see Fig. 7.3. **Object detection** is a fundamental task for situational awareness, to observe which other vessels are steaming nearby [21], see Fig. 7.4. **Object identification** involves determining whether an object stands out from others based on its visual shape and detail. At a fine-grained level, this could mean detecting whether a ship's anchor is raised or lowered. At a coarser level, it may involve estimating the heading of the detected ship [22]. **Object tracking** may combine the

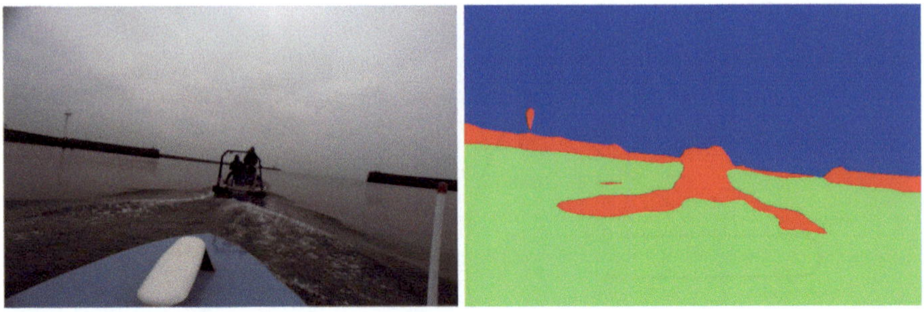

Fig. 7.3 Example of data segmentation and related challenges: for example, the tail waves share the same segment than the boat. Segmentation is done with using AF-deeplab [14]. Image by Tinsae Yehuala, Fugro and University of Twente

Fig. 7.4 Target detection with YOLOv8 results in four ship predictions, of which the two weakest detections landing on are the same ship. Image (1280 × 720) from Aboships dataset [23] cropped to 640 × 320 pixels shows three detectable ships but is affected by glare. The glare with limited dynamical range of the camera causes part of the image to look overly dark

information from the previously mentioned tasks and use, for example, a Kalman filtering-based approach to track the object in image or in other suitable coordinates.

7.2.2 *Sensor Fusion for Safe Operations*

Sensor fusion is the topmost way to enhance the reliability of maritime perception. Just as sea captains cross-check multiple independent sources before making decisions [24], each sensor can ideally be treated as its own independent source of information. In addition to being independent, the sensors must also complement one another so their combined inputs can be integrated into a continuously updated, well-informed awareness of the surrounding situation.

There are two main questions when discussing sensor fusion complementarity: coverage and modality. The sensor setup shown in Fig. 7.2 shows complementary nature in terms of e.g. multiple cameras adding coverage to different directions, but also in that (frontal) targets may be detected in adverse weather even if some of the

Table 7.3 Comparison of relevance of USV situational awareness sensors (adapted from [9])

	RGB camera	Thermal	RADAR	LiDAR	Acoustic
Bandwidth	Optical	LWIR	X,S bands	NIR	Foghorn
Modality	Image	Image	**R,A,I,D	**R,A,I	**A
Spatial accuracy	✓✓	✓	✗	✓✓	✗✗
Sea wave measurement	✓	✗	✗✗(*✓✓)	✗✗	✗✗
FOV	✓	✗	✓✓	✓	✓✓
Distance measurement	✗	✗✗	✓✓	✓✓	✗✗
Object identification	✓✓	✓✓	✗✗	✓	✓
Computational load	✗✗	✗	✓✓	✗	✓
24 × 7, all-weather	✗✗	✓	✓✓	✓ (?)	✓✓(?)
Environmental nemeses	Fog, Rain Night, Glare	Fog, Rain	N/A	Fog, Rain	Wind, Rain

*: Requires inverted polarization from the maritime radar
**: Range, Angle, Intensity, Doppler

modalities are occluded. A data modality refers to the type or domain of information being captured or represented, such as vision or sound, or RADAR, see Table 7.3.

Single-modality sensor fusion is one of the most straightforward methods for combining sensor data. In this approach, only data of the same type is utilized. For example, an object detection algorithm might be extended from a single camera to a two-camera system [25], and even further to a three-camera system [26], to improve reliability and robustness.

Multiple modalities can also be utilized in sensor fusion. For instance, imagery can extend beyond just red-green-blue (RGB) optical bandwidths to include thermal data. This leads to thermal and RGB fusion [27], which combines two different modalities but still fuses the data into a single, joint image frame, allowing for enhanced perception. For multi-modal fusion involving ranging sensors, such as radar and lidar, camera images may be converted into a bird's-eye view representation of the situation (cf. orthophoto). Techniques fusing lidar range measurements with camera data have been explored in the context of autonomous driving [28]. Another variation of this approach involves latent fusion of all ranging modalities, such as lidar, radar, and sonar [29]. In the maritime context, the bird's-eye view approach has been shown to be particularly useful in Arctic waters when tracking ice floes [30].

Further modalities are available from the Inertial Measurement Unit (IMU) that consists of accelerometers, gyroscopes, and magnetometers. A common practice is to fuse the GNSS receiver with the IMU, to obtain position, velocity, and time. At open sea, when waves are present, the IMU may be used to estimate the inclination of the USV, and use this information to guide the algorithm's attention when processing visual data. In coastal areas, where land is visible, sensor fusion techniques may be used between lidar and IMU for localization [31]. These techniques can be extended to include vision data as well [32].

The GNSS data from other ships is available through the Automatic Identification System (AIS). Land stations collect this data as long time series,[2] which are valuable for training AI models. AIS data can be used to train path planning algorithms for long-term decision-making, such as route optimization over extended periods [33]. For short-term decision-making, the USV autopilot should combine AIS information [34] and real-time object detection to support collision avoidance and ensure compliance with the International Regulations for Preventing Collisions at Sea (COLREGs) .

Acoustic sensing for USVs allows for the autopilot to hear the foghorn warning signals and estimate the direction from where the signal is coming from [35]. The autopilot must be able hear them, as the foghorn is an essential IMO regulated safety signal device [9].

Underwater perception presents additional challenges [36]. For vision techniques, these include (1) rapid light attenuation meaning that water absorbs and scatters light much more strongly than air. (2) Color shift: different wavelengths of light are absorbed at different rates. Hence, images become dimmer and more monochromatic with increasing distance or depth, reducing the amount of information that can be captured. (3) More noise and error such as scattering from suspended particles, combined with an overall low contrast. Solutions include artificial illumination in the form of structured light, overall scene illumination, or AI-based image enhancement [37].

The availability of high-quality training data is crucial for deep learning models. While some datasets exist [23, 38], they remain insufficient for diverse tasks and for ensuring a high level of safety. As a result, the use of simulation-generated training data is becoming increasingly important. A simulator can, in principle, provide unlimited training data, but if this data does not correspond to reality, the trained model is useless. In Fig. 7.5, the image generated by AILiveSim is further processed with AI to enhance its realism. For illustration, we present the outputs produced by CycleGAN-turbo, CycleGAN, and CUT. The realism of the generated images can be assessed either through human judgments or quantitative metrics. Ultimately, the decisive test is whether a model trained on these synthetic images performs well on real-world data—success on such benchmark demonstrates true domain adaptation [39].

New research directions for USVs include incorporating the weather station data [40], measuring water turbidity, and fusing these with other modalities, as these areas are currently under-explored. These new sensor fusion techniques could bring advantages in visual and acoustic sensing, because the atmospheric conditions affect the refraction and attenuation of light and acoustic signals. A recent (niche) exteroception challenge is the detection of anchor states of neighboring vessels, see Fig. 7.9 in Con-

[2] HELCOM AIS dataset https://helcom.fi/action-areas/shipping/ais-and-e-navigation/.

Fig. 7.5 Simulated imagery and related challenges (resolution 256 × 256 pixels). Domain adaptation from the simulated image (left) to realistic ones is done using CycleGAN-turbo, CycleGAN, and CUT, respectively. Image by Vigneshwar Karunanithi, Fugro and University of Twente

clusion. This detection task requires close-range vision perception, and is virtually impossible using traditional land radars or satellite imagery. Ultimately, integrating machine learning with domain knowledge—such as the expertise of captains and navigators—would provide valuable insights and improve performance [41].

7.2.3 State of the Robot and Ontology

The state of the robot is typically described mathematically (e.g., by its position and velocity), but it can also be extended to include semantic variables, such as whether the Global Navigation Satellite System (GNSS) positioning is being jammed or whether there is fog limiting visual observations. Therefore, the relevant ontology for the USV includes both external and internal situational awareness. For example, the state of each sensor may be described as shown in Table 7.4.

Updates for the mathematical state are typically handled by a Kalman filter - type of solution. However, a semantic state would require another way of updating. Obtaining a more concise awareness of the situation could include fusing multiple semantic inputs into large language models (LLMs). These inputs could include the internal information but also ontologically structured verbal descriptions of the environment in form of scene graphs [42]. Hence, the internal state of having "degraded vision" could be combined with the environment state "it rains", leading into a better awareness for the autopilot. On the other hand, testing complex deep learning models in the context of safety certificates may prove challenging.

7.3 Who Calls the Shots? Teleoperation Versus Autonomy

Over the years robotic systems have matured from platforms with static movement patterns to platforms that learn optimal movements patterns, thereby increasing the different types of environment in which these platforms are able to move around robustly [43–46]. These new abilities, together with the advances in Artificial Intelligence, were received with great enthusiasm by the scientific community, which started to study, engineer, and develop autonomous robotic systems: robots with the ability to execute complex tasks at *ad hoc* locations in the real world [47–49]. The motivation has been to develop robots that could complete the entire mission autonomously. However, at present, most robots that operate in the real world, such as USVs and UUVs, are still remotely controlled, for various reasons. Two hard-to-overcome bottlenecks are (i) power/energy limitations, which make it difficult to run computationally intensive autonomy stacks for long durations, and (ii) restricted communication bandwidth, whether through low-data-rate acoustic links or physical tethers. Compounding these issues, robots still lack a sufficiently deep understanding of their surroundings to handle the ad-hoc, unfamiliar situations that inevitably arise in open-world deployments [50].

Humans are superior in understanding the real world and thus also the environment in which such robots are operating. Moreover, when they are given the awareness of the situation in which the robot is operating, human operators are well capable of piloting robots [51]. This does require that robots' operators are trained with piloting them, thereby learning the capabilities and performance of robots under various conditions. Still, there are practical limitations for humans to directly control the robots, mostly related to the data that needs to be exchanged between robots and their human operators: control signals, images of the environments, and sometimes forces acting on the robots. Especially underwater, where wireless communication is severely limited, such volumes of data cannot be exchanged [52]. For this reason underwater UUVs are tethered. However, even in configurations where an UUV is tethered to a USV that is piloted onshore via a wireless communication link, data transmission may be delayed or lost. At these moments human operators are not capable of piloting the USV-UUV combination. This means that the tethered USV-UUV combination must determine its remaining operational capabilities, for example, whether to continue the mission autonomously, wait until communication is restored, or return to base, refer to Fig. 7.6.

Communication is not the only trigger for the USV to autonomously decide how to proceed with the mission. Other triggers for activating autonomous decision-making should also be considered, including the underwater use case. These are discussed in the following section, followed by an analysis of how the USV combination might implement such triggering mechanisms.

Table 7.4 Proposition for states for each sensor or subsystem

State	Description
Normal	Functioning optimally, e.g., complying with Table 7.2 requirements
Degraded	Operating but e.g. unable to comply with Table 7.2 requirements
Jammed/Blocked	Sensor fusion (Sect. 7.2.2) with other sensors unsuccessful
Faulty	Sensor malfunctioning or offline, e.g., internal diagnostic error

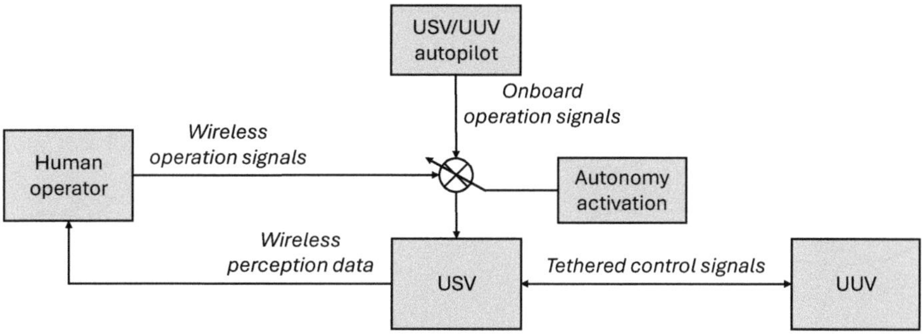

Fig. 7.6 Conceptual architecture of the USV–UUV team with alternating control between the human pilot and the autopilot

7.3.1 Trigger Conditions for Autonomy Activation

Both the human operator and the autopilot are part of a complete system aimed at producing proper inspection results, including measurements and images acquired from the infrastructure that is being inspected by the USV-UUV combination. In order for this inspection functionality to produce its results in a proper manner it has been designed with particular system requirements and environmental conditions in mind. Depending on whether control of the USV–UUV combination is exercised by a human operator or by the autopilot, different requirements and conditions must be satisfied to achieve a given level of inspection performance. The expected performance of these results will degrade, or may even fail entirely, when these requirements and conditions are (partially) not met. Figure 7.7 illustrates this dependency between the inspection performance when the USV/UUV is human-piloted or its inspection performance when it is autonomously piloted.

Before focusing on some key system requirements and environmental conditions that will induce a performance difference between a human-piloted approach and an autonomous approach, it is important to note how the system would be able to make such a (numerical) comparison. To that extent, we define variables that represent the *system state* (Sect. 7.2.3) and the *environmental state* (Sect. 7.2.1). When the corresponding system requirements and environmental conditions are satisfied, these variables induce specific levels of so-called Key Performance Indicators (KPIs), by which the inspection performance can be measured [53–55].

Although there may be many dependencies that affect the inspection performance of the human and the autonomous approach, we discuss some important ones that should at least be taken into account.

System requirements

The overall system is composed of a human operator, the USV/UUV autopilot, the USV and the UUV. In each of these systems faults may arise and their interfaces may not always be reliable. Some key faults and reliability issues are the following:

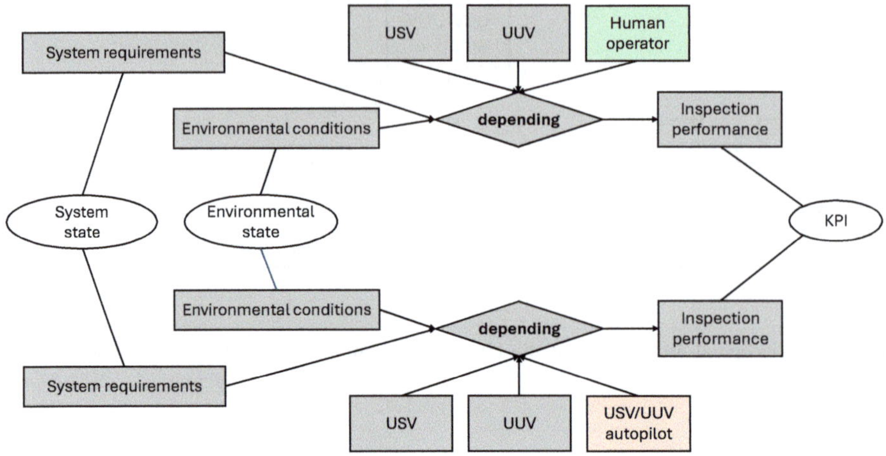

Fig. 7.7 Ontology graph modeling the system performance, where entities are illustrated with rectangular boxes, n-ary relation with diamond boxes and properties with ellipses. Note the symmetry, the human and the autopilot are both facing a similar system

- **Communication Loss**: The USV loses connectivity with the human operator located in the command center, either due to satellite link disruption or interference. Connectivity can be lost completely, or the capacity of the communication link may drop slightly, affecting both the control signal of the operator as well as the ability of the operator to create a situation awareness about the USV-combination.
- **Tether Entanglement** The UUV is tethered to the USV via a cable. While the UUV is inspecting underwater infrastructure, this cable may get stuck on surrounding structures or obstacles located between the UUV and the USV.
- **Propulsion Faults**: The USV is propelled by a combustion engine and the UUV by an electric motor. Either of the two could break or the energy supplied to them (fuel or electric) may reach critical levels.
- **Navigation faults**: The USV and the UUV have an internal navigation unit, in which the USV heavily depends on GNSS signals that may be unreliable, and the UUV's position estimate depends on seabed features and landmarks which may be sparse, ambiguous, or entirely absent in certain operational environments.

Environmental conditions

Both the USV and the UUV must operate in environmental conditions that cannot be controlled, such as weather, surface debris, and underwater obstacles. Sudden shifts in wind, waves, or currents can quickly change the USV's manoeuvrability, sometimes faster than an operator can react. If the USV drifts unexpectedly, the tether to the UUV may get stuck. Likewise, ad-hoc objects (floating logs, fishing gear, uncharted structures) can endanger either vehicle if the perception stack fails to detect them or if sensor data are lost in transmission. These realities make it essential to equip the USV-UUV pair with real-time decision algorithms to evaluate the environmental variables. Such algorithms typically rely on state estimation frameworks, reinforcement learning models, or probabilistic decision-making strategies to determine whether control should be allocated to the human operator or to the onboard autonomy, and to decide when a transfer of control (hand-off) should occur. Only seamless, timely transitions can guarantee reliable behavior under fast-changing environmental conditions.

7.3.2 Potential Solutions

Since most uncertainty will either come from loss in communication or loss in the situation awareness of the robot as perceived and interpreted by its operator, we consider a *layered operation model* in which the human pilot always outranks the UUV's onboard autopilot. This model provides a pragmatic solution for balancing teleoperation and autonomy: the USV-UUV operates autonomously by default, continuously assessing the system state and the environmental state while executing its mission plan. An overview of what the operator may expect from the autonomy of the USV-UUV is shown in the graphical user interface of the operator. The operator may use it as a guideline or disregard it at their choosing. In this approach, if human-issued commands are received, they are intended to override the autonomous control, allowing operators to intervene as necessary.

However, as mentioned, communication latency or temporary communication losses may result in human-issued commands not being received, or arriving late. Unless the system seamlessly resumes autonomous operation, this may result in undesired behavior of the combined USV-UUV system. Therefore, an additional mechanism should be in place to guarantee this seamless transition, i.e., to decide whether the human or the autopilot is more capable of piloting the USV-UUV combination in given circumstances, see Fig. 7.8. These candidate approaches are discussed next.

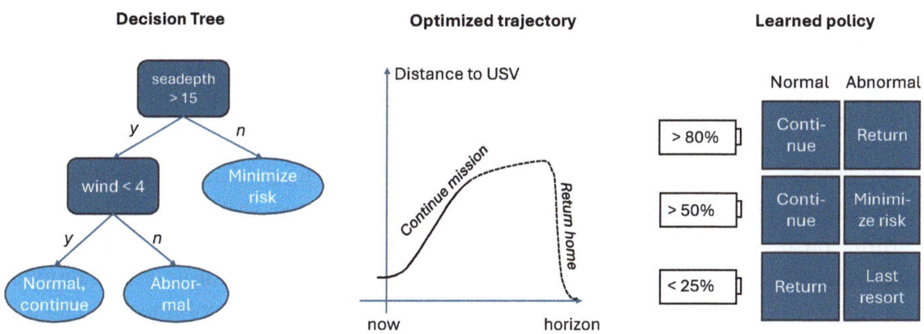

Fig. 7.8 Illustration of planning strategies for the UUV using decision trees (left), model predictive control (center), or the policy after reinforcement learning (right). See Sect. 7.1.2 for terminology

Decision trees

In a decision tree, the decisions are based on a series of (binary) choices and outcomes. The tree structure has nodes representing intermediate outcomes, and branches representing choices. The leaves of the tree represent the final decisions.

Decision trees are commonly used in machine learning for tasks like classification and regression, but have also been used in autonomous decision making [56, 57]. Their advantage in our use-case is that they are easy for a human operator to understand and interpret, they can handle both numerical and categorical data, while also capturing nonlinear relationships between the system and environmental states and the inspection KPIs. An important drawback of decision trees is their sensitivity, i.e., small changes in the input variables may lead to completely different decisions.

Hybrid model predictive control

Model predictive control [58] is an optimization-based control approach in which a prediction model and a (numerical) optimization algorithm are used to determine the optimal control inputs for the given system (in our case the USV-UUV combination) over a given time prediction window in such a way that one or more performance criteria (e.g. mission time, energy consumption) are optimized subject to various constraints (e.g., operational constraints). Next, the computed optimal control inputs are applied to the system, and after a while the state of the system is measured, and the whole optimization process is repeated for a shifted prediction window. In our use case both continuous dynamics (e.g., speed and position of the USV and UUV) and discrete events or actions (e.g., task or mode switching and the occurrence of faults) are present. This combination gives rise to so-called hybrid dynamics [59], which naturally motivates the use of *hybrid model predictive control*. The resulting optimization problem is therefore typically a mixed-integer nonlinear problem, involving both discrete (integer-valued) and continuous (real-valued) decision variables. However, if a piecewise-affine approximation of the nonlinear system dynamics is considered, the problem may be cast into a mixed-integer linear or quadratic form, for which efficient state-of-the-art solvers exist (such as cplex or gurobi).

Reinforcement learning

In reinforcement learning, an agent learns to make decisions by interacting with the system to maximize cumulative performance [60]. The agent takes an informed decision, then observes the consequences (rewards or penalties), and then adjusts its strategy over time to improve its performance.

Reinforcement learning is commonly used in machine learning for tasks like control and decision making [61]. Its advantage in our use-case is that reinforcement-learning-based agents can solve complex decision-making problems and that they may learn from interaction, thereby being able to adapt to dynamic environments. An important drawback of the approach is its reliance on large volumes of data and the limited explainability of the decision-making process by which the AI method arrives at its outcomes.

7.4 Cybersecurity of a Surveilling USV Fleet

USVs may prove highly useful as a patrol fleet for surveillance[3] of offshore and undersea infrastructure, e.g., for liability and insurance applications. We consider a hypothetical situation where commercially operated maintenance USVs are leveraged to collectively acquire a joint situational awareness, which is subsequently shared with multiple parties, such as coast guard and other authorities. This perspective of sharing a situational awareness derived from (autopiloted) USVs calls for an analysis on cybersecurity.

Cybersecurity has become a pillar of modern maritime security. At the global level, the IMO now requires every member flag State, ship operator, and manager to address cyber risk. Its Guidelines on Maritime Cyber Risk Management (MSC-FAL.1/Circ.3) set out high-level principles, while Resolution MSC.428(98) mandates that cyber risk be woven directly into each company's Safety Management System (SMS).[4] The operational context is increasingly cyber-physical: navigation, propulsion, and cargo-handling systems are deeply networked, so weaknesses in access control, interconnection, or networking can expose the vessel to attack.[5]

Within the European Union, the European Maritime Safety Agency (EMSA) supports both the Commission and Member States by pooling expertise and sharing best practice.[6] In parallel, the EU and NATO have pledged—through their 2023 Joint Declaration—to coordinate cybersecurity efforts across ports, shipping companies, and naval forces.[7] Several specialized bodies translate that pledge into practice. NATO Cooperative Cyber Defence Centre of Excellence (CCD COE) supports national decision-makers in strengthening their National Cyber Security Strategies (NCSS).[8] Similarly, the European Union Agency for Cybersecurity (ENISA) works to establish high cybersecurity standards across Europe.[9] Together, these frameworks and institutions provide the layered governance—international, regional, and national—needed to secure today's highly connected maritime infrastructure.

[3] Defense news, Dec 3rd 2024, NATO plans its own drone fleet, https://www.defensenews.com/.

[4] IMO, 2025, The Guidelines on Cyber Security Onboard Ships.

[5] IMO, 2017, Maritime Cyber Risk Management in Safety Management System. Resolution Maritime Safety Committee, MSC 428(98).

[6] European Maritime Security Agency (2022) This is EMSA. Available from: http://www.emsa.europa.eu/about.html.

[7] European Commission. 2023. Joint Declaration on EU-NATO Cooperation, 10 January 2023. https://eur-lex.europa.eu/legal-content/EN/TXT/HTML/?uri=CELEX:52023JC0008.

[8] CCDOE, 2025 The NATO Cooperative Cyber Defence Centre of Excellence is a multinational and interdisciplinary cyber defense hub https://ccdcoe.org/about-us/.

[9] ENISA, 2025 What we do. https://www.enisa.europa.eu/about-enisa/what-we-do.

7.4.1 Surveilling OffShore Infrastructure in the Era of Hybrid Warfare

Hybrid warfare combines multiple warfare types by using conventional and unconventional instruments. NATO has indicated the importance of protecting the maritime environment, especially undersea cables, and pipelines that enable communications and provide energy to NATO and EU countries. The allies collaborate to enhance overall situational awareness by enhancing surveillance mechanisms, and to create strategies that support each other. When deploying these strategies, it is important to do so in all the levels—operational, tactical, and technical [62].

When an USV is deployed above dense cable corridors, its cybersecurity architecture must be specified from the outset. A hybrid attack typically uses a diversion in one domain (e.g. a cyber intrusion) to mask a more destructive act in another (e.g. physically severing a fibre-optic cable). The end-state is the same: to disable critical infrastructure and undermine the basic functions of society. The ISA/IEC 62443 control-system family is the reference standard for hardening such cyber-physical assets [63]. Applied to a USV-UUV system, it mandates protection requirements across four domains: (1) shore-side command-and-control and human operators (2) the USV/UUV autopilot (3) the USV, and (4) the UUV. Interface security is pivotal. A tethered UUV benefits from its physical link, but the radio frequency (RF) or satellite channels that connect the USV to shore are exposed to jamming, spoofing, and *man-in-the-middle* attacks. If the USV loses connectivity with humans in the command center (possibly due to a cyberattack), the autopilot would act as the recovery system. However, capturing these systems is relatively easy.

Navigation may be a prime attack vector, as USVs depend heavily on Global Navigation Satellite Systems (GNSS). Jamming simply blocks the signal; spoofing is more insidious, forging authentic-looking satellite data so the autopilot obeys a falsified track [64, 65]. Resilient navigation therefore requires dissimilar backups such as eLoran[10], magnetometer cross-checks, radar fixes, or even modern celestial navigation.

Man-in-the-middle attacks exploit the two-way data links on which every remotely operated vessel depends, allowing an adversary to intercept, modify, or replay control messages [66]. With even brief administrative-level access to propulsion or power-management subsystems, an attacker can throttle engines, disable steering, or falsify status reports so that an automated safety shutdown is triggered at a moment of their choosing. Once data integrity is lost, the very safeguards designed to protect the vessel can be weaponized to damage equipment or halt the mission [67].

Total automation is neither feasible nor desirable. A safer architecture treats the USV-UUV system as a set of segmented subsystems that can be selectively isolated from a compromised component, while still preserving overall command authority and situational awareness for the operator in the control room. Human oversight is still essential.

---

[10] eLoran is a terrestrial radio-navigation system with an 800–1500 km range.

7.4.2 Shared Hybrid Situational Awareness

Situational awareness relies on a combination of human and machine elements. On the human side, factors such as perception and cognitive processes shape how individuals understand and interact with their surroundings [68]. On the USV side, the autopilot fuses sensor data to build a picture that—under nominal conditions—mirrors the operator's own view (see Fig. 7.7). Once that picture is distributed to actors beyond the AARC-enhanced remote pilot, we refer to it as Hybrid Situational Awareness (HSA).

HSA starts by defining a clear "goal state" for awareness at the strategic, operational, and tactical layers. Reaching and sustaining that goal demands a coherent command-and-control framework able to translate semantic information between humans and machines in real time. Maintaining HSA therefore depends on the joint ability of people and algorithms to sense environmental changes, recognize threats, and adapt to evolving events. Flexible autonomy should let control pass smoothly back and forth between human and system whenever workload, connectivity, or mission phase requires that [69, 70]. Because cyber-physical threats can degrade the situational awareness in seconds, rapid exchange of threat data must be a core service. The command-and-control architecture therefore needs a resilient fall-back path—whether terrestrial, maritime, or space-based—so that a break in the primary link does not collapse decision-making or safety-critical functions.

7.5 Conclusion and Research Directions

AUS present large scale benefits for offshore infrastructure maintenance, removing humans from dull, dirty, and dangerous work. Reaping these benefits requires developing AUS resilience against challenges posed by climate and environment, and cyber-physical factors. The North Sea climate is harsh with strong winds, sudden storms, occasional big waves, and powerful tidal currents. The resilience is achieved by increased autonomy, in the form of Autonomy-Assisted Remote Control (AARC) and an autopilot that can take over and guarantee mission safety. The need for the autonomy is four-fold:

- In case of communication loss, the USV autopilot must take over to guarantee mission safety (in contrast to airline autopilots which are surveilled by pilots).
- Autonomy capabilities are required for over the horizon remote operations because of round-trip delays and limited bandwidth, especially for interaction tasks and when navigating close to structures at the sea.
- Autonomous perception capabilities are needed to reduce the volume of digital situational awareness that must be transmitted; instead of raw sensor data such as video feeds, this awareness can be sent in the more compact form of a digital twin.
- Multi-vehicle operations are best executed with AARC, for efficient and seamless coordination.

Fig. 7.9 Perception AI needs further development. Image of Eagle S ship that was involved in the cable cutting incident at Bay of Finland in Dec 2024 (Image: poliisi.fi). Segmentation is from Segment Everything Everywhere method [71] using "ship anchor" prompt. Although the system can segment people on deck, even though they are smaller than the ship's anchor, reliably recognising the anchor itself will still require additional training

The systems make use of AI in multiple ways. The USV maintains its own situational awareness and performs continuous evaluation in deciding on whether to go on autopilot or not. To this end, the evaluation on whether the operational conditions are normal or abnormal is critical, with the additional understanding about if the objective is to aim for continuing the mission, to minimize risks, or to perform last resort actions. Depending on the system state, the environment, and a list of key performance indicators, the decision for the autopilot to take over can be established (Fig. 7.7).

The latest perception AI techniques are likely to be highly useful. For understanding the surroundings (exteroception), the state of the USV and UUV (interoception), and in formulating the decision-making for triggering between AARC and autopilot. The triggering options need scientific exploration, be it with decision trees, predictive control, or reinforcement learning. Furthermore, after establishing a need to go on autopilot, the system must adapt a suitable control strategy to continue its mission. To this end, joint-embedding predictive architectures offer interesting possibilities [72]. Finally, concerning AARC and autopilot utilization in scenarios that include the use of force, please refer to [73].

References

1. Junji F (2021) Automation levels of automated/autonomous ships. ClassNK Techn J 4:35–50
2. Porathe T, Hoem Å, Rødseth Ø, Fjørtoft K, Johnsen SO (2018) At least as safe as manned shipping? autonomous shipping, safety and "human error". Safety Reliability—Safe Societ Changing World 417–425
3. Autonomous and remotely operated ships. Standard, Sept (2018)

4. Ericson II CA (2011) Hazard analysis techniques for system safety. Hoboken, Wiley and Sons, incorporated
5. Aven T (2019) The science of risk analysis: foundation and practice. Abingdon, UK: Routledge
6. Wróbel K, Montewka J, Kujala P (2017) Towards the assessment of potential impact of unmanned vessels on maritime transportation safety. Reliability Eng Syst afety 165:155–169
7. Valdez Banda OA (2024) MASS and safety. Springer Nature Switzerland, pp 67–91
8. Fan C, Montewka J, Zhang D (2022) A risk comparison framework for autonomous ships navigation. Reliability Eng Syst Safety 226:108709
9. Thombre S, Zhao Z, Ramm-Schmidt H, Garcia JM, Malkamäki T, Nikolskiy S, Hammarberg T, Nuortie H, Bhuiyan MZ, Särkkä S, Lehtola VV. (2020) Sensors and AI techniques for situational awareness in autonomous ships: a review. IEEE Trans Intell Transport Syst
10. Burmeister HC, Bruhn W, Walther L, Moræus JA, Sage-Fuller B (2016) D8.6: final Report: Autonomous Bridge. Technical report, The Fraunhofer Center for Maritime Logistics and Services
11. Liu Z, Zhang Y, Xiang Y, Yuan C(2016) Unmanned surface vehicles: an overview of developments and challenges. Annual Rev Control 41:71–93
12. Humphries F, Horne R, Olsen M, Dunbabin M, Tranter K (2023) Uncrewed autonomous marine vessels test the limits of maritime safety frameworks. WMU J Maritime Affairs 22(3):317–344
13. Liu RW, Guo Y, Nie J, Hu Q, Xiong Z, Yu H, Guizani M (2022) Intelligent edge-enabled efficient multi-source data fusion for autonomous surface vehicles in maritime internet of things. IEEE Trans Green Commun Netw 6(3):1574–1587
14. Cheng L, Xiong R, Wu J, Yan X, Yang C, Zhang Y, He Y (2024) Fast segmentation algorithm of USV accessible area based on attention fast deeplabv3. IEEE Sensors J 24(15):24168–24177
15. Li W, Zhang R, Lin H, Guo Y, Ma C, Yang X (2024) Saccadedet: a novel dual-stage architecture for rapid and accurate detection in gigapixel images. In: Joint European conference on machine learning and knowledge discovery in databases. Springer, pp 392–408
16. Dosovitskiy A, Beyer L, Kolesnikov A, Weissenborn D, Zhai X, Unterthiner T, Dehghani M, Minderer M, Heigold G, Gelly S et al (2021) An image is worth 16x16 words: transformers for image recognition at scale. An image is worth 16x16 Words: transformers for image recognition at scale, international conference on learning representations (ICLR), https://openreview.net/forum?id=YicbFdNTTy
17. Kai H, Yunhe W, Hanting C, Xinghao C, Jianyuan G, Zhenhua L, Yehui T, An X, Xu C, Xu Y et al (2022) A survey on vision transformer. IEEE Trans Pattern Anal Mach Intell 45(1):87–110
18. Zhu L, Liao B, Zhang Q, Wang X, Liu W, Wang X (2024) Vision mamba: efficient visual representation learning with bidirectional state space model. In: Forty-first international conference on machine learning
19. Dao T, Gu A (2024) Transformers are ssms: generalized models and efficient algorithms through structured state space duality. Proceedings of the 41st International Conference on Machine Learning (ICML)
20. Awais M, Naseer M, Khan S, Anwer RM, Cholakkal H, Shah M, Yang M-H, Khan FS (2025) Foundation models defining a new era in vision: a survey and outlook. IEEE Trans Pattern Anal Mach Intell
21. Er MJ, Zhang Y, Chen J, Gao W (2023) Ship detection with deep learning: a survey. Artif Intell Rev 56(10):11825–11865
22. Gorad A, Hassan S, Särkkä S (2023) Vessel bearing estimation using visible and thermal imaging. In: Scandinavian conference on image analysis. Springer, pp 373–381
23. Bogdan I, Jesper W, Valentin S, Johan L (2023) A benchmark for maritime object detection with Centernet on an improved dataset, Aboships-plus. J Marine Sci Eng 11(9):1638
24. Lehtola VV, Montewka J, Salokannel J (2020) Sea captains' views on automated ship route optimization in ice-covered waters. The J Navigat 73(2):364–383
25. Yifan S, Yu W, Guangmiao Z, Huihui L, Wu Y (2024) Stereoyolo: a stereo vision-based method for maritime object recognition and localization. J Marine Sci Eng 12(1):197
26. Kotaro Y, Shigehiro Y, Takeshi H (2025) On-ship trinocular stereo vision: an experimental study for long-range high-accuracy localization of other vessels. J Marine Sci Eng 13(1):115

27. Wan Z, Zhang P, Wang Y, Yong S, Stepputtis S, Sycara K, Xie Y (2025) Sigma: Siamese mamba network for multi-modal semantic segmentation. IEEE/CVF Winter conference on applications of Computer Vision (WACV)
28. Liu Z, Tang H, Amini A, Yang X, Mao H, Rus DL, Han S (2023) Bevfusion: multi-task multi-sensor fusion with unified bird's-eye view representation. In: 2023 IEEE international conference on robotics and automation (ICRA). IEEE, pp 2774–2781
29. Balemans N, Anwar A, Steckel J, Mercelis S (2024) Lidar-bind: multi-modal sensor fusion through shared latent embeddings. IEEE Robot Automat Lett 9(11):9159–9166
30. Sandru A, Hyyti H, Visala A, Kujala P (2020) A complete process for shipborne sea-ice field analysis using machine vision. IFAC-PapersOnLine 53(2):14539–14545
31. Xu W, Cai Y, He D, Lin J, Zhang F (2022) Fast-lio2: fast direct lidar-inertial odometry. IEEE Trans Robot 38(4):2053–2073
32. Zheng C, Xu W, Zou Z, Hua T, Yuan C, He D, Zhou B, Liu Z, Lin J, Zhu F et al (2024) Fast-livo2: fast, direct lidar-inertial-visual odometry. IEEE Trans Robot
33. Nguyen D, Fablet R (2024) A transformer network with sparse augmented data representation and cross entropy loss for AIS-based vessel trajectory prediction. IEEE Access
34. Sonntag V, Perrusquia A, Tsourdos A, Guo W (2025) A colregs compliance reinforcement learning approach for USV Manoeuvring in track-following and collision avoidance problems. Ocean Eng 316:119907
35. Gorad A, Zhao Z, García JMV, Lehtola V, Hammarberg T, Ramm-Schmidt H, Islam S, Thombre S, Särkkä S (2025) Bearing estimation using foghorn sounds. Appl Acoust 231:110560
36. Huy DQ, Sadjoli N, Azam AB, Elhadidi B, Cai Y, Seet G (2023) Object perception in underwater environments: a survey on sensors and sensing methodologies. Ocean Eng 267:113202
37. Raveendran S, Patil MD, Birajdar GK (2021) Underwater image enhancement: a comprehensive review, recent trends, challenges and applications. Artif Intell Rev 54:5413–5467
38. Zhenfeng S, Wu W, Zhongyuan W, Du W, Chengyuan L (2018) Seaships: a large-scale precisely annotated dataset for ship detection. IEEE Trans Multimedia 20(10):2593–2604
39. Kay J, Haucke T, Stathatos S, Deng S, Young E, Perona P, Beery S, Van Horn G (2025) Align and distill: unifying and improving domain adaptive object detection. Trans Mach Learn Res. Featured Certification
40. Lehtola VV, Mäkelä M, de Oliveira Marques T, Montloin L (2022) Tropospheric wet tomography and PPP: Joint estimation from GNSS crowdsourcing data. Adv Space Res 70(8):2399–2411
41. Ilioudi A, Dabiri A, Wolf BJ, De Schutter B (2022) Deep learning for object detection and segmentation in videos: towards an integration with domain knowledge. IEEE Access 10:34562–34576
42. Xiaojun C, Pengzhen R, Xu P, Zhihui L, Xiaojiang C, Alex H (2021) A comprehensive survey of scene graphs: generation and application. IEEE Trans Pattern Anal Mach Intell 45(1):1–26
43. Haarnoja T, Ha S, Zhou A, Tan J, Tucker G, Levine S (2019) Learning to walk via deep reinforcement learning. Proceedings of robotics: science and systems
44. Li T, Lambert N, Calandra R, Meier F, Rai A (2020) Learning generalizable locomotion skills with hierarchical reinforcement learning. In: 2020 IEEE international conference on robotics and automation (ICRA), IEEE, pp 413–419
45. Shi H, Zhou B, Zeng H, Wang F, Dong Y, Li J, Wang K, Tian H, Meng MQ-H (2022) Reinforcement learning with evolutionary trajectory generator: a general approach for quadrupedal locomotion. IEEE Robot Automat Lett 7(2):3085–3092
46. Jiachen Y, Jingfei N, Meng X, Jiabao W, Yang L (2022) Intelligent path planning of underwater robot based on reinforcement learning. IEEE Trans Automat Sci Eng 20(3):1983–1996
47. Sørensen AJ, Ludvigsen M (2015) Towards integrated autonomous underwater operations. IFAC-PapersOnLine 48(2):107–118
48. Petillot YR, Antonelli G, Casalino G, Ferreira F (2019) Underwater robots: from remotely operated vehicles to intervention-autonomous underwater vehicles. IEEE Robot Automat Magazine 26(2):94–101

49. Edward M, Ignacio C, William A, Corina B (2022) Autonomous underwater manipulation: current trends in dynamics, control, planning, perception, and future directions. Current Robot Reports 3(4):187–198
50. Martins A, Almeida C, Pereira R, Sytnyk D, Soares E, Matias B, Peixoto PA, Ferreira A, Machado D, Almeida J (2024) Robotic field experiments for critical infrastructure protection in the repmus 2023 military exercise. In: OCEANS 2024-Singapore. IEEE, pp 1–6
51. Adetunji FO, Ellis N, Koskinopoulou M, Carlucho I, Petillot YR (2024) Digital twins below the surface: enhancing underwater teleoperation. In OCEANS 2024-Singapore (pp. 1–8). IEEE
52. Van der Werf I, Dol H, Blom K, Heusdens R, Hendriks RC, Leus G (2024) On the equivalence of osdm and otfs. Signal Process 214:109254
53. Sifat AH (2024) A safety-performance framework for computational awareness in autonomous robots
54. Verhagen RS, Neerincx MA, Tielman ML (2024) Meaningful human control and variable autonomy in human-robot teams for firefighting. Front Robot AI 11:1323980
55. Sisman B, Djedilbaev T (2024) Awareness in robots. In: European robotics forum. Springer, pp 98–102
56. Peddi R, Bezzo N (2023) A decision tree-based monitoring and recovery framework for autonomous robots with decision uncertainties. In: 2023 IEEE/RSJ international conference on intelligent robots and systems (IROS). IEEE, pp 9722–9728
57. Dubslaff C, Klös V, Päßler J (2024) Template decision diagrams for meta control and explainability. In: World conference on explainable artificial intelligence. Springer, pp 219–242
58. Rawlings JB, Mayne DQ (2009) In: Model predictive control: theory and design. Nob Hill Publishing, Madison, Wisconsin
59. Lunze J, Lamnabhi-Lagarrigue F (eds) (2009) In: Handbook of hybrid systems control: theory, tools, applications. Cambridge University Press
60. Sutton RS, Barto AG (2018) In: Reinforcement learning: an introduction. MIT Press
61. Carlucho I, De Paula M, Wang S, Petillot Y, Acosta GG (2018) Adaptive low-level control of autonomous underwater vehicles using deep reinforcement learning. Robot Autonomous Syst 107:71–86
62. Pöyhönen J, Lehto M (2020) Cyber security: trust based architecture in the management of an organization security. In: Proceedings of the 18th European conference on cyber warfare and security ECCWS2020, 25–26 June 2020, Chester, UK. Academic Conferences and Publishing International Limited Reading, pp 304–313
63. Hollerer S, Sauter T, Kastner W (2022) Risk assessments considering safety, security, and their interdependencies in ot environments. In: Proceedings of the 17th international conference on availability, reliability and security, pp 1–8
64. Borhani-Darian P, Li H, Wu P, Closas P (2024) Detecting GNSS spoofing using deep learning. EURASIP J Adv Signal Process 2024(1):14
65. Rødningen T, Hauglin H, Rødningsby A (2023) Over-the-air jamming and spoofing tests of gnss timing devices. In: 2023 Joint conference of the European frequency and time forum and IEEE international frequency control symposium (EFTF/IFCS). IEEE, pp 1–3
66. Simola J, Takala A, Lehkonen R, Frantti T, Savola R (2024) Improving detection capabilities in ot environments through multisource data sensors. In: European conference on cyber warfare and security, vol 23. pp 496–505
67. Denzler P, Hollerer S, Frühwirth T, Kastner W (2021) Identification of security threats, safety hazards, and interdependencies in industrial edge computing. In: 2021 IEEE/ACM symposium on edge computing (SEC). IEEE, pp 397–402
68. Endsley MR (1995) Measurement of situation awareness in dynamic systems. Human Factors 37(1):65–84
69. Endsley MR (1988) Design and evaluation for situation awareness enhancement. In: Proceedings of the human factors society annual meeting, vol 32. Sage Publications Sage CA, Los Angeles, CA, pp 97–101
70. Endsley MR (2015) Autonomous horizons: system autonomy in the air force-a path to the future. In: United States air force office of the chief scientist, AF/ST TR, vol 15(6). pp 1–34

71. Zou X, Yang J, Zhang H, Li F, Li L, Wang J, Wang L, Gao J, Lee YJ (2024) Segment everything everywhere all at once. In: Advances in neural information processing systems, pp 36
72. Dawid A, LeCun Y 2024) Introduction to latent variable energy-based models: a path toward autonomous machine intelligence. J Statist Mechan: Theory Experim 2024(10):104011
73. Verdiesen EP (2024) Comprehensive human oversight over autonomous weapon systems, Ph.D. dissertation, TU Delft, The Netherlands

Chapter 8
A Research Portfolio to Advance the Resilience of Energy Systems: The Resilient Energy Systems Mission Campaign

Alexandria Haddad, Shamina Hossain-McKenzie, Jamie Thorpe, Holly Eagleston, Laura Biedermann, Matthew Reno, Michael Ropp, Kyle Skolfield, and Jack Flicker

Abstract Energy security is a pressing global concern, as nations dependent on electric grids and critical infrastructure face vulnerabilities from various threats, including electromagnetic pulse, cyber-physical attacks, kinetic attacks, and increasingly frequent natural disasters. For the U.S. and its allies, these threats represent significant security risks, as compromised infrastructures can undermine essential mission functions. Securing these systems is a top priority, and the international community acknowledges the need for cross-border cooperation and partnerships. In response, Sandia National Laboratories has initiated a Mission Campaign focused on Resilient Energy Systems, aimed at advancing innovative science and technology to enhance energy security and the resilience of critical infrastructure. While the campaign primarily targets U.S. energy systems, its research outcomes are applicable globally and align with NATO's energy security objectives. This chapter provides an overview of the Mission Campaign's three research thrusts: (1) Science of Vulnerabilities, which develops tools to assess risks and vulnerabilities; (2) Materials, Device, and Cyber Innovation, which seeks new high-energy materials and cost-effective solutions for physical and cyber threats; and (3) System-Level Threat-Informed Computational Science, which analyzes critical elements of the electrical energy system to create a more resilient architecture. The chapter also describes an integrated portfolio of research across four applied areas: a cyber vulnerabilities toolset, resilience measurement and optimization, solid-state transformers, and Electromagnetic Pulse mitigation. Finally, the chapter highlights specific projects and tools under development, and outlines plans for translating research into real-world applications.

A. Haddad (✉) · S. Hossain-McKenzie · J. Thorpe · L. Biedermann · M. Reno · M. Ropp · K. Skolfield · J. Flicker
Sandia National Laboratories, Albuquerque, NM, USA
e-mail: awhadda@sandia.gov

H. Eagleston
Firescape Inc., Albuquerque, NM, USA

M. Huber et al. (eds.), *Critical Infrastructure Systems and Extreme Threat Nexus: Preparing for Resilience and Security*, NATO Science for Peace and Security Series C: Environmental Security, https://doi.org/10.1007/978-3-032-13170-6_8

Keywords Vulnerabilities · Materials · Devices · Cyber · Innovation · System-level · Threat-informed · Computational science · Resilience · Optimization · Solid-state transformers · Electromagnetic pulse

8.1 Introduction

The NATO alliance confronts significant energy resilience challenges shaped by its military operations and the complex geopolitical landscape. Key issues include ensuring reliable energy supplies for military forces, protecting critical energy infrastructure from various threats, and adapting to an evolving energy landscape characterized by climate change and technological advancements. NATO's reliance on civilian energy networks exposes it to vulnerabilities from natural disasters, cyber-attacks, and industrial accidents, necessitating a robust operational energy strategy.

Cyber-attacks are of increasing concern. A notable example is the CRASHOVERRIDE attack on the Ukrainian power grid in 2015 [1] that left 230,000 customers without power. But recent examples of large-scale cyber-attacks such as the attack on Danish critical infrastructure in May of 2023 [2]–during which 22 companies were compromised—reinforce the need for resilience against these intentional threats. These threats combined with threats from increasing weather events [3] and aging infrastructure are exacerbated by the fact that many countries' infrastructure is interconnected across borders leading to the need for a holistic, international view of infrastructure resilience.

The U.S., serving approximately 72% of its electricity customers through investor-owned utilities [4], shares many of these energy resilience challenges. The U.S. energy landscape is characterized by an aging grid infrastructure, cybersecurity threats, and the need for modernization to accommodate increasing energy demands. Both NATO and the U.S. recognize the necessity of international cooperation and public–private partnerships to enhance energy security and resilience.

This chapter aims to leverage lessons learned and best practices from U.S. energy resilience research, conducted within the Resilient Energy Systems Mission Campaign (RES MC) at Sandia National Laboratories, to bolster NATO's energy security. By examining the commonalities and unique challenges faced by both entities, we can identify strategies for improving energy resilience in military operations and critical infrastructure protection. Ultimately, enhancing NATO's energy security will require a multifaceted approach that integrates innovative solutions, robust infrastructure investments, and collaborative efforts among member states and private sector stakeholders.

At inception, the RES MC organized around three research thrusts, described in Sect. 8.2 of this chapter. After executing more than 45 projects, the research portfolio is culminating in four Marquee Product areas, described in Sect. 8.3, intended to provide a framework for demonstrating real world applications of research. The RES investment roadmap presented in Sect. 8.2 displays the link between the research thrusts and Marquee Project areas as the research has progressed over the course of the seven-year Mission Campaign lifecycle. Following the timeline in the MC roadmap, Sect. 8.4 provides overviews of completed projects and current research projects in Sect. 8.5. Section 8.6 provides the current research strategy for the MC moving forward as well as thoughts about life after the MC and conclusions in Sect. 8.7.

8.2 Overview of the RES MC

The U.S. Department of Energy's Laboratory Directed Research and Development (LDRD) program invests in high-risk, potentially high-payoff activities that support national security missions and advance science and engineering. As Sandia's primary source of discretionary R&D funding, LDRD enables research not tied to ongoing program spending, addressing the needs of the Department of Energy (DOE), National Nuclear Security Administration (NNSA), and the nation. This flexibility allows for quick responses to emerging R&D needs and opportunities, exploring disruptive science and technology for future missions. Mission Campaigns are critical components of the LDRD program and are agile, time-limited (5–7 years), and strategically focused research portfolios intended to evolve to meet critical mission needs for U.S. national security.

Within the LDRD program and in response to the increasing threats to the energy system and the need for a more resilient and secure energy system, in 2020 Sandia National Laboratories (Sandia) funded the Resilient Energy Systems Mission Campaign (RES MC), a 7-year, $50 M + effort focused on increasing the resilience of the electrical energy system to intentional threats. Although focused on U.S. energy security and critical infrastructure it aligns broadly with the international community and NATO's energy security strategy, in particular. The RES MC leverages Sandia's unique Science and Technology (S&T) competencies to address the resilience, survivability and recovery of national energy systems, particularly electrical, against high-impact intentional threats. This research portfolio is delivering:

- An understanding/assessment of current vulnerabilities,
- Innovative materials, device technologies, and cyber solutions that toughen equipment nodes and increase system resilience, and
- Advanced system-level algorithmic, topological, and modeling approaches to enable defensive postures, create system flexibility and responsiveness, limit consequences, and accelerate recovery.

The RES MC's work to advance energy systems' resilience is intricately linked to the nexus of cyber-physical-social infrastructures and climate change, encompassing the interplay of technological, physical, social, and environmental factors that influence energy stability. Cybersecurity plays a vital role in safeguarding energy infrastructure from digital threats, while physical components, such as power plants and transmission lines, must be resilient to various challenges, including extreme weather events exacerbated by climate change. Social factors, including community engagement and public awareness, are essential for fostering resilience and adapting to shifting energy needs. Furthermore, energy systems are interconnected with other critical infrastructures, making it crucial to understand these interdependencies to prevent cascading failures. The adaptive capacity of energy systems to respond to evolving threats and demands is key to maintaining resilience. Additionally, adapting to climate change impacts and ensuring the security of renewable energy technologies are critical for long-term viability. An integrated approach that addresses these interconnected elements is necessary to ensure a reliable and secure energy supply in the face of ongoing challenges from climate change and cyber threats.

Given the range of existing and emerging threats to the nation's energy infrastructure, the RES MC primarily targets intentional threats, including cyber-attacks, cyber-physical attacks, and electromagnetic pulses (EMPs) from high-altitude nuclear explosions. While the grid is a central element of the energy system, the funded R&D aims to provide value across multiple energy domains and related critical infrastructures. The RES MC seeks to enhance knowledge of vulnerabilities, advance technical capabilities, and test novel system architectures to improve energy infrastructure resiliency.

For the RES MC, intentional threats against energy systems were deemed the highest priority, since these systems underpin every other critical infrastructure—water, food, transportation, communications, finance, and military operations—and have complex failure modes that can start with the loss of a few components but end in a cascading collapse that propagates across the country. This prioritization is reinforced by a 2017 Congressional Study that identified cyber-physical attacks and EMPs from a high-altitude nuclear explosion, and Geomagnetic Disturbances (GMDs) as urgent national security concerns [5].

Substation attacks, and the vulnerability of today's transformers, represent a significant threat to the integrity and reliability of electrical grids because these facilities serve as critical nodes in the distribution of electricity from generation sources to consumers. Such attacks can take various forms, including physical intrusions, cyber intrusions, or a combination of both, with the potential to disrupt power supply and compromise public safety. Notable attacks such as the 2013 sniper attack on a California substation [6] or the 2022 North Carolina substation attack [7], highlight the vulnerabilities of these infrastructures to coordinated assaults that can cause extensive damage and prolonged outages.

The RES MC is organized around three research thrusts. The **Science of Vulnerabilities** thrust supports development of new tools to assess vulnerabilities and risk in energy and critical infrastructure systems, including potential physical failure mechanisms of node electrical hardware and control equipment for intentional and sophisticated kinetic, EMP, GMD, and cyberattacks. The **Materials, Devices, and Cyber Innovation** thrust focuses on the discovery and development of new high-energy materials, cost-effective physical devices, and cyber solutions. The last thrust, **System-Level Threat-Informed Computational Science**, considers the critical elements of the electrical energy system, mechanisms of cascading failures, and potential approaches for a more resilient/secure architecture. The goal is to create

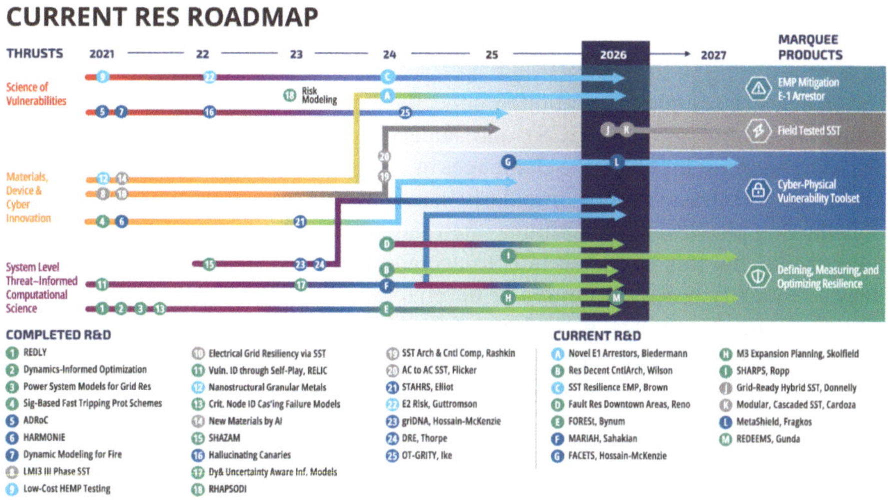

Fig. 8.1 RES MC roadmap

tools to analyze and assess how to optimize and deploy solutions for maximum benefit in terms of economics, security, and resilience. Together, these three thrust areas create an integrated portfolio of R&D to answer fundamental questions relevant to improving resilience among US energy systems and other critical infrastructures.

To address these vulnerabilities, the RES MC team developed an annually revolving call for projects, starting in FY21, seeking proposals in three research thrust areas. Figure 8.1 captures the completed and active RES MC projects.

8.3 Marquee Research Products

During the initial stages of the RES MC, selecting projects aligned with the research thrusts provided a diverse portfolio. However, to maximize impact, it was essential to focus on portfolio integration, project cohesion and demonstratable research outcomes. To enable this, Sandia leadership defined four Marquee Product areas. Marquee Product areas would allow for demonstration of real-world impact and meaningful contributions to energy and national security. Those four areas are described below. Research funded in the later years of the MC is intentionally focused on the Marquee Product areas and builds on prior research to provide demonstratable products in each of the four Marquee areas by the end of the RES MC in September of 2027.

8.3.1 Cyber-Physical Vulnerability Tools

The increasing complexity and interconnectivity of Operational Technology (OT), industrial control systems (ICS) and supervisory control and data acquisition (SCADA) systems highlight the challenges in assessing and mitigating vulnerabilities within critical infrastructure, particularly in the energy sector. Traditional approaches to evaluating physical vulnerabilities often rely on damage thresholds for individual components, but this approach fails to account for the stochastic nature of failure and is frequently based on limited or unreliable testing. Furthermore, the evolving landscape of cyber threats, including manipulation of network-control systems and malicious actions by embedded devices, exacerbates these vulnerabilities. Further complicating OT network situational awareness, protocols are typically poorly documented with a diverse array of ICS devices, necessitating a more robust analytical framework.

In this area, the RES MC called for proposals with novel approaches for advanced analysis of components and the timely ingestion of system information into existing emulytics and modeling-and-simulation (mod-sim) capabilities. Additionally, the team encouraged research that enabled rigorous cyber-physical analysis of energy and critical infrastructure systems, integrating information technology and operational technology, using network and process-layer data. The team also required work centered in secure data exchange between existing cybersecurity tools and power-system analytics, along with the development of cyber-physical data fusion, and machine learning (ML) pipelines using clearly defined implementation requirements.

8.3.2 Electromagnetic Pulse (EMP) Mitigation

The effects of an EMP, such as those resulting from a high-altitude detonation of a nuclear weapon can be damaging to the electric grid and critical grid components such as transformers [8]. The time regime for EMPs is in the nanoseconds and grid operators currently do not have devices that can mitigate such fast energy pulses as they couple with grid components. For EMP mitigation, our research objectives focus on developing compact nanosecond-response-time megavolt pulse arrestors for rapid rise time voltage spikes on the order of nanoseconds, EMPs to enhance the resilience of energy systems against high-altitude electromagnetic pulse (HEMP) effects. HEMP effects can vary based on system configurations, the vulnerability of individual devices, and existing protective measures. We seek to contribute to the development of protective solutions that not only safeguard energy systems from immediate HEMP impacts but also ensure their continued operation in the face of such challenges. Focusing on these areas, we seek to develop effective strategies and technologies to mitigate the risks posed by HEMP, thereby enhancing the overall reliability and robustness of the electric grid.

8.3.3 Solid State Transformers (SST)

Traditional transformer technology is antiquated and vulnerable to numerous threats such as EMPs and cyber attacks. In addition, they are ill-suited to the dynamic demands of the modern grid which rely on unidirectional power systems [9]. Solid State transformers represent a significant advance in transformer resilience and technical capability. For solid-state transformers (SSTs), our research objectives focus on developing innovative solutions to replace outdated power system components to enhance resilience against emerging threats. We aim to create smaller, less expensive, controllable, modular solid-state power converters and substations that incorporate advanced circuit designs and control strategies to improve functionality and modularity within distributed architectures. Key areas of interest include the exploration of materials and components that ensure high reliability and longevity, with a particular emphasis on novel thermal management approaches. This work encompasses the development of compact, low-loss, high-frequency capacitors, and inductors, as well as high-voltage, high-current semiconductor switches capable of operating at high switching frequencies and fast slew rates.

Our objectives also include the creation of programmable and responsive transmission-scale SSTs that are significantly lighter and more compact compared to traditional systems. Ultimately, our aim is to enable paradigm shifts in energy transfer methods, increasing flexibility and resilience while achieving ambitious targets of approximately 10/kVA, 20 W/cm^3 power density, and efficiency levels of 99% or greater.

8.3.4 Defining, Measuring, and Optimizing Resilience

From the Presidential Policy Directive *Critical Infrastructure Security and Resilience* (PPD-21): "The term "resilience" is defined as the ability to prepare for and adapt to changing conditions and withstand and recover rapidly from disruptions. Resilience includes the ability to withstand and recover from deliberate attacks, accidents, or naturally occurring threats or incidents" [8]. From this we can determine that we must provide [9]:

- Rigorous and recursive planning
- Hardened infrastructure
- Efficient redundancies
- Adaptability to differing states
- Threat neutrality

Current planning models for power-grid systems often suffer from low temporal fidelity and limited geographic scale, primarily due to computational constraints. To address these challenges, new paradigms are needed to facilitate accurate and tractable algorithmic strategies capable of modeling relevant dynamics across a wide range of temporal and physical scales. Our research seeks to develop scalable modeling and analysis capabilities that provide rigorous solutions for power-grid planning and operational applications, with a focus on quantifying resilience. This work includes the implementation of scalable discrete-optimization methods, potentially leveraging heuristics and ML techniques, to assess resilience measures and optimization objectives that can quantify grid resilience for utilities, particularly in evaluating the impacts of proposed modifications. The framework should account for grid dynamics and consider various threat scenarios while encouraging the consideration of multiple competing objectives. The vision is to create highly flexible, modular, self-healing, and smart reconfigurable systems that support a transition from centralized to decentralized energy architectures, enabling asynchronous, fractal, hybrid, and multifrequency operational paradigms while effectively modeling large-scale failure modes of non-linearly interconnected electrical and other systems.

8.4 Examples of Completed Projects

At the current stage of the RES MC, we have completed over 37 R&D projects. The following subsections provides a sample of the work funded over the last five years. For a complete list of RES MC funded projects with corresponding success stories, please visit our website, www.sandia.gov/resilience/.

8.4.1 Harmonized Automatic Relay Mitigation of Nefarious Intentional Events (HARMONIE)

The HARMONIE special protection scheme (HARMONIE-SPS) project was designed to deploy adaptive cyber-physical response to mitigate unpredictable disturbances within the electric grid [10]. Central to the HARMONIE-SPS is a ML classification framework, using graph neural networks and transformer models, that evaluates real-time cyber-physical data to identify the operational state of the system—whether it is functioning normally, experiencing a cyber disturbance, a physical disturbance, or a combination of both. This classification informs appropriate responses, which may include cyber-physical corrective actions.

The HARMONIE-SPS incorporates innovative strategies, such as a consensus algorithm-based relay voting scheme, an automated pairing algorithm for triggering conditions and corrective actions that leverages clustering and support vector machines, and a cyber network traffic routing optimization algorithm. Tested within a sophisticated emulation environment that combines a real-time digital simulator (RTDS) and SCEPTRE™ [11], the platform is equipped to handle the complexities introduced by smart-grid technologies and the increasing threat landscape, including EMPs and extreme weather events. By processing both cyber and physical

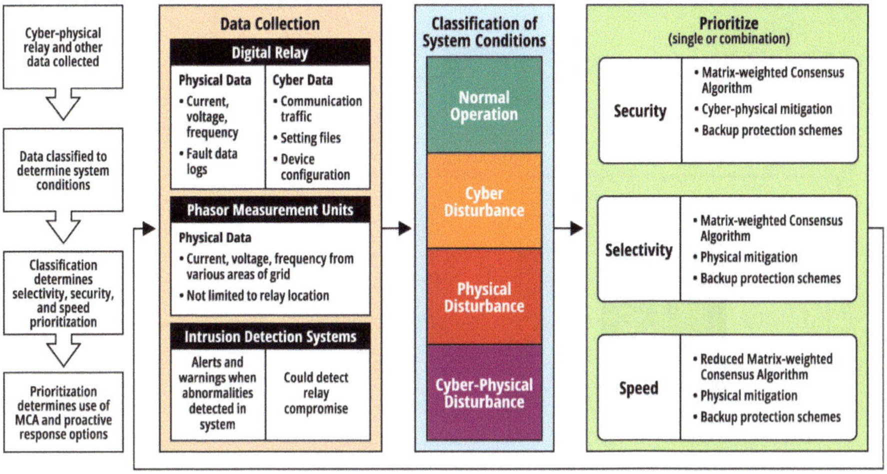

Fig. 8.2 HARMONIE-SPS enables real-time analysis of grid cyber-physical data and increased situational awareness

data from protective relays and out-of-band measurements, HARMONIE enhances situational awareness and enables proactive responses to disturbances, significantly bolstering the resilience of the electric grid against both intentional and inadvertent threats. Figure 8.2 Includes a diagram depicting the data collection and operational classification system enabling real-time cyber-physical data analysis.

The final report [10] details the methodologies employed, including the ML framework and the collaborative efforts with the ADROC project [12], while showcasing HARMONIE-SPS's effectiveness through various experimental scenarios. The HARMONIE model has also been an instrumental capability for other Sandia projects, including RES MC projects: ADROC [12], griDNA [14, 15], FACETS, DRE [16]. For additional information, see [17–19].

8.4.2 Advancing Resilience of Control Systems (ADROC)

The ADROC platform [12] is an innovative cyber experimentation workflow designed to conduct resilience assessments for cyber-physical and industrial control systems. With the growing complexity of these systems, ADROC addresses a critical challenge: How to effectively conduct cyber experimentation and gather the necessary data for quantitative cyber resilience analyses. Existing cyber resilience frameworks are often qualitative or driven by expert opinion, making them difficult to implement consistently and validate. Other frameworks are highly generic or difficult to adapt for specific systems. Experimentation with high-fidelity system modeling can be a useful tool to overcome some of these limitations, but the experimentation

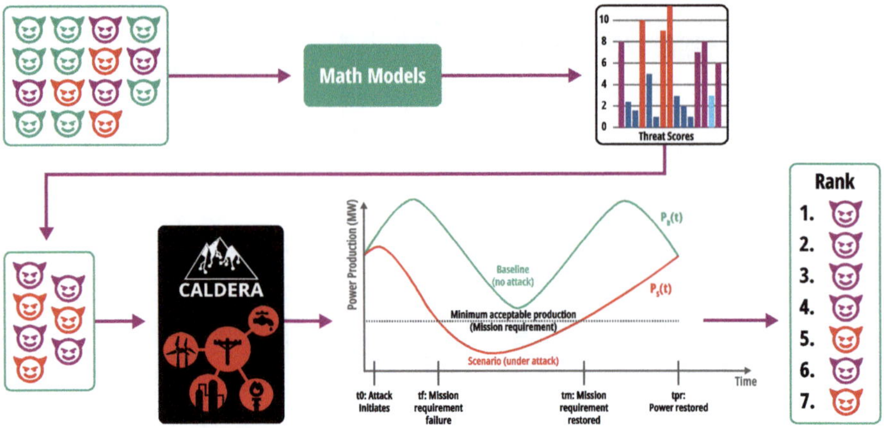

Fig. 8.3 ADROC platform workflow

process is often time-consuming and not scalable due to lack of automation, consistency, and repeatability. These limitations are a particular challenge when complex and dynamic threats require many experiments to characterize. ADROC addresses these challenges by combining two modeling and experimentation approaches—computationally efficient math modeling and high-fidelity system modeling—into a single, multi-fidelity workflow.

Figure 8.3 presents ADROC's platform workflow. ADROC's first phase, math modeling, allows for test scenarios of lower concern to be quickly discounted by calculating broader metrics on lower-fidelity models that can be evaluated very quickly [20]. The second phase, high-fidelity system modeling, supports more detailed, real-time scenario analysis on a more precise representation of the real-world system [21]. Data from the higher-fidelity models provide useful resilience metrics that can be used to rank scenarios and motivate prioritization of limited resources for improved cyber resilience.

ARDOC advances efficient uncertainty modeling in cyber systems, high-fidelity threat modeling, cyber resilience metrics development, and a workflow for mixed-fidelity cyber-physical experimentation. This approach enables more scalable and repeatable data-driven analysis for cyber-physical systems, ultimately making efficient, quantitative cyber resilience experimentation a more tractable option for threat prioritization on cyber-physical systems.

8.4.3 *Dynamic Modeling for Fire*

Wildfires, whether ignited naturally, accidentally, or maliciously, pose a significant threat to critical electric-grid infrastructure in the western U.S., often resulting in large-scale blackouts. This project developed a novel framework for using satellite imagery and ML techniques to identify hazardous fuel conditions and dynamically generate a wildfire hazard map [22]. The map highlights challenging areas for fire suppression, while geospatially indicating the most at-risk portions of the electric grid. This data-driven, on-demand algorithmic approach enhances existing fire spread modeling capabilities through the characterization of spatially continuous and uncertainty-aware weather and fuel drivers. The framework uses satellite imagery and Remote Automated Weather Station data to update fuel models regularly, incorporating process-constrained ML techniques to refine live- and dead-fuel moisture estimates. Figure 8.4 Includes a process flow of the fire modeling analytical framework.

Using the historical Bobcat Fire, which burned over 113,000 acres in 2020, the methodology was tested and demonstrated the effectiveness of the custom inputs in simulating wildfire behavior. The wildfire impacts were integrated across Southern California with a commonly studied electric-grid model, employing stochastic ignitions informed by the framework's custom fuel-characterization layers to assess the impacts on grid infrastructure. The analyses also show the components with the highest probability of wildfire risk and could help decision-makers prioritize resiliency strategies [9].

This project's innovations led to the wildfire risk and mitigation intelligence startup, Firescape, in 2025.

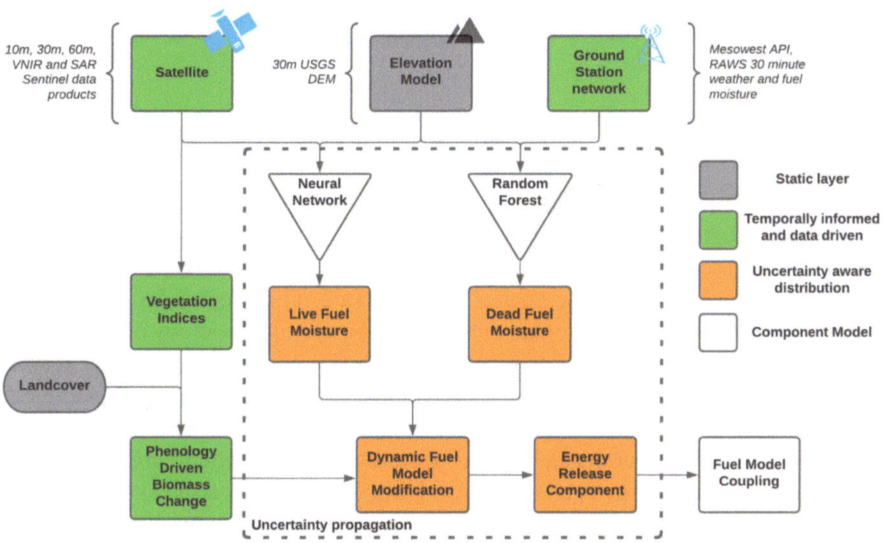

Fig. 8.4 Integrated power systems outputs with fire spread contours and model mitigation strategies

8.4.4 Innovative Nanostructural Granular Metals for Improved Electromagnetic Pulse (EMP) Protection

While the probability of a high-altitude nuclear detonation is low, such a detonation would generate EMP events that have the potential to disrupt or damage critical electrical grid infrastructure. EMPs result in dangerous high-voltage, high current transients that couple to grid transmission lines, compounding over long distances. These resulting EMP-induced transients include early-time (E1) transients with nanosecond rise-times, as well as slower E2 and E3 transients. No grid-scale protection against these ultrafast E1-transients exists today. This team is developing innovative high voltage-strength materials for future grid protection devices [66].

The team investigated the electrical, structural, and chemical properties of granular metals for use in future E1-EMP protection devices. Granular metals, consisting of metal nanoparticles within an insulating matrix, have complementary tunneling and capacitive conduction paths to safely shunt harmful nanosecond transients to ground. The team discovered that, through optimization of the granular metal material family, growth conditions, and post-growth thermal annealing, they could create high-voltage strength granular metals that demonstrate the robust filter performance necessary for protection devices.

Research accomplishments, which were highlighted in *Nanotechnology* [24] and *Journal of Physics: Condensed Matter* [25], led to the receipt of follow-on funding from the RES MC and DOE's Advanced Research Projects Agency–Energy (ARPA-E) (see Sect. 5.03NOVEL E1 ARRESTORS).

8.4.5 Signal Based, Fast Tripping Protection Schemes

Power-system protection is essential for the reliable operation of electrical grids, particularly in detecting and isolating faults to ensure resilience and national security. The ability to accurately identify and swiftly remove electrical faults is critical for minimizing impacts on defense-critical infrastructure. This project developed a new fast-tripping protection scheme that enhances the speed and accuracy of distribution system protection, thereby increasing grid stability during disturbances such as EMP events, lightning strikes, and switching operations. The primary challenges addressed by the researchers include achieving high precision in fault location to mitigate impacts and ensuring that the algorithms operate quickly enough to prevent damage to the broader electrical grid, all without relying on communication systems.

Modern power systems face significant stability issues, as conventional overcurrent and impedance-based protection schemes are often too slow to isolate faulted equipment within the necessary milliseconds [26]. While communication-assisted protection schemes, such as current differential and permissive transfer trip methods, can isolate faults more rapidly, they depend on communication infrastructure, which

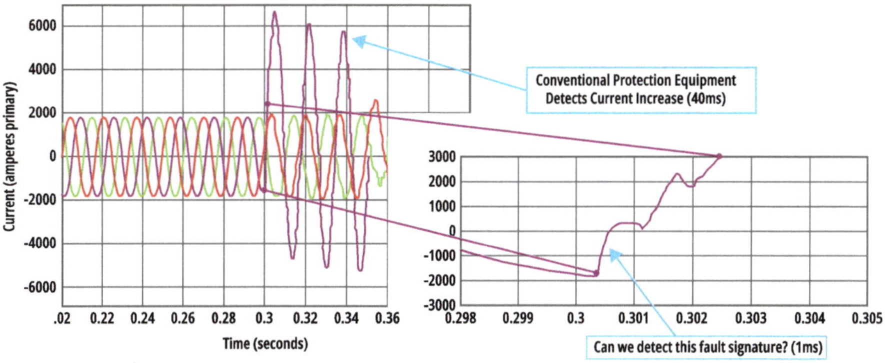

Fig. 8.5 Conventional methods take 10–40 ms to detect a fault in the system

can fail during critical events. Figure 8.5 presents the detection delay of conventional fault detection methods. To overcome these limitations, this project developed a communication-free adaptive protection module for integration into power system intelligent electronic devices [27]. This innovative protection scheme is designed to maintain reliable operation under extreme conditions, including potential cyberattacks that may compromise communication networks.

These protection schemes leveraged the traveling wave signature characteristics to differentiate fault currents from normal load currents. By utilizing high-frequency measurements exceeding 100 kHz, the signal-based fast tripping protection mechanisms can detect faults locally within 1 ms and identify characteristics about the fault, such as type and location, using artificial intelligence (AI). This rapid response capability is achieved through an understanding of the underlying physics of the grid, effectively addressing cyber risks and enhancing the overall protection of the electrical grid. This advanced protection scheme represents a significant advancement in power system resilience, ensuring that critical infrastructure remains secure and operational during adverse events.

The project received initial follow-on funding from DOE's Solar Energy Technologies Office (SETO), Office of Science, and the Advanced Research Projects Agency—Energy (ARPA-E), resulting in increased capability. Recently, this capability has been applied to the reduction in wildfire ignition by utilizing the fault location to quickly detect and isolate an incident before it leads to a full-blown wildfire. This application has been validated using hardware in the loop and field testing in Sandia's microgrid [28].

8.4.6 Self-healing Adaptive Zeta-Alpha Microgrid (SHAZAM)

The SHAZAM project [29] focused on advancing the performance of self-healing power systems (SHePS) that operate solely on local measurements, eliminating the need for data-sharing communication systems [30]. This innovative approach aimed to enhance the performance and robustness of power systems, particularly in resilience-critical scenarios, such as cyberattacks and natural disasters, where communications may not be available. A SHePS must be able to perform (a) protection, or detection and isolation of a fault; and (b) restoration, or the re-energizing of all the healthy system parts [29]. This project yielded a collection of techniques designed to address three key challenges associated with traditional SHePS:

1. Reliance on data sharing via high-speed communication networks [31–32].
2. Inability to perform self-assembly or form ad-hoc networked microgrids [33].
3. Scalability issues because of the large amounts of data from multiple endpoints.

The significance of SHePS lies in their ability to automatically detect faults and restore normal operations, making them critical for enhancing electric power resilience. However, traditional SHePS often depend on high-speed communication networks, which can be costly, unreliable during emergencies, and vulnerable to cyber threats. By eliminating the need for communication, SHAZAM's approach allows for the formation of ad-hoc networked microgrids, improving resilience and operational flexibility. This project not only addresses the limitations of existing systems but also paves the way for scalable, efficient power systems that can adapt to the challenges posed by geographically distributed inverter-based resources (IBRs).

Results demonstrate that self-healing power systems can achieve surprisingly high performance and exceptional robustness while relying exclusively on local measurements. The project developed numerous transient simulation tools and successfully implemented fundamental functions, such as automatic load-generation balancing and fault isolation. Notably, the system can perform black-start and post-fault self-assembly across various scenarios, ensuring maximum load energization and effective fault isolation. Additionally, techniques were established to prevent unintentional self-assembly into closed loops and to manage thermal overloads during load transfers, further enhancing system reliability.

SHAZAM has two follow-on LDRD projects, discussed in Sect. 8.5.6. Additional information [34–38].

8.5 Examples of Current Projects

8.5.1 Flexible, Autonomous, and Cyber-Physical Ecosystem for Trusted Security (FACETS)

The FACETS Vulnerability Toolset Marquee Project intends to enhance the monitoring and defense of both legacy and modern power systems through interoperable solutions. It will deliver comprehensive cyber-physical situational awareness (CPSA) across interconnected systems, allowing for a flexible capability that can adapt to emerging needs. FACETS will support grid planning, operation, and response decision-making, thereby improving resilience and security through a defense-in-depth approach. The primary outcome of FACETS is a tool-agnostic and plug-and-play cyber-physical security, operation, automation, and response (SOAR) capability. This system will integrate distributed cyber-physical data, facilitate adaptive detection and response using federated learning, and provide aggregate alerting and visualization for enhanced situational awareness.

The main output of FACETS will be a comprehensive ecosystem that defines necessary capabilities, integration requirements, and secure implementation strategies, along with a cyber-physical integrated security operation center (CP-ISOC) for unified data collection, alerting, and response. Figure 8.6 presents the vision for a cyber-physical integrated security operation center. The project will leverage various RES-MC capabilities, including:

1. **griDNA**: For distributed data collection and CPSA fusion analysis [14–15, 39].
2. **OT-GRITY**: For detecting cyber host execution integrity compromises [23, 40].
3. **Hallucinating Canaries (HC)**: For detecting cyber field device compromises [41–42].
4. **HARMONIE-SPS**: For adaptive cyber-physical remedial action schemes [77].
5. **STAHRS**: For wide-area measurement-based remedial action schemes [43].
6. **DRE**: For rigorous cyber-physical emulation experimentation [16].

The FACETS project focuses on five main objectives:

1. **Architecture**: Develop a tool-agnostic architecture using software wrappers and containers for flexible integration and user-friendly interfaces, allowing for seamless deployment across various platforms.
2. **Federated Learning (FL) Approach**: Implement FL to enhance data privacy and enable multi-modal analysis across decentralized devices, integrating results from various detection tools while ensuring transparency through Explainable AI.
3. **CPSA-focused Correlation and Fusion Analysis**: Address challenges in aggregating and fusing diverse data types and time domains from multiple tools to enhance situational awareness.
4. **Secure Implementation Strategy** [45]: Establish a secure data-sharing framework that protects both open and private channels, leveraging technologies like

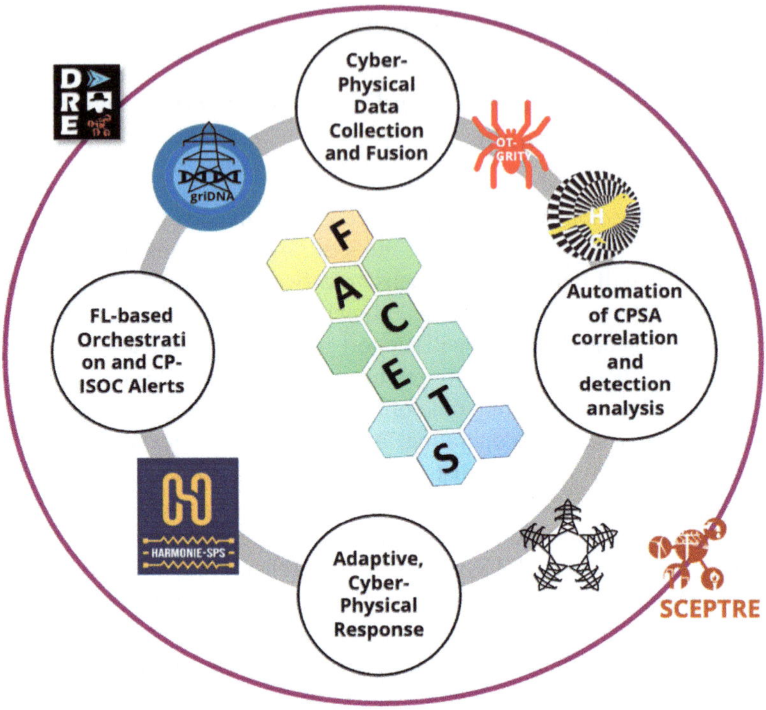

Fig. 8.6 Vision: integrated security operation center (SOC)

distributed ledger systems and cryptographic services to ensure resilience and confidentiality.

5. **Testing Approach**: Validate the integrated ecosystem against cyber-physical attack scenarios using existing emulation environments and frameworks like MITRE ATT&CK, assessing performance and robustness through uncertainty quantification (UQ) metrics.

This project will result in customizable software that can integrate with any other commercial and/or government off-the-shelf tools, as well as novel CPSA capabilities. Ultimately, FACETS will serve as a foundational capability for a next-generation CP-ISOC, allowing for flexible addition and removal of detection and mitigation tools while enhancing operational and security decision-making across cyber-physical systems.

8.5.2 AC to AC Solid State Transformer (SST)

In the Solid State Transformer Marquee Project, SSTs represent a transformative technology for the future of electrical grids, offering enhanced operational flexibility through the integration of advanced power electronics. By decoupling power system dynamics, SSTs significantly mitigate the risk of catastrophic failures that can cascade throughout the grid. The architecture of conventional SSTs involves complex configurations with multiple full-bridge and half-bridge stages, usually custom-designed for a specific location, that have long lead times and are extremely expensive to replace. Modular SSTs, unlike traditional transformers and custom-built SSTs, can be adapted to fit various power systems, promoting electrical infrastructure resiliency and efficiency. The widespread adoption of SSTs is expected to improve power quality and grid reliability while keeping costs manageable.

The project has three parallel tracks: Bidirectional switch (BiDFET) design and fabrication, passive materials development, and SST design and prototyping, described below.

1. The design, development, and production of a monolithically integrated bidirectional BiDFET is an extension from an existing and proven MOSFET design. The use of the newly developed BiDFET will significantly reduce the needed component count, providing a reduction in overall topology size and complexity of SSTs. Additionally, the innovative BiDFET design provides reduced low voltage losses and reverse-transfer capacitances.
2. Passive material development for both high frequency (HF) magnetic materials use innovative magnetic core materials, developed at Sandia. The newly developed materials will provide long-term stress reliability compared to typical materials currently used in SST HF transformers.
3. The project is developing a modular AC:AC block topology that, provided the use of bidirectional switch elements, can be implemented using only six switches and no direct current bus. This innovation simplifies SST design and enhances efficiency, making it more viable for modern grid applications. Additionally, the block topology allows the combination of multiple SSTs in parallel and series for flexible usage in a variety of power system environments.

This research has led to the development of a start-up company partnering with Sandia, Public Service Company of NM, and Mitsubishi on an upcoming DOE-funded project focusing on utilizing the SST as a drop-in replacement for a single-phase pole-mount transformer in a fielded lighting circuit. Additional information [46–47].

8.5.3 Novel E1 Arrestors

A high-altitude nuclear detonation generates EMP events, resulting in dangerous high-voltage, high current transients that couple to grid transmission lines,

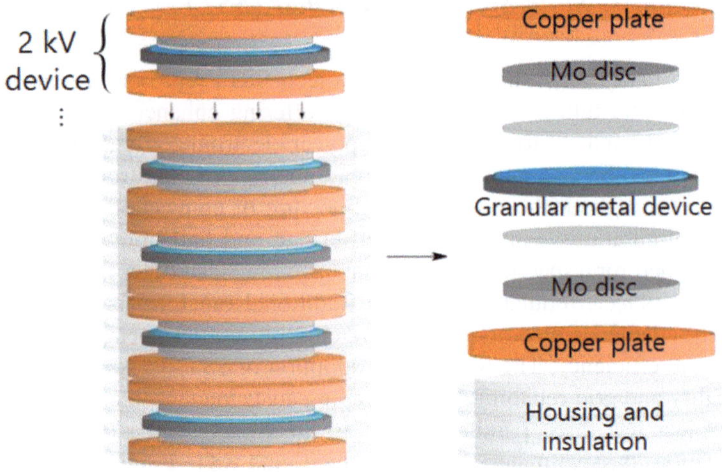

Fig. 8.7 Prototype arrester design showing series-stacked elements, designed by VA Tech collaborator

compounding over long distances. No grid-scale protection against these ultrafast E1-transients exists today. The EMP Mitigation Marquee Project's team is developing self-breaking, nanosecond responsive E1 arresters that would shunt E1 transients to ground. High-voltage strength, high-current density granular-metals comprising conductive molybdenum nanoparticles in an insulating silicon nitride matrix are the active materials in these E1 arresters. These silicon nitride-based granular metals exhibit unsurpassed high-pass filter performance, a key property for self-breaking shunt arresters. High-voltage breakdown strengths of 1 MV/cm will enable compact, low self-inductance E1 arresters operating at grid voltages. The project's accomplishments are summarized in two *Journal of Applied Physics* publications [48–49]. Figure 8.7 includes a visual representation of the design for a prototype E1 arrestor device.

8.5.4 Multi-hazard, Multi-objective, Multi-investment Optimal Transmission Infrastructure (M3)

The increasing frequency and severity of threats, both from intentional actors, such as cyber-attacks, and natural disasters like wildfires and floods, pose significant challenges to global power systems. As these events become more common, the risk of simultaneous occurrences rises, highlighting the vulnerabilities within current power infrastructure. It is crucial to strategically site infrastructure investments to withstand the impacts of multiple natural disasters.

Current methodologies for power system infrastructure investment often rely on expansion planning models that oversimplify the complexities of resilience. These models typically focus on singular risks or fail to account for the interplay of various hazards, which complicates decision-making processes. The challenge is further exacerbated by the need to consider diverse investment options—such as batteries, solar energy, and undergrounded lines—across different time scales. Although recent efforts have employed decomposition techniques to improve solution algorithms, progress has been limited in effectively addressing the multifaceted nature of resilience in power systems.

The 3 M project [50], under the Defining, Measuring and Optimizing Resilience Marquee Project, aims to advance the field by developing enhanced models and methods for expansion planning that explicitly incorporate multiple sources of risk, whether isolated or compounded. By balancing risk mitigation with expected capital and operating costs, the project seeks to create a resilient expansion planning framework. This framework will utilize a multi-objective generalized disjunctive programming optimization approach to jointly address operational constraints and discrete investment decisions, facilitating a comprehensive analysis of resilience trade-offs. This project will leverage previously developed Sandia and RES MC tools, such as IDAES-GTEP (GitHub, "IDAES/idaes-gtep, online), (IDAES, online), [53] and FORESt.

To validate the proposed framework, the project will apply its flexible formulation and solution acceleration techniques to realistic test systems in California, leveraging geolocated datasets that include population demographics and natural disaster risks, particularly wildfire ignition. The outcomes will demonstrate how targeted infrastructure investments can bolster grid resilience against various natural disasters simultaneously, providing valuable insights into the trade-offs associated with individual and overlapping risks. Importantly, the framework developed will be adaptable for use at regional or national levels, equipping decision-makers with the tools necessary to respond effectively to a wide range of potential threats.

8.5.5 Fault Resilience for Downtown Areas and Critical Customers in Low-Voltage Meshed Networks

This Defining, Measuring and Optimizing Resilience Marquee Project [46] aims to develop a next-generation protection device for low-voltage meshed distribution networks, facilitating the integration of distributed energy resources (DERs) for backup generation during resilience events. By transforming these networks into microgrid building blocks, the device will significantly enhance grid reliability and resilience for critical infrastructure, such as hospitals and urban centers, especially during widespread transmission outages.

Current Institute of Electrical and Electronics Engineers (IEEE) standards limit DER interconnections within secondary networks, which provide redundancy but remain vulnerable to transmission system outages during disasters or cyber-attacks. Microgrids can mitigate this vulnerability by allowing DERs to operate independently until the transmission system is restored. However, the capabilities of existing network protector relays often hinder DER integration, creating barriers to microgrid adoption.

The innovative protection device will eliminate constraints on microgrid adoption in secondary networks and improve the resilience of critical infrastructure across the U.S. [55]. This project will create a network protector unit that differentiates between normal operations and fault currents, using both communication-based and local measurement methods. Testing will involve hardware-in-the-loop simulations to ensure effectiveness [56]. Successful implementation of this device will enhance reliability and resilience for critical facilities, reducing the public liability during system disruptions, and bolstering national security.

The project team is working with utilities for device verification and validation, scoping potential locations for network protector replacements and cybersecurity testing of communication-based protection algorithms. Since one of the major limitations are around utility process and IEEE standards for not allowing modification of the network protector, efforts are ongoing to update the interconnection screening criteria and IEEE 1547.6 standard, funded by DOE SETO [57].

8.5.6 Self-healing Adaptive Resilient Power Systems (SHARPS)

SHePS are essential for ensuring resilience in critical infrastructure, such as defense, emergency response, and communications, especially during severe natural or man-made disruptions to the power grid. Current SHePS concepts often depend on high-speed communication networks, which can be costly, cyber-vulnerable, and unreliable during crises. Previous research at Sandia, as described in Sect. 8.4.6 [29], has shown that significant improvements in power system performance can be achieved without relying on these high-speed communications.

The assessment of power-system reliability typically excludes Major Event Days (MEDs) [58], which are crucial periods when the most severe impacts on society occur. Enhancing power system resilience during these MEDs is vital, and SHePS can provide this resilience through automated self-organization. However, existing concepts often still rely on impractical communication networks.

This project will tackle the following SHAZAM-identified challenges:

1. Enabling system components to function without explicit knowledge of isolation switch statuses
2. Locating faults without causing widespread outages
3. Accommodating both inverter-based and rotating generation
4. Isolating faults to mitigate wildfire risks; and

5. Ensuring that grid-following assets can operate autonomously without causing instability.

The SHARPS initiative, under the Defining, Measuring and Optimizing Resilience Marquee Project, aims to build upon the successes of previous work to create affordable SHePS that can deliver threat-agnostic resilience during MEDs without the need for high-speed, real-time communications. A key objective of the SHARPS project is to advance the technology readiness level (TRL) of SHAZAM from approximately TRL 3 to TRL 7 [59]. The SHARPS team will collaborate closely with industrial partners to create a comprehensive demonstration plan for SHARPS + SHAZAM at the IDEAL Center at New Mexico State University. This initiative aims to showcase the practical applications of SHePS in enhancing grid resilience, ultimately contributing to a more robust and reliable power infrastructure capable of withstanding diverse threats.

8.6 Moving Forward

For reference, Table 8.1 provides a comparative analysis of projects highlighted during this chapter. Our goal by the end of the RES MC is to not only have developed a significant portfolio of successful fundamental capabilities that can continue to be built upon moving forward, but to have demonstratable products for each Marquee area as follows.

For the Cyber-Physical Vulnerabilities Toolset Marquee Product, the FACETS framework will not only integrate internally developed vulnerability-assessment and mitigation tools but also provide a framework for integration of any Commercial Off-the-Shelf (COTS) or Government Off-the-Shelf (GOTS) tools. This framework will provide support for grid planning, operation, and response decision-making, thereby improving resilience and security through a defense-in-depth approach. Additionally, the mission campaign has funded close to a dozen cyber-physical vulnerability assessment and mitigation tools.

For the EMP Mitigation Marquee Product, we highlighted two projects that have developed new materials for faster EMP arrestor response times and will develop an innovative device, respectively. Additionally, the mission campaign has funded work to understand/measure the risk to electric power system infrastructure from a HEMP/E2 pulse and an EMP-resistant SST.

For the SST Marquee Product, the mission campaign's marquee product will be a field demonstrated SST. Although the AC:AC SST that we highlighted in Sect. 8.5 is in its final year, we are currently assessing two additional projects to bring this goal over the finish line. Additionally, we have funded several other successful SST projects, specifically a Solid-State Transformer Architecture, and Control Compensation for Common Mode Currents.

The Defining and Measuring Resilience Marquee area has been perhaps the most challenging due to the complexity, multi-faceted nature of the system and range of

Table 8.1 Comparative summary of the RES MC highlighted projects

Project title	Summary	Marquee area
HARMONIE	Develop an adaptive, cyber-physical special protection scheme (SPS) that uses machine learning to process real-time grid data and inform an automated response to unpredictable disturbances	Cyber-physical vulnerability tools
ADROC	Provides a platform for quantitative, repeatable cyber resilience assessment and threat prioritization for industrial control systems (ICS) using a hybrid modeling approach	Cyber-physical vulnerability tools
Dynamic modeling for wildfire	Develop a tool to determine wildfire risk to critical infrastructure in near-real-time and understand resulting impacts to the loadshed that could lead to cascading failure	Defining, Measuring, and Optimizing resilience
Innovative nanostructureal granular metals for improved electromagnetic pulse	Develop novel materials (nanostructural granular metals) for nanosecond-responsive arresters to protect the electrical grid against the early-time (E1) component of a HEMP	Electromagnetic pulse (EMP) mitigation
Signal based, Fast tripping schemes	Improves the speed and accuracy of grid protection systems by using high-frequency measurements (> 100 kHz) and signal processing to detect and locate faults in under 1 ms	Defining, Measuring, and Optimizing resilience
Self-healing adaptive zeta-alpha microgrid (SHAZAM)	Creates a toolbox for self-healing power systems (SHePS) that can form self-assembling microgrids using only local measurements, eliminating the need for vulnerable high-speed communications	Defining, Measuring, and Optimizing resilience
Flexible, Autonomous, and Cyber-physical ecosystem for trusted security (FACETS)	Develop customizable software that can integrate with any other commercial and/or government off-the-shelf tools for imprived cyber-physical situational awareness	Cyber-physical vulnerability tools

(continued)

Table 8.1 (continued)

Project title	Summary	Marquee area
AC to AC solid state transformer (SST)	Focuses on developing a more efficient and reliable single-stage SST topology, which requires fabricating custom bidirectional semiconductor devices (BiDFETs) that are not commercially available	Solid state transformers (SST)
Novel E1 arrestors	Develop a fast rise time high voltage (E1) EMP prototype arrestor device	Electromagnetic pulse (EMP) mitigation
Multi-hazard, Multi-objective, Multi-investment optimal transmission infrastructure (3 M)	Develop improved mathematical models and methods for solving generation and transmission expansion planning problems that explicitly capture multiple (isolated or compound) sources of system risk	Defining, Measuring, and Optimizing resilience
Fault resilience for downtown areas and critical customers in low-voltage meshed networks	Develops next-generation protection schemes that enable the use of distributed energy resources (DER) for backup power on low-voltage (LV) meshed networks in dense urban areas	Defining, Measuring, and optimizing resilience
Self-healing adaptive resilient power systems (SHARPS)	Creates a toolbox for self-healing power systems (SHePS) that can form self-assembling microgrids using only local measurements, eliminating the need for vulnerable high-speed communications	Defining, Measuring, and Optimizing resilience

threats. Depending on the analysis tool combined with the threat the analysis and solution space can contain millions of potential data points that are required to be measured. Scalability has proven to be a significant challenge.

For this Marquee area our researchers have been focused in three independent, but similar tracks:

1. **Optimization and Expansion Planning**: Our Marquee Project for this area is a combination of several optimization projects focusing on expansion planning. The project described in Sect. 8.5 Multi-hazard, Multi-objective, Multi-investment Optimal Transmission Infrastructure (M3) is one of those projects.
2. **Self-Healing Microgrids**: The SHAZAM and follow-on SHARPS projects will be the Marquee project for this track.

3. **Resilient Control Systems for the Future Grid**: We have funded several projects that have developed resilient control systems for the future grid. We highlighted two in this chapter: Signal Based, Fast Tripping Protection Schemes [46]. In addition, we have funded several other projects such as Resilient Decentralized Control Architectures for Robust Operations and Design to Manmade Threats that are still in the middle of R&D.

The LDRD program at Sandia is designed to foster innovative, fundamental, high-risk, high-reward R&D projects that typically, but not always, finish at a lower TRL. The "Golden Path" for an LDRD project and the research team is to finish the project with a clear path to sponsorship that will lead to the eventual completion with the work in the hands of the end user, be that government, military, or industry. This path requires a good deal of partnering across Sandia, as well as with academia, the intended end user. Sometimes these innovations lead to capabilities licensable through Sandia's Licensing and Technology Transfer program (for more information https://ip.sandia.gov/).

Additionally, the RES MC leadership team actively engages with each principal investigator (PI) to assist with their "Life after LDRD" plan. One of the ways the RES MC leadership has engaged externally, has been through establishing an external advisory board. Our external advisory board is comprised of top national laboratory, academic, government, and industry, leaders. Cooperative Research and Development Agreements (CRADAs) with two electric utilities have been initiated to address real-world research gaps and to partner for verification and validation of LDRD work. The RES MC also partnered with two university and NSF research foundations, UIUC's CITES and Georgia Tech's AI4Opt.

During the final years of the RES MC the leadership team has been focused on delivering the four marquee demonstration products and how we will engage with our internal program managers in Grid Security and Communications, Cyber-Physical Mission groups Complex Systems for National Security, Electrical Sciences, and other groups and departments across the labs.

8.7 Conclusion

Advancing S&T and R&D to enhance the resilience of energy systems is vital in the context of interconnected cyber-physical-social infrastructures and climate change challenges. As energy systems increasingly integrate with digital technologies, they become more vulnerable to cyber threats and extreme weather events. Focusing on innovative solutions strengthens the resilience of energy infrastructures, ensuring their reliability and adaptability while safeguarding critical resources. This holistic approach supports societal stability, enabling communities to thrive amidst disruptions.

Bridging the gap between S&T, R&D, and real-life action to enhance energy systems resilience presents several challenges. These include integrating knowledge across disciplines, securing adequate funding for R&D initiatives, and navigating regulatory barriers that hinder technology adoption. Effective stakeholder engagement is complicated by differing interests, and data availability issues can impede informed decision-making. Public skepticism may limit acceptance of innovative solutions, while the complexity of energy systems makes predicting impacts difficult. Balancing long-term resilience goals with short-term needs, addressing workforce skill gaps, and managing climate change uncertainty further complicate planning and implementation. Overcoming these challenges requires coordinated efforts, investment in education, and flexible policies that promote innovation.

The selection and execution of RES MC projects aimed at enhancing energy systems resilience has revealed key lessons. A holistic approach is essential, considering the interconnectedness of technical, environmental, and social factors. Interdisciplinary collaboration fosters innovation, while proactive risk management and adaptive practices improve preparedness and responsiveness. Investment in smart technologies enhances monitoring capabilities, and community involvement ensures alignment with local needs. Strong policy support and workforce development are critical, as is promoting data sharing to build trust. Continuous learning and improvement are vital for long-term success, enabling stakeholders to effectively strengthen resilience against cyber threats and climate change.

Key strategies for improving energy systems resilience include fostering interdisciplinary collaboration, developing integrated risk assessment frameworks, and implementing adaptive management practices. Promoting smart technologies, encouraging public–private partnerships, and engaging local communities in planning are essential. Advocating for supportive policies, utilizing scenario planning tools, investing in education and training, and promoting data sharing can enhance situational awareness and coordination. By adopting these practices, stakeholders can effectively advance S&T and R&D efforts to ensure the reliability and adaptability of energy systems.

In relation to NATO, advancing S&T and R&D can significantly enhance security and resilience by strengthening critical infrastructure, improving situational awareness, and enhancing cybersecurity. Resilient energy systems support effective joint operations and facilitate climate adaptation strategies. Collaboration among member states fosters innovation and informs policy development, while education and training enhance personnel capabilities. Research into sustainable energy solutions can also help NATO transition to sustainable practices, ultimately creating a robust network of interconnected energy systems that bolster collective security and readiness in times of crisis.

Acknowledgements The authors would like to thank the following for their assistance with content and editing: Cynthia Phillips, Rossitza Homan, Karen Tafoya, Craig Lawton, William Hart, and Robert Kaplar.

References

1. Pollard M (2024) A case study of Russian cyber-attacks on the Ukrainian power grid: implications and best practices for the United States. [Online]. Available: https://digitalcommons.pepperdine.edu/cgi/viewcontent.cgi?article=1216&context=ppr
2. SektorCERT (2023) The attack against Danish critical infrastructure. TP Clear. [Online]. Available: https://sektorcert.dk/wp-content/uploads/2023/11/SektorCERT-The-attack-against-Danish-critical-infrastructure-TLP-CLEAR.pdf
3. Harrabin R (2025) NASA data reveals dramatic rise in intensity of weather events. The Guardian. [Online]. Available: https://www.theguardian.com/world/2025/jun/17/nasa-data-reveals-dramatic-rise-in-intensity-of-weather-events
4. U.S. Energy Information Administration (2019) Investor-owned utilities served 72% of customers in 2017" 2019. [Online]. Available: https://www.eia.gov/todayinenergy/detail.php?id=40913
5. Coats D (2017) Worldwide threat assessment of the US intelligence community: senate select committee on intelligence. [Online]. Available: https://www.dni.gov/files/documents/Newsroom/Testimonies/SSCI%20Unclassified%20SFR%20-%20Final.pdf
6. Memmott M (2014) Sniper attack on calif. Power Station Raises Terrorism Fears, NPR. [Online]. Available: https://www.npr.org/sections/thetwo-way/2014/02/05/272015606/sniper-attack-on-calif-power-station-raises-terrorism-fears
7. Sullivan B, Canal N (2022) What we know about the attack on two North Carolina power substations. npr. [Online]. Available: https://www.npr.org/2022/12/05/1140775417/north-carolina-substation-attack-what-we-know.
8. The White House Office of the Press Secretary (2013) Presidential policy directive—critical infrastructure security and resilience. [Online]. Available: https://obamawhitehouse.archives.gov/the-press-office/2013/02/12/presidential-policy-directive-critical-infrastructure-security-and-resil
9. Bindewald G (2022) A snapshot of select U.S. DOE resilience activities. DOE Office of Electricity. [Online]. Available: https://www.energy.gov/sites/default/files/2022-07/DOE-OE_Presentation_SEAB_July26.pdf
10. Hossain-McKenzie S, Jacobs N, Summers A, Kolaczkowski B, Goes C, Fasano R, Mao Z, Homoud LA, Davis K, Overbye T (2022) Harmonized automatic relay mitigation of nefarious intentional events (HARMONIE)—special protection scheme (SPS). Sandia National Laboratories. [Online]. Available: https://www.osti.gov/biblio/1890265
11. Sandia National Laboratories (2016) SCEPTRE. [Online]. Available: https://www.osti.gov/servlets/purl/1376989
12. Thorpe JT, Fasano RE, Livesay ML, Sahakian MA, Reinbolt HM, Vugrin ED (2022) ADROC: an emulation experimentation platform for advancing resilience of control systems. Sandia National Laboratories. [Online]. Available: https://www.osti.gov/biblio/1891703
13. R&D WORLD (2024) R&D 100 finalists for 2024 are announced. Aug 2024. [Online]. Available: https://www.rdworldonline.com/rd-100-finalists-for-2024-are-announced/
14. Cordeiro P, Chavez A, Hossain-McKenzie S, Stenger A, Bayless S, Clark R, Behrendt S, Hawkins J, Davis K (2023) Considerations for secure data exchange to achieve cyber-physical situational awareness in the electric grid. In: 2023 IEEE power and energy conference at Illinois (PECI), Champaign, IL, USA
15. Blakely L, Fragkos G, Hossain-McKenzie S, Goes C, Summers A, Haque KA, Davis K (2024) A comparison study of feature extraction and data fusion techniques for improving cyber-physical situational awareness. In: 2024 IEEE/PES transmission and distribution conference and exposition (T&D), Anaheim, CA, USA
16. Thorpe JE, Krakauer NJ, Stephens JA, Williams HSB (2024) Toward rigorous experimentation and analysis in cyber-physical system emulation. In: 2024 IEEE Kansas power and energy conference (KPEC), Manhattan, KS, USA

17. Hossain-McKenzie SS, Cordeiro PG, Chavez AR, Adams R, Stenger A, Bayless S, Clark R, Behrendt S, Hawkins J, Davis K (2023) Considerations for secure data exchange to achieve cyber-physical situational awareness in the electric grid. In: Power and energy conference, Urbana, Illinois, United States of America, March
18. Chavez AR, Hossain-McKenzie SS, Cordeiro PG (2023) Considerations for secure data exchange to achieve cyber-physical situational awareness in the electric grid. In: Power and energy conference, urbana-champaign, Alabama, United States of America, March
19. Hossain-McKenzie SS, Chavez AR, Mayes DJ (2023) Autonomous solutions for advancing electric grid cybersecurity. In: DOE cyber resilience R&D symposium, golden, Colorado, United States of America, March
20. Livesay M, Vugrin E, Thorpe J (Unpublished) Adding diffusion in POMDP learning for uncertainty in cyber-physical systems
21. Thorpe J, Fasano R, Sahakian MG, Gonzales A, Hahn A, Morris J, Ortiz T, Reinbolt H, Vugrin ED (2022) A cyber-physical experimentation platform for resilience analysis. In: Sat-CPS '22: proceedings of the 2022 ACM workshop on secure and trustworthy cyber-physical systems. [Online]. Available: https://dl.acm.org/doi/10.1145/3510547.3517916
22. Eagleston H (2020) Dynamic monitoring to analyze grid vulnerability to fire. Sandia national laboratories. [Online]. Available: https://www.sandia.gov/resilience/clone/
23. Sandia National Laboratories (2024) RESEARCH—laboratory directed research and development FY23 annual report. [Online]. Available: https://www.osti.gov/servlets/purl/2429995
24. Gilbert SJ, Mererson ML, Kotula PG, Rosenberg SG, Kmieciak TG, McGarry MP, Siegal MP, Biedermann LB (2023) Granular metals with SiNx dielectrics. Nanotechnology 34(41):415706
25. Gilbert SJ, Rosenberg SG, Kotula PG, Kmieciak TG, Biedermann LB, Siegal MP (2022) The effect of metal–insulator interface interactions on electrical transport in granular metals. J Phys: Condens Matter 34(20):204007
26. Wilches-Bernal MJ-AF, Reno MJ (2023) Local, single-ended, traveling-wave fault location on distribution systems using frequency and time-domain data. IEEE Access
27. Jimenez-Aparicio M, Hernandez-Alvidrez J, Montoya AY, Reno MJ (2023) Micro random forest: a local, high-speed implementation of a machine-learning fault location method for distribution power systems. In: IEEE PES general meeting
28. Paruthiyil S, Bidram A, Jimenez-Aparicio M, Hernandez-Alvidrez J, Dow ARR, Reno MJ, Bauer D (2025) Traveling wave-based fault detection and location in a real low-voltage DC microgrid. IET Smart Grid
29. Ropp ME, Lavrova O, Reno MJ, Silva E, Densel MA, Kassabian LN, Biswal M, Ramoko A, Ranade SJ (2023) Self-healing, self-assembling islanded power systems using only local measurements. OSTI.gov, Sep 2023. [Online]. Available: https://www.osti.gov/biblio/2430242
30. Ghosh D, Sharman R, Rao H, Upadhyaya S (2007) Self-healing systems-survey and synthesis. Elsevier Decis Support Syst 42:2164–2185
31. Lagos D, Papaspiliotopoulos V, Korres G, Hatziargyriou N (2021) Microgrid protection against internal faults. IEEE Power and Energy Magazine, May/June, pp 20–35
32. Maurer L, Stevens A, Reder W (2012) Tales from the frontline: keys to successful self-healing distribution projects. IEEE Power Energ Mag 10(2):100–106
33. Yang L, Xiao HCF, Lai Y, Chollot Y (2019) The experiences of decentralized self-healing grid. In: IEEE international conference on advanced power system automation and protection (APAP), October
34. Densel M, Ropp M, Silva E, Lavrova O (2024) A technique for thermal overload mitigation in a self-healing intentional island system using only local measurements. IEEE ISGT-NA
35. Silva E, Ropp M, Lavrova O (2023) Demonstration of a self-healing power system concept using manufacturer specific inverter models In: CIGRE grid of the future conference, Kansas City, MO, USA
36. Ropp M, Silva E, Lavrova O (2023) Avoiding asynchronous connections of adjacent micro grids in self-healing power systems. In: CIGRE grid of the future, Kansas City, MO, USA
37. Ropp ME, Reno MJ, Biswal M (2023) Detection and prevention of unintentional formation of loops in self-healing power systems and microgrids. IEEE Trans Power Delivery 38(4):2682–2691

38. Silva E, Lavrova O, Ropp M (2022) Protection elements for self-healing microgrids using only local measurements. In: 2022 North American power symposium (NAPS), Salt Lake City, UT, USA
39. Haque KA, Davis K, Blakely L, Hossain-McKenzie S, Fragkos G, Goes C (2024) Multimodal learning in cyber-physical system: a deep dive with WSCC 9-bus system. In: 2024 IEEE power and energy society innovative smart grid technologies conference (ISGT), Washington, DC, USA
40. Alrawi O, Ike M, Pruett M, Kasturi RP, Barua S, Hirani T, Hill B, Saltaformaggio B (2021) Forecasting malware capabilities from cyber attack memory images. In: 30th USENIX security symposium (USENIX Security 21)
41. Sandia National Laboratories (2022) Hallucinating canaries: instrumented firmware re-hosting for cyber resilience in energy systems. [Online]. Available: https://www.sandia.gov/app/uploads/sites/77/2022/02/20220105_RES_8.5x11_Factsheet_MC_Abraham-Clements.pdf
42. Clements AA, Gustafson E, Scharnowski T, Grosen P, Fritz D, Kruegel C, Vigna G, Bagchi S, Payer M (2020) Hallucinator: firmware re-hosting through abstraction layer emulation. USENIX Association, pp 1201–1218. [Online]. Available: https://www.usenix.org/conference/usenixsecurity20/presentation/clements
43. Choi H, Elliott RT (2023) Sizing energy storage systems to mitigate variability of renewable generation for grid stability using inverse uncertainty propagation. In: 2023 IEEE power and energy society innovative smart grid technologies conference (ISGT), Washington DC, District of Columbia, United States of America, January
44. Thorpe JE, Krakauer NJ, Stephens JA, Stroble H, Williams B (2024) Toward rigorous experimentation and analysis in cyber-physical system emulation. In: 2024 IEEE Kansas power and energy conference (KPEC), Manhattan, KS, USA
45. Cordeiro P, Chavez A, Reyna A, Hossain-McKenzie S, Fragkos G, Collins T, Summers A, Haque A, Davis K (2025) Design for secure and resilient data exchange across distributed cyber-physical sensors and analytics in decentralized energy systems. In: Hawaii international conference on system sciences (HICSS), Honolulu, HI, USA
46. Mueller JA, Flicker J, Dow A, Rodriguez LG, Palacios F (2024) Isolated three-phase AC-AC converter with phase shift modulation. In: IEEE applied power electronics conference and exposition (APEC), Long Beach, CA, USA
47. Mueller J, Kaplar R, Flicker J, Rodriguez LG, Binder A, Ropp M, Gill L et al (2022) System integration analysis for modular solid-state substations, SAND2022–13124. Sandia National Laboratories, Albuquerque, NM (United States)
48. Gilbert S, McGarry M, Meyerson M, Kotula P, Yates L, Sharma P, Siegal M, Biedermann L (2025) Structural, chemical, and electronic control in Co-SiNx granular metal high-pass filters. J Appl Phys 137:065101. https://doi.org/10.1063/5.0238917
49. Bowman T, Kmieciak T, Biedermann L (2024) Nanosecond transient validation of surge arrester models to predict electromagnetic pulse response. IEEE Trans Electromagnet Compatibility 67(1). https://doi.org/10.1109/temc.2024.3486980
50. Skolfield JK, Escobedo AR (2022) Operations research in optimal power flow: a guide to recent and emerging methodologies and applications. Eur J Oper Res 300(2):387–404
51. GitHub, "IDAES/idaes-gtep," [Online]. Available: https://github.com/IDAES/idaes-gtep
52. IDEAS (2022) The Institute for Design of Advanced Energy Systems (IDAES) [Online]. Available: https://idaes.org/about/
53. Bynum LM, Hackebeil GA, Hart WE, Laird CD, Nicholson BL, Siirola JD, Watson J-P, Woodruff DL (2021) Pyomo—optimization modeling in Python. 3rd edn. Springer
54. Mueller JA, Flicker J, Dow A, Rodriguez LG, Palacios F (2024) Isolated three-phase AC-AC converter with phase shift modulation. In: 2024 IEEE applied power electronics conference and exposition (APEC), Long Beach, CA, USA
55. Azzolini JA, Reno MJ, Ropp ME, Cheng Z, Udren E, Holbach J (2023) Increasing DER hosting capacity in meshed low-voltage grids with modified network protector relay settings. In: IEEE innovation smart grid technologies Latin America (ISGT LA), San Juan, PR, USA

56. Joshi M, Snow M, Bidram A, Reno MJ, Ropp M, Azzolini AJ (2024) Hardware-in-the-loop testing of direct transfer trip for network protector units in the presence of distributed energy resources. In: North American power symposium (NAPS)
57. Azzolini JA, Ropp M, Reno MJ (2024) Options for upgrading low-voltage spot network protection to increase DER interconnection capacity. Sandia National Laboratories; SAND2024-11102R
58. IEEE Std 1366-2022 (2022) IEEE guide for electric power distribution reliability indices
59. Government Accountability Office (GAO) (2020) Technology readiness assessment guide: best practices for evaluating the readiness of technology for use in acquisition programs and projects. [Online]. Available: https://www.gao.gov/products/gao-20-48g
60. Bodeau D, Graubart R, McQuaid R, Woodill J (2018) Cyber resiliency metrics and Scorind in practice—use case methodology and examples. MITRE. [Online]. Available: https://www.mitre.org/sites/default/files/2021-11/prs-18-3375-cyber-resiliency-metrics-and-scoring-in-practice-use-case-methodologies-and-examples.pdf
61. Li H, Shetye K, Overbye T, Hossain-McKenzie S, Davis K (2021) Towards the automation of remedial action schemes design. In: Hawaii international conference on system sciences. [Online]. Available: https://www.researchgate.net/publication/349153016_Towards_the_Automation_of_Remedial_Action_Schemes_Design
62. Hossain-McKenzie S, Calzada D, Jacobs N, Goes C, Summers A, Davis K, Li H, Mao Z, Overbye T, Shetye K (2021) Adaptive, cyber-physical special protection schemes to defend the electric grid against predictable and unpredictable disturbances. IEEE explore: 2021 resilience week (RWS). [Online]. Available: https://ieeexplore.ieee.org/document/9611801/authors#authors
63. Summers A, Goes C, Calzada D, Jacobs N, Hossain-McKenzie S, Mao Z (2022) Towards cyber-physical special protection schemes: design and development of a co-simulation testbed leveraging SCEPTRE. In: IEEE explore: 2022 power and energy conference at Illinois (PECI). [Online]. Available: https://ieeexplore.ieee.org/document/9744043
64. Calzada D, Hossain-McKenzie S, Mao Z (2022) Deep learning architecture for processing cyber-physical data in the electric grid. In: IEEE explore: 2022 power and energy conference at Illinois (PECI). [Online]. Available: https://ieeexplore.ieee.org/document/9744015
65. Jacobs N, Summers A, Hossain-McKenzie S, Calzada D, Li H, Mao Z, Goes C, Davis K, Shetye K (2022) Next-generation relay voting scheme design leveraging consensus algorithms. IEEE Explore: 2021 IEEE power and energy conference at Illinois (PECI). [Online]. Available: https://ieeexplore.ieee.org/document/9435201
66. Sandia National Laboratories (2024) Institute for the Design of Advanced Energy Systems (IDAES). Center for Computing Research (CCR), [Online]. Available: https://www.sandia.gov/ccr/project/institute-for-the-design-of-advanced-energy-systems-idaes/

Additional References

67. Industrial Control Systems Cyber Emergency Response Team (2016) Recommended practice: improving industrial control system cybersecurity with defense-in-depth strategies. Department of Homeland Security
68. Office of Cybersecurity, Energy Security, and Emergency Response, "Cybersecurity Capability Maturity Model (C2M2)," Department of Energy (2022)
69. "SOAR versus SIEM: What is the Difference?" Cyberpedia, Palo Alto Networks, [Online]. Available: https://www.paloaltonetworks.com/cyberpedia/what-is-soar-vs-siem
70. Johnson J, Jones CB, Chavez A, Hossain-McKenzie S (2023) SOAR4DER: security orchestration, automation, and response for distributed energy resources. In: Power systems cybersecurity: methods, concepts, and best practices. Springer International Publishing
71. Office of Intelligence and Analysis (2024) Homeland threat assessment. Department of Homeland Security

72. National Cybersecurity Strategy (2023) The White House
73. Office of Electricity Delivery and Energy Reliability (2018) Multiyear Plan for Energy Sector Cybersecurity. Department of Energy
74. Cordeiro P, Chavez A, Hossain-McKenzie S, Stenger A, Bayless S, Clark R, Behrendt S, Hawkins J, Davis K (2023) Considerations for secure data exchange to achieve cyber-physical situational awareness in the electric grid. In: 2023 IEEE power and energy conference at Illinois (PECI)
75. Ike M, Phan K, Valme R, Bui D, Sadoski K (2024) OT-GRITY: detecting ICS integrity attacks in OT via physical world driven characterization of SCADA execution and memory-space behaviours. SAND2024–00490, Sandia National Laboratories
76. Thomas CZ, Wilson J, Wright C (2023) Hallucinating canaries: security field devices through instrumented firmware re-hosting. In: SAND2023–11473R, Sandia National Laboratories
77. Hossain-McKenzie S, Jacobs N, Summers A, Kolaczkowski B, Goes C, Fasano R, Mao Z, Al Homoud L, Davis K, Overbye T (2022) Harmonized automatic relay mitigation of nefarious intentional events (HARMONIE)-special protection scheme (SPS). SAND2022–13427, Sandia National Laboratories
78. Afosa-Baffoe J, Hill B, Trudnowski D, Elliott R (2024) Initial results in PMU testing and modeling of GPS timing errors. In: SAND2024–04184PE, North American Synchrophasor initiative work group meeting, Salt Lake City, UT
79. Thorpe JE, Krakauer NJ, Stephens JA, Stroble H, Williams B (2024) Toward rigorous experimentation and analysis in cyber-physical system emulation. In: Kansas power and energy conference [Accepted]
80. Boettiger C (2015) An introduction to Docker for reproducible research. ACM SIGOPS Operating Systems Review
81. Burns B, Beda J, Hightower K (2022) In: Kubernetes: up and running. O'Reilly Media, Inc.
82. Bhattacharya (2022) Applied machine learning explain ability techniques: make ML models explainable and trustworthy for practical applications using LIME, SHAP, and more. Packt Publishing Ltd
83. "Store and manage your data on IPFS", [Online]. Available: https://ipfs.tech/
84. Van Dam R, Dinh T-N, Cordi C, Jacobus G, Pattengale N, Elliott S (2019) Proteus: a dlt-agnostic emulation and analysis framework. In: 12th USENIX workshop on cyber security experimentation and test (CSET 19)
85. Mayle, Stubbs J, Jacobus G, Elliot S (2021) Distributed ledger technology based communication architectures for physical security systems. Sandia National Laboratories, SAND2021–7270A'
86. Dinh T-N (2022) The synchronic web: primer. https://doi.org/10.2172/1862729
87. Berry J, Birch G, Dixon K, Ganti A, Goss K, Mayer C, Onunkwo U, Phillips C, Saia J, Shead T (2023) Cicada: player-scalable, fault-tolerant secure multiparty computation. Sandia National Laboratories, SAND2023–12004C
88. "MITRE ATT&CK: ICS Matrix," The MITRE Corporation, [Online]. Available: https://attack.mitre.org/matrices/ics/
89. "MITRE CALDERA," The MITRE Corporation, [Online]. Available: https://caldera.mitre.org/

Part IV
Cyber-Physical Process Characterization

Chapter 9
Diversified Reservoir Management in Enhancing Water Security and Climate Resiliency

I. T. Okioga, J. Zhu, J. Morris, S. S. Shirley, M. T. McEvoy, S. Baumgardner, L. Stoltz, and S. Lofaro

Abstract In response to the escalating challenges of extreme weather patterns from climate change, water scarcity, and increasing water demands, diversified reservoir management has become essential for building resilient water systems and supporting community growth and development. This chapter presents a case study from the Western Virginia Water Authority, a regional water utility located in southwest Virginia, USA, to highlight resiliency planning though diversified reservoir management supported by regional cooperation, source water quality protection, supplemental water sourcing, and comprehensive monitoring and emergency action planning. Three distinct reservoir types, off-stream, in-stream and hybrid, each playing a distinct role in sustaining and stabilizing source water supplies, supplementing natural flows, supporting ecosystem needs, and providing environmental, community and economic benefits, are discussed. The off-stream reservoir sits completely outside natural waterways, offering controlled filling and reduced environmental disruption. The reservoir can potentially function as a high-flow storage system, which reduces flooding through flow diversions, and provides a resilient water supply solution during low-flow periods. While drawbacks include limited natural recharge, the need for active refill management, as well as capacity constraints during extended droughts, these challenges are mitigated by the presence of the other two reservoirs. The in-stream reservoir, which directly harnesses both water and hydroelectric power from river flows, provides consistent water availability and renewable energy. The hybrid reservoir, which combines surface runoff and flow diversion from nearby waterways, balances environmental impact with reliable water supply. This chapter highlights the complementary benefits of integrating all three reservoir types within a single utility. This multi-mode approach enhances water resource resiliency, mitigates climate risks, and promotes sustainable, renewable energy production, offering an adaptable solution to water and energy demands. Emphasis is also placed on

---

I. T. Okioga (✉) · J. Zhu · J. Morris · S. S. Shirley · M. T. McEvoy · S. Baumgardner · L. Stoltz · S. Lofaro
Western Virginia Water Authority, Roanoke, VA, USA
e-mail: irene.okioga@westernvawater.org

M. Huber et al. (eds.), *Critical Infrastructure Systems and Extreme Threat Nexus: Preparing for Resilience and Security*, NATO Science for Peace and Security Series C: Environmental Security, https://doi.org/10.1007/978-3-032-13170-6_9

regional cooperation and integrated strategies that incorporate considerations of surrounding ecosystems.

Keywords Regional cooperation · Water scarcity · Diversified water supply · Integrated reservoirs · Resiliency planning · Climate change · Reservoir types (off-stream · in-stream and hybrid)

9.1 Background and Objectives

Prompted by the 2002 East Coast drought, which impacted several U.S. states and parts of Canada, the Western Virginia Water Authority (Authority) was established in July 2004, when the City of Roanoke and the Roanoke County Utility Departments in Virginia, USA, consolidated to address water scarcity concerns and the increasing cost of wastewater treatment by integrating management and operations. The Authority became the first regional water utility in the Commonwealth of Virginia to manage both water and wastewater services across multiple localities.

In November 2009, Franklin County was integrated into the Authority, followed by Botetourt County in July 2015. The Town of Boones Mill, Town of Vinton and Ferrum systems were acquired more recently in December 2021, July 2022, and August 2024 respectively, creating a utility that serves the largest number of distinct member jurisdictions in the Commonwealth (Fig. 9.1). The consolidation enabled the Authority to implement an integrated approach to regional water resources management, including the coordinated management of its drinking water system, supplied mainly by the Spring Hollow Reservoir, the Smith Mountain Lake Water system (co-owned by the Bedford Regional Water Authority), the Carvins Cove Reservoir, the Beaverdam Creek and Falling Creek Reservoirs, Crystal Spring, and Muse Spring, among multiple other groundwater systems throughout the Roanoke Valley [26].

Combined, the Authority's main water treatment facilities can currently treat over 51 million gallons of water per day (MGDs), not including smaller ground water systems, and the Beaverdam Creek and Falling Creek Reservoirs (Table 9.1). The Beaverdam Creek Reservoir, which is currently undergoing dam improvements, is not a stand-alone reservoir, but instead supplies water to the Falling Creek Reservoir through a man-made tunnel. Both reservoirs are currently not in operation. The Authority's current annual water production averages 19.6 MGDs, suggesting a substantial reserve to accommodate future growth, address emergency demands, and allow certain interconnected portions of the system to be taken offline for maintenance and improvements, as needed. The water distribution system currently comprises approximately 1350 miles of pipe and 70,000 customer connections, including industrial, commercial, institutional, and residential customers.

The objective of this chapter is to describe best practices for water supply reservoir management from the Authority's experience, which includes integrating multiple reservoir types within a single utility. The article begins by categorizing reservoirs based on their location relative to the water source, and then continues to provide

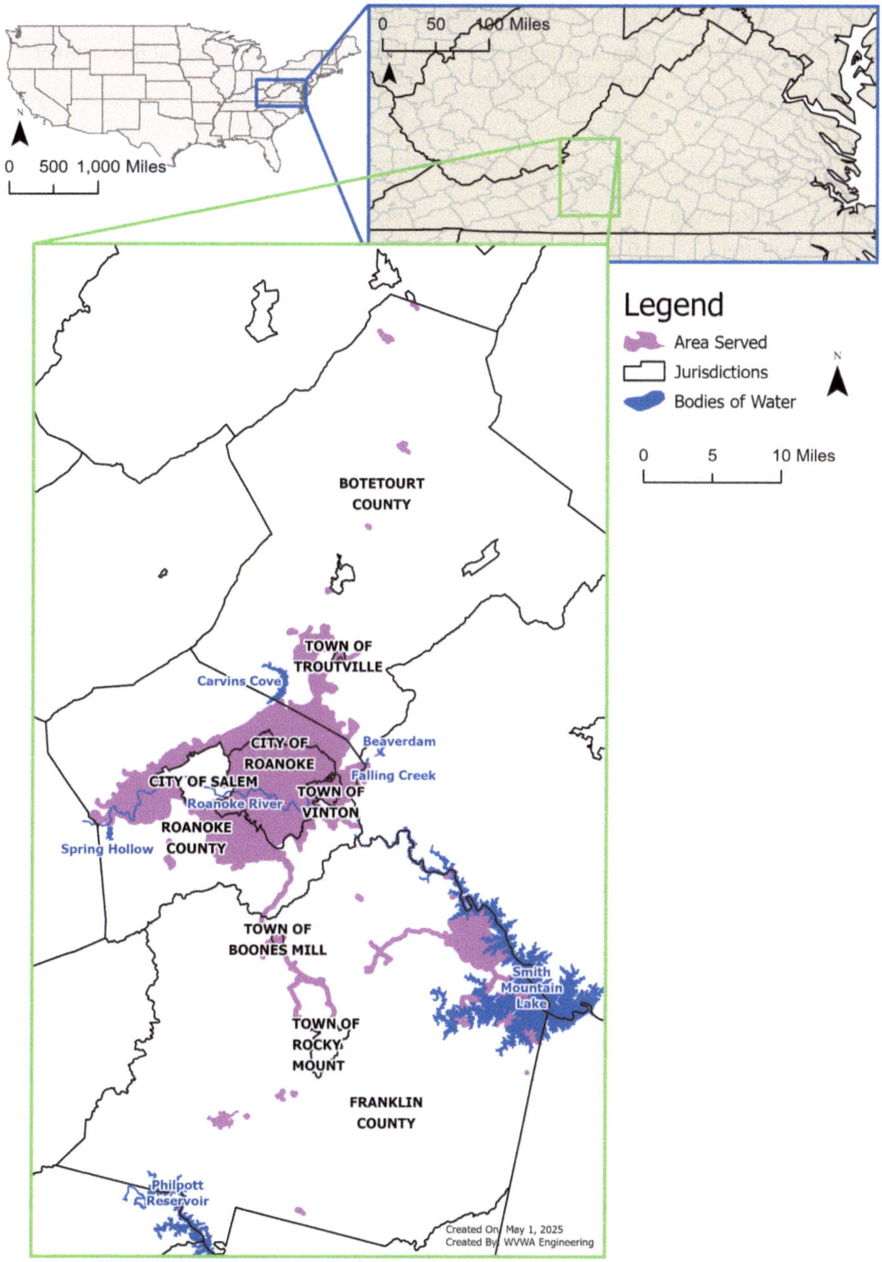

Fig. 9.1 Western Virginia water authority service area map

Table 9.1 Authority main drinking water system average production and plant capacities

Main water supply sources	Annual average production (MGDs)	Plant capacity (MGDs)	Percent production (%)
Spring Hollow Reservoir	3.4	18	17
Smith Mountain Lake	1.4	3	7
Carvins Cove Reservoir	9.5	24	48
Beaverdam/Falling Creek Reservoirs[a]	0.0	2	0
Crystal Spring	4.2	4.7	21
Muse Spring	1.1	2	6
Total	19.6	51.7[b]	100

[a]Falling Creek Reservoir is currently not in operation
[b]Total does not include Falling Creek Plant capacity

an overview of the Authority's reservoir systems, specifically reviewing operations of the Spring Hollow, the Smith Mountain Lake, and the Carvins Cove reservoirs, to emphasize their complementary advantages and balanced limitations. Finally, the article concludes by exploring additional considerations for resilient reservoir planning.

9.2 Reservoir Types

Water utilities rely on a variety of reservoir types to ensure a stable and resilient water supply, each with unique advantages and challenges. For this chapter, three reservoir types are identified based on their location relative to their water source to highlight their operational dynamics, advantages, and limitations in managing water storage and distribution across varying environmental and operational conditions. The reservoir types are as follows:

Off-stream reservoirs, which receive water through flow diversions and artificial conveyance systems including pumps and pipelines, based on being located entirely outside rivers, streams, or other natural waterways;

In-stream reservoirs, which are constructed directly within rivers, streams, and other natural waterways; and Hybrid reservoirs, which integrate in-stream and off-stream characteristics to facilitate flow impoundment from natural waterways as well as artificial flow diversions.

9.2.1 *Off-Stream Reservoir—Spring Hollow*

The Spring Hollow Reservoir in Roanoke County, Virginia, USA, is an off-stream reservoir, whose water is pumped from the Roanoke River. The reservoir, which reached full pond for the first time in 1995, holds 3.2 billion gallons at a full pond elevation, providing about 17% of the current water production, as a percentage of the Authority's main water supply totals. Full pond is achieved when the water level in a reservoir reaches the spillway crest elevation, which is the ideal upper limit of its normal operating range (Fig. 9.2).

The Spring Hollow Reservoir is primarily designed to manage water resources as a supply reservoir while also addressing environmental impacts, with one of the key mechanisms being the use of a regulated flow bypass system that accommodates spawning of the Roanoke Log Perch, a formerly protected fish species that was delisted in July 2025 (U.S. Fish & Wildlife Service 2025). While regulated flows help minimize ecological impacts, the reservoir reliance on external transfers becomes a concern during extended dry periods when bypass volumes are maintained at the expense of drinking water supply. The reservoir advantages are instead realized during high flow periods during which time selective pumping becomes important to prevent sediment-laden flows into the reservoirs. Depending on flow quality and quantities, off-stream reservoirs can potentially function as high-flow storage systems with the potential of reducing flooding through flow diversion.

The Texas Water Conservation Association (TWCA) Flood Response Committee [19] distinguishes between three reservoir types characterized by their primary functions as: (1) water supply; (2) flood control; and, (3) dual purpose reservoirs, noting that while water supply reservoirs are ideally maintained at full pond levels to ensure reliable supply, they can also provide "unintended flood mitigation benefits" and serve

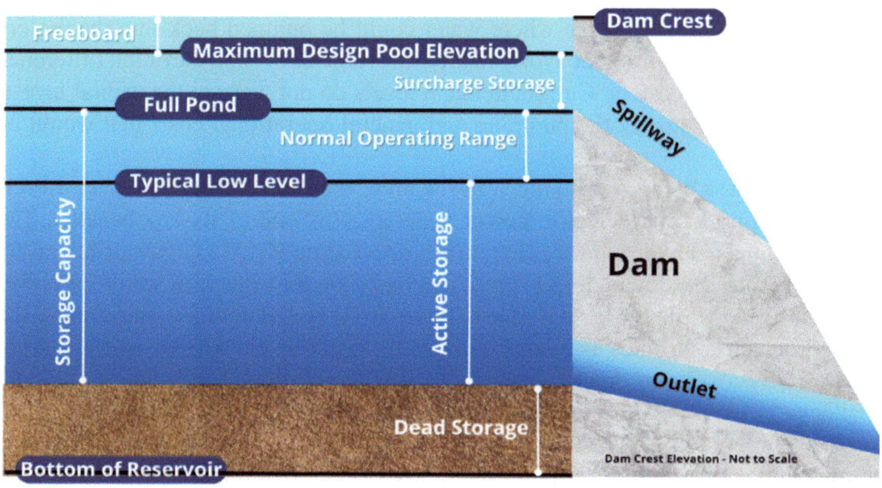

Fig. 9.2 Reservoir elevations

as dual purpose reservoirs based on the additional capacity (surcharge and freeboard storage) above their normal operating full pond elevation (Fig. 9.2). Therefore, theoretically, the Spring Hollow reservoir could potentially function as a dual-purpose high-flow storage system to reduce flooding through flow diversions. However, in reviewing this possibility, there are concerns regarding use of reservoir surcharge and freeboard capacities, related to compromise in structural integrity—the surcharge and freeboard zones are typically not designed to withstand water pressure over extended periods, and this can lead to structural failure. With no safety margins in place, there is also more risk to structural damage caused by overtopping the dam crest. In addition, the existing raw water pumps would not be able to keep up with required flow capacities to mitigate flooding. Lastly, as flood flows are often sediment-laden, operating the Spring Hollow reservoir as a dual-purpose reservoir would negatively impact water quality and potentially clog pumps or lead to excess reservoir sedimentation. Overall, the required operations would be complex and associated maintenance costly.

9.2.2 In-Stream Reservoir—Smith Mountain Lake

The Smith Mountain Lake Reservoir is an in-stream reservoir, with primary inflows from the Roanoke and the Blackwater River. Located in Franklin, Bedford and Pittsylvania Counties in Virginia, USA, the reservoir capacity is approximately 367 billion gallons at full pond elevation. The Smith Mountain Lake Reservoir reached full pond elevation for the first time in 1966. The reservoir currently accounts for approximately 7% of the current water production, as a percentage of the Authority's main water supply totals. However, the Smith Mountain Lake is by far the largest of the three reservoirs. If the water resources within the Roanoke Valley were not able to meet the area's drinking water demand, then this resource has been identified as a long-range supply source where water could be pumped back into the region. However, this is currently not hydraulically favorable due to the elevation difference of approximately 300 feet. Current demand levels do not warrant the capital investments required or justify the pumping and general operational costs that would be needed.

In addition to being a drinking water source, the Smith Mountain Lake Reservoir also supports hydroelectric power generation as one of its key benefits. The average annual net hydropower generation is approximately 342 Gigawatt-hours [6]. Reservoir water levels can drop 1.5 to 2 feet based on the release of water for power generation, and while the level is able to return to full pond within a short time using the hydroelectric turbines as pumps, fluctuating water levels are considered disruptive to local ecosystems that are sensitive to water level shifts, highlighting a broader challenge in hydropower reservoir management.

The Three Gorges Dam, the world's largest hydropower project, offers a compelling case study of the complexities of in-stream hydropower reservoir management. The most notable success of this project is the massive output of clean energy of over 100 terawatt-hours annually, which significantly reduced China's reliance on coal and spurred economic growth in the area due to increased availability of electricity. The dam is also instrumental in mitigating floods downstream, improving navigability along the Yangtze River, further boosting inland shipping and regional economic development [29]. However, despite the remarkable engineering and economic benefits, the project has triggered significant ecological challenges in the Yangtze River Basin, many of which continue to persist even after two decades of project construction. Sedimentation became acute a few years after the impoundment. It was estimated that the dam had retained 150 million tons of sediment per year since 2003 [14]. The retention of sediment led to loss of Yangtze delta and erosion of the riverbeds and wetlands which were critical breeding grounds for fish and birds [30]. Various strategies have been used to manage downstream sedimentation starvation including sediment flushing, re-evaluation of dam operations, real-time monitoring and forecasting and dredging. Another well-noted challenge is eutrophication and frequent algal blooms downstream from stagnation of water in the reservoir, creating water quality issues for downstream water users. The dam also severely disrupted the natural flow regime and the habitat for aquatic life, causing the devastating loss in biodiversity in the Yangtze River basin.

In 2020, the International Union for Conservation of Nature officially declared extinction of the iconic Chinese Paddlefish, which reignited international attention on the ecological consequences of the Three Gorges Dam and other similar large dam projects in general [18]. To address the ecological degradation issue, the Yangtze River Protection Law was passed in 2021 and a 10-year fishing ban was introduced to provide better enforcement of measured aimed to restore and protect the river ecology from the negative impacts from the Dam. Overall, the ecological consequences of the Three Gorges Dam highlight the interconnectedness of the reservoir and the river systems and the wide- and far-ranging impacts of in-stream reservoirs. It offers a cautionary tale to balance the engineering design for human use with comprehensive environmental stewardship.

Environmental challenges associated with the Smith Mountain Lake Reservoir, and associated dam, mainly include nutrient accumulation in its two main branches, the Roanoke River arm and the Blackwater River arm. A study conducted by the Virginia Polytechnic Institute and State University (Virginia Tech), Virginia, USA, in 1972 indicated concerns surrounding the rapid rate of deterioration of water quality in the Roanoke River arm when the Smith Mountain Lake Reservoir was first brought into operating level in 1965 [17]. The lake, functioning as a nutrient trap, was identified as a contributing factor of eutrophication. More recently, during the summer of 2023, elevated levels of cyanobacteria blooms (a naturally occurring bacteria, more commonly known as blue-green algae) were confirmed in the Blackwater River arm. While cyanobacteria are generally harmless at low levels, some species create toxins (cyanotoxins) when they overgrow and form a "harmful algal bloom" (HAB). The toxins created can cause gastroenteritis, liver damage, neurological symptoms, and skin conditions that can affect both animals and humans [9, 21]. Factors contributing to the bloom include overabundant nutrients, elevated water temperatures, slow-moving water, and sunlight [9, 16].

While the cyanotoxins levels that occurred in 2023 were within state advisory levels, the Virginia Department of Environmental Quality (DEQ) noted water samples with cyanobacteria above advisory levels [16]. The Blackwater arm of Smith Mountain Lake was therefore placed under a swimming advisory by the Virginia Department of Health (VDH), from June 6 to August 24, 2023 which coincided with peak recreational season, therefore prompting significant public interest and concern.

The Virginia Water Resources Research Center at Virginia Tech in collaboration with VDH, DEQ, the Smith Mountain Lake Association and other advisory committees from various institutions in the Commonwealth of Virginia, USA, are currently conducting a study that aims to provide short- and long-term strategies to prevent future cyanobacteria formation of HABs in Smith Mountain Lake [9]. Lessons learned in the Smith Mountain Lake incident, which may potentially be highlighted in the study, include placing central emphasis on nutrient reduction at the source to improve raw water quality. This, in turn, plays a key role in protecting public health, and reducing the need for costly reactionary treatment measures. In addition to surface water quality concerns, HABs have a direct impact on the quality of treated (finished) drinking water and treatment cost.

A study by the Unites States Environmental Protection Agency (US EPA) [21] reported that cyanotoxins in raw water can break through conventional water treatment processes, therefore undermining the quality of finished water. In addition, utilities that use chlorine for disinfection may increase the dosage of chlorine disinfectant to break down cyanotoxins, leading to increased risk of disinfection by-products (DBP) [2]. The higher chemical usage leads to increased treatment costs. Therefore, in addition to focusing on source water quality, emphasis must also be placed on balancing effective cyanotoxin removal with minimizing costs, and most importantly, minimizing DBP formation, and building system redundancies for both raw and finished water.

Although the Spring Hollow Reservoir and Carvins Cove Systems are interconnected via the Authority's North Loop Transmission Main (Fig. 9.3), allowing finished water from the respective water treatment facilities (WTFs) to be conveyed to shared service areas, the Smith Mountain Lake System is currently not connected with any other system. This limits the Authority's ability to fully benefit from an integrated water supply network. However, long-term planning includes demand-driven initiatives to connect the Smith Mountain Lake System to the Carvins Cove and Spring Hollow distribution systems. This integration will improve service redundancy and system resilience in the overall service area, in addition to reducing supply disruptions specific to each individual reservoir type.

9.2.3 Hybrid Reservoir—Carvins Cove Reservoir

The Carvins Cove Reservoir, which first reached full pond elevation in 1946, is situated within the Carvins Cove Natural Reserve, a 12,672-acre watershed in Botetourt

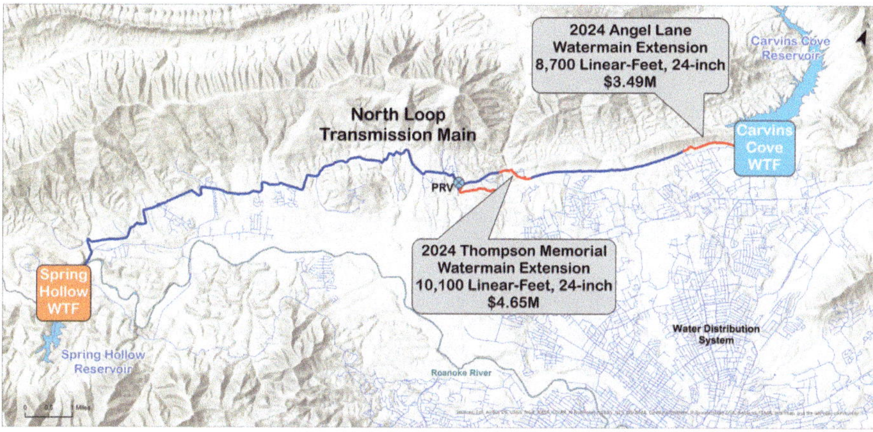

Fig. 9.3 Spring Hollow Water Treatment Facility (WTF) and Carvins Cove WTF Interconnection to shared service areas via the north loop transmission main (Loop recently closed by the two Watermain Extension Projects shown)

and Roanoke Counties, Virginia, USA. The reservoir holds approximately 6.42-billion gallons of water at full pond. During normal production, the Carvins Cove Reservoir supplies about 48% of the current water production, as a percentage of the Authority's main water supply totals. Spring Hollow Reservoir and Carvins Cove Reservoir are managed in a way that attempts to keep the remaining volume of the reservoirs within 5% of each other. Demand is shifted towards the fuller reservoir if this 5% spread is breeched. Stand-by wells, which are not normally in use, are considered for service if the Carvins Cove Reservoir level significantly drops.

The land in the Carvins Cove Natural Reserve above the 1200-foot contour is owned and managed by the City of Roanoke. The land below this elevation, and the reservoir, are owned and operated by the Authority. The Carvins Cove Reservoir is considered a hybrid reservoir based on the water supply regime that combines both natural rainfall runoff and flow diversions. In addition to receiving runoff from the watershed, the reservoir is fed from two underground tunnels with diversion gates that can convey flow from Tinker Creek and Catawba Creek which is in a different watershed, resulting in an interbasin transfer, as illustrated in Fig. 9.4.

At almost 13,000 acres, the Carvins Cove watershed is substantial. If a significant rain event falls over this catchment, the reservoir can, and has, gained up to four feet of water without supplemental flows from the tunnels. Because of significant runoff potential from the watershed, the standard operating procedure includes maintaining additional required freeboard.

The Carvins Cove Reservoir is the threshold indicator for water restrictions during droughts. Declining reservoir levels in the Carvins Cove Reservoir during a particular month of the year trigger specific water use restrictions based on the severity of drought conditions. Initial triggers prompt early response actions such as voluntary conservation. Subsequent triggers indicate an eminent water shortage, and strict

Fig. 9.4 Schematics of the Carvins Cove reservoir water supply regime

water rationing may eventually be necessary. The Authority has three stages of water use restrictions as follows: Stage 1: Voluntary Conservation, Stage 2: Mandatory Restrictions and Stage 3: Emergency Restrictions [27].

Climate change has impacted the Carvins Cove Reservoir in multiple ways. Increased windstorms have toppled weak, damaged or dead trees within the watershed. This contributes to fuel on the forest floor, increasing the risk of forest fires. Additionally, more frequent and intense rainfall events often wash minerals and nutrient-laden sediments into the reservoir. The excess nutrients can encourage algal blooms, while the minerals can deplete oxygen levels in the reservoir water column. Under low-oxygen (anoxic) conditions, iron and manganese can become soluble, creating challenges for water treatment using filtration. Lastly, rising air temperatures create favorable conditions for algae growth. Elevated temperatures increase algal metabolic rates and extend their growing season. In addition, algae have access to nutrients trapped in stratified layers caused by warm water, therefore encouraging their growth and proliferation of cyanobacteria. Warm waters also enhance the release of nutrients from sediments, especially phosphorous, further fueling blooms.

While the Carvins Cove Reservoir watershed includes more than 11,000 acres of conservation easements, which protect the reservoir by restricting land usage that could contaminate water supply, its forested landscape increases vulnerabilities related to fires, resulting in impacts on source water quality and infrastructure damage. Generally, forest fires within the catchment areas of reservoirs present significant threats, including water contamination and the associated costs for mitigation efforts.

Two notable instances of fire occurrences on water supply catchments include the 1996 Buffalo Creek Fire and the 2002 Hayman Fire, both in Colorado, USA. The latter being regarded as Colorado's largest recorded wildfire that destroyed 137,760 acres, including 7500 of the 8500 forested acres of the Cheesman Reservoir watershed [4]. Following both the 1996 and 2002 fires, impacts including erosion and sedimentation problems from subsequent rains have resulted in a total investment in restoration projects of more than $27 million, including water quality treatment, sediment and debris removal, reclamation techniques and infrastructure improvements [4]. More recently, in 2022, Las Vegas, New Mexico, USA experienced the largest fire in the state's history, the Hermit's Peak-Calf Canyon fire, which burned more than 341,735 acres of land over four months including the upper reaches of the Gallinas River watershed, to the extent that the resulting sediment loading overwhelmed the water treatment system. Federal assistance in the tune of $140 million was set aside to replace the water treatment facility [24]. In August 2024, the City of Las Vegas was approved $98 million of the funding as the first installment [7]. The section that follows details forestry management at the Carvins Cove Reservoir as one "soft-path" management approach that encompasses surrounding ecosystems to support more resilient and integrated strategies.

9.3 Other Hard- and Soft-Path Considerations for Resilient Reservoir Planning

The Authority benefits from both hard- and soft-path reservoir management solutions. Hard-path solutions involve building or modifying physical infrastructure, including construction and modifications of reservoir systems, preferably with a diverse portfolio of reservoir types, as well as alternative redundant water supply sources. These solutions are often capital-intensive, and potentially present challenges, including balancing the need for major infrastructure investments against limited financial resources. Securing funding for large-scale redundant upgrades may compete with other priorities, such as aging infrastructure that demand more urgent attention. This highlights the need for more cost-effective approaches that can complement traditional infrastructure upgrades.

Soft-path solutions emphasize careful planning and management that prioritizes social and individual well-being, while focusing on lower-cost approaches such as technology application, environmental protection and sustainable operations, among other complementary approaches [3]. While engineering solutions are required to implement actual reservoir systems, governance and political support structures are necessary to foster regional water cooperation to boost resiliency.

A critical component of implementing soft-path solutions is incorporating diverse stakeholder perspectives with the aim of delivering services that "match users' needs" rather than simply "delivering quantities" [3]. Engaging diverse stakeholders, such as local communities, environmental groups, and regulatory agencies, ensures that water supply and management strategies are efficient, and at the same time responsive to varying needs, priorities and viewpoints.

While this inclusive approach presents challenges for the Authority, such as balancing conflicting interests and coordinating across multiple governance levels, considering the Authority's multiple regional jurisdictions, it opens significant opportunities, including fostering public trust, enhancing transparency and supporting equitable strategies.

The sections that follow discuss both hard-path and soft-path solutions, considering climate change as a catalyst for regional cooperation, water quality concerns as a driver for forest management and integrated water supply systems, alternative source water supply for diversification, **advanced technologies** for monitoring, and emergency action planning for preparedness.

9.3.1 Planning for Extreme Weather Patterns Through Regional Water Cooperation and Supply Management

Planning for resilient reservoirs and water supply systems is becoming increasingly critical as water utilities grapple with the challenges posed by climate change, evolving environmental conditions, and shifting societal demands. More frequent extreme weather events—such as changing precipitation patterns, wildfires, and prolonged heatwaves—are straining existing reservoir capacity and operational management. Extreme weather patterns may therefore continue to be catalysts for regional water supply cooperation. For example, the Colorado River Basin, a major watershed in western USA that spans seven states and northwestern Mexico, experienced declining flows alongside rising demand, compelling water planners in the USA to adopt regional water supply cooperation strategies and implement drought response initiatives [20]. The Authority, having been established because of drought, has particularly benefited from regional cooperation. The 2002 drought exposed vulnerabilities in the Roanoke Valley water supply as it relates to lack of emergency preparedness and lack of redundancies due to fragmented and isolated water systems. In forming the Authority, a unified, system-based approach to water supply planning was made possible, and the consolidated infrastructure from each member jurisdiction was also made available to further facilitate redundancy planning and therefore improve the overall system resiliency. The Authority currently maintains a master plan to ensure its ability to meet the water demand needs of its current and potential future customer base. The plan considers water demand by forecasting future needs, which drives infrastructure improvements and evaluation of options for expanding existing resources or developing alternative water sources. In addition to these capital investments, the plan also incorporates soft measures such as water conservation programs, public education initiatives, and demand management strategies. This comprehensive approach helps ensure a reliable and sustainable water supply for our growing populations and evolving needs.

9.3.2 Managing Emerging Water Quality Concerns Through Forest Management and Integrating Reservoir Systems

Beyond quantity, maintaining source water quality is an increasingly pressing concern, as both natural disasters and anthropogenic activities introduce new contaminants into water supplies. The growing intensity and frequency of wildfires, for instance, have been shown to significantly increase the concentrations of chemicals, heavy metals, and sediments in receiving waters—often by orders of magnitude compared to pre-fire conditions [13].

In 2007 a committee was convened to develop the Carvins Cove Natural Reserve Park Management Plan. The Carvins Cove Advisory Committee comprised a diverse group of stakeholders, including the Authority; local city and county officials; parks, trails, greenways and recreation experts; representatives from federal, state and educational institutions; and other environmental advocates. The plan lays out the Authority's and City of Roanoke's land ownership rights, the park limits and management goals, guidelines and actions, including the recommendations for a Forest Stewardship Plan, and a Fire Response Plan, with the indication that fire fuel loading is at "extreme danger levels in some areas of the watershed" [15]. Fuel loading is the amount of combustible material (fuels in weight per unit area) present in a forest, including both live vegetation (dense growth of inflammable-leafed shrubs) and dead vegetation (forest litter, downed trees etc.) that can feed a wildfire. The Plan also notes that existing roads may not be adequate for fire/emergency vehicle access.

With the understanding of the risks associated with forest fires, in December 2024 the Authority, in collaboration with the City of Roanoke, initiated a Forest Stewardship Plan for the Carvins Cove Natural Reserve. This is currently being implemented in two phases. Phase I comprises a Forest Assessment, to identify potential stressors in the Carvins Cover Reservoir watershed and determine mitigation measures associated with forest fires. This initial nine-month effort will also develop a Wildfire Incident Checklist aimed at providing immediate actions to prevent, respond to, and recover from a wildfire at the Carvins Cove Natural Reserve.

Resources developed in Phase I will be used to complete a comprehensive Forest Stewardship Plan in Phase II. The Forest Stewardship Plan will cover general long-term best management practices for both source water protection and recreation, including wildfire considerations, aimed at ensuring source water protection and improving watershed water quality.

Concurrently, emerging contaminants such as per- and polyfluoroalkyl substances (PFAS) and microplastics have raised public health concerns, prompting regulatory action. In April 2024, the US EPA established the first-ever enforceable maximum contaminant levels for a subset of PFAS compounds, necessitating advanced treatment technologies or alternative water sourcing for many utilities. On May 14, 2025, US EPA rescinded enforceable limits for four PFAS compounds, and announced its intent to keep and extend compliance deadlines for Perfluorooctanoic Acid (PFOA) and PFOS (Perfluorooctane Sulfonic Acid) from 2029 to 2031 [23].

In 2020, the Authority detected PFAS contamination in one of its reservoirs. Investigations revealed that the contamination originated from an upstream water treatment equipment manufacturer plant's discharge to a local wastewater treatment facility that had a discharge permit for the source river. While the Authority responded swiftly—identifying the contamination source and evaluating treatment options—there was an immediate need to secure an alternative water supply to mitigate risks. Fortunately, the system's reservoir interconnection enabled continued service to the affected areas through shifting supplies. This demonstrates the value of redundancy and flexibility in water supply planning.

9.3.3 Supplemental Sources

The Authority drinking water system is supplied by several groundwater springs and wells in addition to the reservoir systems previously discussed. Crystal Spring provides approximately 21% of the current water production, as a percentage of the Authority's main water supply totals and Muse Spring provides approximately 2%. A series of wells that provide less optimal groundwater, due to either quality or quantity, are kept in an operational state in the event of drought conditions. These wells may be brought online if the interconnected system starts to feel the effects of a drought. In addition, the Authority also can purchase water from nearby City of Salem's water department. There exists an interconnection that allows water to flow in either direction, as needed during emergencies or under drought conditions.

As water scarcity and extreme weather patterns put increasing pressure on utilities, to strengthen water supply redundancy and resilience, utilities are going beyond conventional governance and political frameworks to find innovative and synergistic solutions. A prominent example is the Bay Area Regional Reliability Partnership (BARR) California, USA, which is a collaborative, voluntary initiative among the eight large water utilities in the Bay Area of Northern California, aimed at building resilience to droughts and climate change through strengthened regional water supply planning and coordination. Among its achievements is the development of a Shared Water Access Program that explore interagency water transfers. The framework lays out options to bank surplus supplies in wet years and evaluates how water can be shared with drier service areas through existing reservoirs and pipes [1].

9.3.4 Utilizing Advanced Monitoring Technologies

The use of advanced technologies in water supply reservoir management is an emerging practice that is transforming how utilities monitor, operate, and optimize reservoir systems. Remote sensing technologies, such as satellite imagery and drones, provide real-time data on reservoir levels, sedimentation, and surrounding watershed conditions, enabling proactive decision-making. For example, the US EPA Cyanobacteria Assessment Network (EPA CyAN) is a multi-agency initiative that collaborates with several US federal agencies including the National Aeronautics and Space Administration (NASA), the National Oceanic and Atmospheric Administration (NOAA), the United States Geological Survey (USGS) and the US Amy Corps of Engineers (USACE) to explore the use of remote sensing in developing early warning systems for detecting algal blooms in water supply sources [22]. Ground-based sensors, including internet of things (IoT)-enabled water quality probes and flow meters, which can continuously track parameters such as turbidity, temperature, contaminant levels, water flow, and reservoir depth, improve operators' understanding of reservoir conditions, and therefore enhance operational efficiency.

The unprecedented capabilities to generate large amounts of data is propelling the development of predictive modeling and AI-driven analytics that allow reservoir managers to forecast reservoir water quality and hydraulic conditions, and assess the impacts of climate variability and extreme weather events. As such, the Authority has partnered closely with research institutes to pilot the use of sensors and advanced modeling to anticipate and prepare for water quality changes in reservoirs [5, 25]. In addition, to remain aligned with technological advancements, the Authority promotes continuous learning, partners with leading technology vendors, and actively engages in knowledge sharing with industry leaders and professional organizations. This includes collaboration with the Water Information Sharing and Analysis Center (WaterISAC), a U.S. based nonprofit that provides security and threat information specifically for the water and wastewater sector to help utilities protect critical infrastructure. While the use of advanced monitoring technologies in water treatment systems can significantly boost operational efficiency and data analytics, they also have the potential to increase the risk of cyber-incidents.

In order to safeguard the Authority's data and information, to ensure business continuity, and preserve trust with employees, customers and vendors, the Authority has prioritized cybersecurity. The Authority has partnership with the US Department of Homeland Security (DHS)'s Cybersecurity and Infrastructure Security Agency (CISA), utilizing many of the services it offers, including training and vulnerability testing. Since 2023, the Authority has also outsourced its Cyber Security Monitoring and Response to a qualified third-party vendor to help mitigate the risk of data breaches, financial losses and reputational damage. The Authority's internal Information Technology (IT) and Operational Technology (OT) staff is focused on staying abreast of the current threat landscape, while also educating the rest of the organization through quarterly training on cyber security risks to build a culture of security and to enhance the Authority's overall security posture. The IT and OT teams also formed an internal Cyber Security Committee that meets regularly, to ensure coordination between IT and OT for the protection of both IT and OT networks and data against unauthorized access, damage, or disruption.

Overall, the Authority reduces cyber security risk by following industry standards and applying best practices, including those developed by the National Institute of Standards and Technology (NIST), a US federal agency under the Department of Commerce which supports cybersecurity frameworks. The NIST Special Publication 800-82r3, Guide to Operational Technology Security [10] is an industry standard resource for securing OT systems typically found in water treatment facilities. The principles in NIST SP 800-82r3, highlight the importance of defense-in-depth (multiple layers of security controls) since no one security control is perfect. The following defense-in-depth concepts can be used to develop layers of protection to securely integrate the growing need for advanced monitoring technologies:

Asset Identification and Asset Management: The first step is to create a complete list of all hardware and software components in the Supervisory Control and Data Acquisition (SCADA) ecosystem. This list includes the various sensors, data acquisition systems, communication equipment, servers, and workstations used to collect real-time data for the monitoring, controlling, and managing of industrial processes and allowing operators to make informed decisions and control operations remotely. Understanding the significance and value of each asset will help establish a complete picture giving the ability to focus security efforts and recognize anomalies [NIST SP 800-82r3, Section 6.1.1—Asset Management (ID.AM)].

Network Segregation and Architecture: NIST recommends segmenting the OT network from the enterprise IT network. This segment becomes the first line of defense if the IT network is compromised. These connections should be well documented and easily disconnected. Additional segmentation is recommended within the OT network following the Purdue Model [28]. This model creates zones increasing the attack complexity of the most critical assets [NIST SP 800-82r3, Section 5.2.3.1—Network Architecture].

Access Control: This concept encompasses strong authentication based on roles and responsibility (least privilege) and systematic review of user accounts and permissions. Remote access to OT systems, crucial for monitoring wide-area systems, require implementing strong access controls like Virtual Private Networks (VPNs) with the latest authentication and encryption methods [NIST SP 800-82r3, Section 6.2—Protect, and Section E.3.6 Remote Access].

System Hardening: Not knowing the source point or the destination target of the attack, securing each individual device on the OT network becomes important. This practice involves monitoring cybersecurity alerts and advisories from government agencies, such as CISA and device manufactures. Mitigations might include patching vulnerabilities, turning unnecessary services and ports off, and implementing system- or event- specific secure configurations recommended [NIST SP 800-82r3, Section 5.2.5.2 Patching, and Section 5.2.5.4 Configuration Management].

Intrusion Detection and Prevention: Implementing intrusion detection systems (IDPS) geared towards industrial control system and associated protocols provides a valuable window into detecting and stopping malicious activity. In addition, these systems must be fed the proper threat intelligence [NIST SP 800-82r3, Section 5.2.3.3 Network Monitoring, and Section 6.3.2.1 Network Monitoring (DE.CM-1)].

Security Monitoring and Logging: Comprehensive and auditable logging of all system events including access attempts, configuration changes, and network traffic is essential. These logs can be used for incident detection, response, and forensic analysis. [NIST SP 800-82r3, Section 5.2.3.2 Centralized Logging].

Vulnerability Management: According to NIST SP 800-82r3 (DE.CM-8), vulnerability scanning in the OT environment should be an ongoing process that uses both automated and manual means to find vulnerabilities. Common techniques include performance and penetration testing, regular security audits, assessments, peer reviews, and continuous monitoring with passive or active scanning. [NIST SP 800-82, Section 6.3.2.4 Vulnerability Scanning (DE.CM-8)].

By taking advantage of the principles and guidelines outlined in NIST publications, water treatment facilities can strategically implement advanced monitoring technologies while significantly mitigating the associated cybersecurity risks. A proactive and NIST-informed approach, or similar framework, is fundamental to ensure the continued reliability and security of this critical infrastructure.

9.3.5 Proactive Maintenance and Emergency Action Plans

There has been increasing recognition of dams and reservoirs as critical infrastructures that have the capability to cause catastrophic damage when they fail from natural or man-made causes. In May 2020 after several days of heavy rainfall, the Edenville Dam in Michigan, USA failed, triggering the failure of the downstream Sanford Dam. The cascading failures led to widespread flooding in Midland County, displacing nearly 10,000 residents and causing extensive property damage [12]. Aging infrastructure and the lack of proper oversight from governing bodies were at the core of the failing dams. As the essential role of water utilities to manage this critical infrastructure continues to gain recognition, there is an increasing emphasis on proactive rehabilitation and maintenance of these critical structures to prevent future failures. In addition, developing robust emergency response plans is an integral component of water reservoir resilience management. Having a well-developed Emergency Action Plan (EAP) provides reservoir managers with a structured response for potential failures or hazardous events. EAPs identify the impacts and the resources needed to address emergency events, and to coordinate with other stakeholders like first responders, local authorities, and impacted communities.

The Authority conducts inundation studies for all its reservoirs to determine the time-to-location and depth of water in the downstream channel in the event a dam breach was to occur. The inundation studies are generated by modeling the downstream river channel from the dam. The model looks at several scenarios, such as a sunny-day dam breech, that occurs when there is no additional precipitation, thus no storm-related contributions to flow, as well as response during a precipitation event that adds more volume to the flood wave. Inundation zones captured in shapefiles are shared with Emergency Response Coordinators and the National Weather Service (NWS). Parcels that will likely be impacted by a dam breach are highlighted in these maps. In the event of a dam emergency, the NWS is contacted and asked to issue cell phone alerts to property owners of the impacted parcels.

9.3.6 Economic Impacts of Integrated Reservoir Management Approach

Since reservoirs are the intersection of multiple use cases including water supply source, recreation, power generation, irrigation and ecological conservation, and reservoir infrastructure projects are typically expensive and complex, it is imperative to employ an objective and unified approach to plan and justify the impacts and uses of the various planned uses. Conducting a comprehensive economic analysis to quantify the benefits and costs of the multiple uses is essential to govern existing reservoir infrastructure planning, as well as development of future projects. In a study done by the Institute for Nature Resources at Oregon State University that evaluates the tradeoff economics of reservoirs that serve the dual purposes of water storage and flood control in the Willamettee River Basin [11], the loss in annual recreational benefits are expected to increase to over 12 million towards the end of the century while the current annual flood control benefits are estimated to be more than 1 billion and are expected to triple by 2100. Based on the economic analysis, there is a strong economic justification to keep the reservoir level low during wet season when the flood risk is high before considering filling for storage. The Oak Ridge National Laboratory conducted an economic benefits analysis of multipurpose hydropower reservoirs and concluded that while power generation was a critical component, other use cases like recreation and irrigation provided the most economic benefits, with significant revenue stream from recreation tourism spending and the value of irrigated crops [8]. Multipurpose reservoirs are not only sole water supply sources, they hold significant environmental, economic and social consequences that extends beyond local and regional economies. Prudent economic analysis should be incorporated in integrated reservoir planning and management to inform reservoir operation for optimal shared benefits.

9.4 Conclusion

In conclusion, the Authority's reservoir management addresses the complex challenges posed by extreme weather patterns. The Authority's unique position in managing a wide portfolio of interconnected water supply systems that include off-stream, in-stream, and hybrid reservoirs, facilitates water quality control, supports ecosystems, and potentially reduces flood risks, all while ensuring reliable and sustainable water supply, water security and climate resiliency.

The Authority's multi-mode reservoir approach developed organically, and fortuitously, through the consolidation of infrastructure from its member jurisdictions when the Authority was formed. As the Authority continues to grow, additional redundancies are being realized. However, there is a strong rationale for deliberately adopting planned measures that consider multiple reservoir types, based on the complementary advantages and balanced limitations. Future research recommendations therefore include developing optimization frameworks or models that balance operational dynamics, advantages, and limitations of different reservoir types, for strategic multi-mode reservoir implementation under varying environmental and socio-economic conditions.

The Authority's Spring Hollow Reservoir, an off-stream reservoir, and the Carvins Cove Reservoir, a hybrid reservoir, currently, in combination, provide 66% of the system demands, as a percentage of the Authority's main water production. Planning efforts have therefore focused on building redundancy and resiliency between these two systems, including allowing finished water from the respective water treatment facilities to be conveyed to shared service areas.

Creating or closing "loops" in distribution lines is common practice that facilitates maintenance flexibility, improved reliability and improved water quality based on eliminating dead-ends that are a water age concern. Loops in transmission mains move larger volumes of water from sources and provide alternate supply routes to distribution systems. Transmission main loops require more planning due to construction complexities and higher cost of the larger transmission lines. The Authority, therefore, took a phased project implementation approach in constructing resiliency improvements to connect the two systems, with the most recent construction completed in 2024.

While none of the Authority reservoirs currently utilize surcharge and freeboard storage for flood mitigation, future studies are recommended to assess the feasibility of retrofitting or implementing structural upgrades to support safe and effective dual-purpose reservoir operations under evolving climate and land use conditions. Off-stream reservoirs, such as the Spring Hollow reservoir, which have controlled filling capabilities, could therefore be evaluated for their potential to serve as high-flow storage systems, reducing flood risks through flow diversions, while enhancing water supply resilience during low-flow periods.

The Smith Mountain Lake reservoir is an in-stream reservoir, which directly harnesses both water and hydroelectric power from river flows. The reservoir currently accounts for approximately 7% of the current water production, as a percentage of the Authority's main water supply totals.

While the lake is owned by Appalachian Power, the Smith Mountain Lake Water Treatment Facility is jointly owned by the Authority and the Bedford Regional Water Authority. As such, all three entities share an interest in the lake's sustainable management. However, interests can compete during droughts or peak power demands, if water supply needs and hydropower generation are not in full alignment. Coordinated reservoir releases that carefully balance energy production while maintaining lake levels are therefore important.

Cyanotoxins are a water quality concern of the Smith Mountain Lake, and while traditional chemical treatment such as chlorine is typically used in cyanotoxin removal in water treatment processes, more research on additional advance processes that reduce the risk of DBP formation is recommended, in addition to water quality studies exploring strategies that prevent future cyanobacteria formation.

Research on water treatment should also include a review of efficient methods for removing ash, sediments, and other debris resulting from forest fires, with careful consideration of their impacts on treatment infrastructure to prevent damage and maintain system reliability. This is especially critical for forest-protected water sources, such as the Carvins Cove Reservoir, which are highly vulnerable to the effects of wildfires.

While the Authority builds new redundancies as it continues to grow and acquire new systems, including ground water supplies, it also faces capital improvement expenses to fully integrate some systems and bring them up to current Authority regional design and construction standards. These hard-path solutions are often necessary, but costly. Therefore, best practices that include soft-path measures such as water quality and environmental management, consideration of existing supplemental water supplies, monitoring, proactive maintenance, and emergency action planning, among other measures, play a vital role in long-term sustainability of the Authority's water resources, as these measures do not require extensive physical infrastructure investments.

References

1. Bay Area Regional Reliability Partnership (2020) Retrieved from https://www.bayareareliability.com/
2. Foreman K, Vacs Renwick D, McCabe M, Cadwallader A, Holsinger H, Kormondy C, Albert R (2021) Effects of harmful algal blooms on regulated disinfection byproducts: findings from five utility case studies. American Water Works Association (AWWA) Water Science
3. Gleick PH (2003) Global freshwater resources: soft-path solutions for the 21st century. Science 302(5650):1524. https://doi.org/10.1126/science.1089967
4. Hall S (2017) The legacy of Colorado's largest wildfire. Retrieved from Denvar Water: https://www.denverwater.org/tap/legacy-colorados-largest-wildfire
5. Hammond NW, Birgand F, Carey CC, Bookout B, Breef-Pilz A, Schreiber ME (2023) High-frequency sensor data capture short-term variability in Fe and Mn concentration due to hypolimnetic oxygenation and seasonal dynamics in a drinking water resservoir. Water Research
6. Hydropower Reform Coalition (n.d.) Smith Mountain P-2210. Retrieved from https://hydroreform.org/hydro-project/smith-mountain-p-2210/?utm_source=chatgpt.com
7. Lohmann P (2024) Las Vegas to get $98 million to replace water treatment facilities after 2022 wildfire damage. Retrieved from Source New Mexico: https://sourcenm.com/briefs/las-vegas-n-m-to-get-98-million-to-replace-water-treatment-facilities-after-2022-wildfire-damage/

8. Marisol Bonnet AW (2015) The economic benefits of multipurpose reservoirs in the United States-Federal Hydropower Fleet
9. Muras A (2025) Collaborative algal bloom study underway at Smith Mountain Lake. Retrieved from Virginia Water Resources Research Center at Virginia Tech: https://www.vwrrc.vt.edu/2025/04/08/collaborative-algal-bloom-study-underway-at-smith-mountain-lake/?utm_source=chatgpt.com
10. NIST (2023) Guide to Operational Technology (OT) security: NIST Publishes SP 800-82, Revision 3. Retrieved from Computer Security Resource Center (CSRC): https://csrc.nist.gov/News/2023/nist-publishes-sp-800-82-revision-3
11. Oregon State University Insitute for Natural Resources (2016) Reservoir Economics
12. Panda A (2020) Case study: America's dam disaster. Retrieved from University of Michigan Center for Social Solutions
13. Paul M, LeDuc S, Lassiter M, Moorhead L, Noyes P, Leibowitz S (2022) Wildfire induces changes in receiving waters: a review with considerations for water quality management. Water Resources Res 58:e2021WR030699. https://doi.org/10.1029/2021WR030699
14. Reuters (2007) Three Gorges dam causes downstream erosion-study
15. Roanoke City and Western Virginia Water Authority (2007) Carvins Cove natural reserve park management plan. Retrieved from Roanoke Parks and Recreation Department: https://www.playroanoke.com/wp-content/uploads/2023/12/Carvins-Cove-Natural-Reserve-Park-Management-Plan-12.17.2007.pdf
16. Smith Mountain Lake Association (n.d.) Be HAB aware at Smith Mountain Lake. Retrieved from Smith Mountain Lake Association: https://smlassociation.org/hab/
17. Sparger RH (1972) A productivity study on Smith Mountain Lake. Retrieved from Virginia Polytechnic Institute and State University : https://vtechworks.lib.vt.edu/items/a66c47db-742c-423a-8e14-a3687e7c2f95?utm_source=chatgpt.com
18. Striebel B (2020) Sadly, it is not the year of the Sturgeon in China. Retrieved from https://www.wwf.org.co/?359593/Chinese-paddlefish-extinct
19. TWCA (2018) Flooding in Texas: preparation and response. Retrieved from Texas Water Association: https://texaswaterassociation.org/page-18080/9274071
20. U.S. Climate Resilience Toolkit (2016) A new generation of water planners confronts change along the Colorado River. Retrieved from Climate.gov: https://toolkit.climate.gov/case-study/new-generation-water-planners-confronts-change-along-colorado-river
21. US EPA (2016) Water treatment optimization for cyanotoxins. Retrieved from US EPA: https://www.epa.gov/ground-water-and-drinking-water/water-treatment-optimization-cyanotoxins-document
22. US EPA (2024) US EPA Cynaobacteria Assessment Newtwork (CyAN). Retrieved from https://www.epa.gov/water-research/cyanobacteria-assessment-network-cyan
23. US EPA (2025) Per- and Polyfluoroalkyl Substances (PFAS). Final PFAS National Primary Drinking Water Regulation. Retrieved from US EPA: https://www.epa.gov/sdwa/and-polyfluoroalkyl-substances-pfas
24. Walton B (2023) New Mexico's largest fire wrecked this city's water source. Era of megafires endangers water supplies in American West. Retrieved from Circle of Blue: https://www.circleofblue.org/2023/supply/water-management/infrastructure/new-mexicos-largest-fire-wrecked-this-citys-water-source/
25. Wander HL, Thomas RQ, Moore TN, Lofton ME, Breef-Pilz A, Carey CC (2023) Data assimilation experiments inform monitoring needs for near-term ecological forecasts in a eutrophic reservoir. Ecosphere
26. Western Virginia Water Authority (2025) Water quality reports. Retrieved from Western Virginia Water Authority: https://www.westernvawater.org/learn/water-quality-reports
27. Western Virginia Water Authority (n.d.) Drought contingency plan. Retrieved from Western Virginia Water Authority: https://www.westernvawater.org/water/water-sources-and-treatment/drought-contingency-plan
28. Williams TJ (1992) The purdue enterprise reference architecture: a technical guide for CIM planning and implementation. Instrument Society of America, Research Triangle Park, NC

29. Yang L, Zeng S, Xia J, Wang Y, Huang R, Chen M (2022) Effects of the three Gorges Dam on the downstream streamflow based on a large-scale hydrological and hydrodynamics coupled model. J Hydrol Reg Stud
30. Yongqiang Zhou J (2016) Impacts of three Gorges Reservoir on the sedimentation regimes in the downstream-linked two largest Chinese freshwater lakes. Sci Rep

Chapter 10
Agro-Environmental Security Under Climate Change (Central Europe): Climate-Borne Risk and Their Evaluation

Hana Středová, Petra Fukalová, Emma Vacková, and Tomáš Středa

Abstract While the agro-sector currently remains less impacted by direct cyber threats due to the incomplete application of advanced IT methods, the increasing adoption of Agriculture 4.0 technologies will soon expose this sector to a broader array of digital vulnerabilities. However, in terms of overall security, climate change poses even more serious risks for agriculture compared to other sectors, as it is inherently dependent on interactions with nature and the environment. Our study addresses the escalating impact of climate-driven abiotic threats on agro-environmental security, focusing on Central Europe, with particular emphasis on Czechia. Through a detailed analysis, we identify and evaluate key climate risks, including drought, extreme temperature variations, and flooding, assessing their probability and potential consequences for the resilience of the agro-sector. Our analysis integrates multiple data sources, including risk assessments conducted by first-year students of the Agroecology MSc program at Mendel University, who represent emerging professionals and future stakeholders in the field. In contrast, national risk assessments typically address climate threats from a broader environmental security perspective. By synthesizing data from various sources, our findings reveal critical insights into the alignment and gaps between agro-environmental and broader environmental security assessments. This work underscores the pressing need for adaptive strategies in agro-environmental security, emphasizing the urgency of enhancing resilience in the Czech agricultural landscape against climate-induced risks.

Keywords Threats · Analyses · Environmental security · Resilience · Adaptation

H. Středová (✉) · P. Fukalová · E. Vacková · T. Středa
Faculty of Agrisciences, Mendel University in Brno, Brno, Czechia
e-mail: hana.stredova@mendelu.cz

T. Středa
Czech Hydrometeorological Institute, Branch Brno, Brno, Czechia

M. Huber et al. (eds.), *Critical Infrastructure Systems and Extreme Threat Nexus: Preparing for Resilience and Security*, NATO Science for Peace and Security Series C: Environmental Security, https://doi.org/10.1007/978-3-032-13170-6_10

10.1 Introduction

10.1.1 The Security of the Agro-Sector: Challenges from Both Environmental and Cyber Threats

Environmental security encompasses the protection of natural ecosystems and human societies from environmental hazards, including climate change, natural disasters, and resource depletion. Agro-sector security focuses on safeguarding agricultural systems against all threats while ensuring sustainable food production. Increased environmental pressure driven by climate change forms the foundation of the concept of agro-environmental security, as well as the increasing reliance on digital infrastructure within agro-sector represents emerging concept of agro-cyber security. Since this book focuses on the Cyber-Physical and Social Infrastructure–Climate Change (CPSICC) Nexus, we note at the outset that our chapter primarily centers on agro-environmental security and climate-related threats. However, toward the end of our study, we broaden the perspective by introducing the "cyber" dimension of agriculture and associated cyber-physical risks. This addition situates our findings within the larger context of digital transformation and evolving risk landscapes. The integration of advanced technologies, such as the Internet of Things (IoT), artificial intelligence (AI), and big data underpins the emergence of Agriculture 4.0 and 5.0 [5], which are reshaping agricultural systems. Agriculture 5.0, in particular, emphasizes smart innovations that increase productivity while minimizing environmental impact and addressing social and political challenges in food systems [5, 13]. While our main focus remains on climate-driven agro-environmental risks, we include cyber-physical risks in our concluding section to provide a more comprehensive understanding of emerging threats and resilience strategies.

10.1.2 Agro-Environmental Security and Climate Change Nexus

Climate change has a profound impact on extreme weather events with negative effect on agriculture. These events intensify their frequency and duration, disrupting the environment and the lives of residents, especially in vulnerable areas. In Czechia (central Europe), the main agro-environmental threats described in the Environmental Security Concept 2021–2030 with a view to 2050 and in the Climate Change Adaptation Strategy for the Czech Republic for 2021–2030, which is in line with the EU Adaptation Strategy are: long-term drought, floods, heavy rainfall, extremely high temperatures, extreme wind, frost, slope instability and wildlife fires. As these threats are often interconnected [17], Gill and Malamu [11, 36, 42] urgently call for comprehensive strategy addressing so-called "multi-hazards". As documented by

[3] the number of articles on multi-hazards in Europe is increasing. Many publications focus on the link between drought and insects, especially in temperate forests. The combination of high temperatures and prolonged drought has an impact on the increase in the occurrence of destructive fires in Europe (Bowman et al. 2009). In Europe, areas most prone to fires are located in the Mediterranean region [12, 28].

10.1.3 Primary Threats

From an agro-environmental security viewpoint, weather manifestations such as heavy rainfall, extreme temperatures, wind, abnormal frost and snow cover represent primary threats mostly fostered by climate change. It is estimated that while extremes related to low temperature will occur less than once per decade in Europe, high temperatures are expected to occur multiple times annually [1]. High temperatures thus represent a new challenge, especially for food security, due to their impact on crop production (food, fodder, and other commodities) and livestock production (lack of fodder). Increased temperatures also affect soil fertility, the risk of soil erosion, reduce water quality or availability and accelerate the spread of diseases and pests of agricultural crops.

A positive benefit to the contrary can be e.g. earlier sowing, accelerating the phenological phase of plants and growing heat-loving varieties at higher altitudes [16]. Scientists estimate that by 2050, the ripening of agricultural crops in central Europe will be accelerated by up to 10–14 days in lowlands and by up to 15–20 days in higher elevations [26]. This positive effect of faster vegetation development contradicts with increasing spring frosts, which occur after the germination and emergence of herbs/vegetables. Vineyards, fruit trees, rape, potatoes and sugar beet are most affected. In France, for example, up to 80% of wine-growing areas were damaged in 2021. This causes considerable ecological and economic damage to agriculture and forestry and future scenarios predicts that more than one third of European, Asian and North American forest will be damaged this way. Raising awareness of the effects of late spring frosts, assessing the risks in terms of frequency and severity for each region, and breeding hardy varieties could help to address this problem [21]. We now believe that global change is also leading to a reduction in snow cover in central Europe as documented already by [40]. The incidence of frost days with insufficient rainfall or snowfall will also increase, leading to an increase in demand for irrigation water. Crop rotation practices, vegetation growth and crop yields will also be affected.

Considerable damage of agricultural crops can be caused by strong winds. The phenomenon of wind erosion in Czechia is described in detail by Středová et al. [37] and Kučera et al. [19]. Eroded wind-borne soil particles damage plants, strong wind may also make difficult an application of pesticides or fertilizers. On June 24, 2021, several serious thunderstorm incidents took place in Central Europe, with losses estimated at several billions of euros. A tornado, that occurred at the same time in Czechia, was one of the most powerful incidents in its history [41], forming primarily due to a hot and unstable air mass of tropical origin. Since the tornado that swept through southeastern Czechia in 2021, there has been ongoing speculation about whether climate change played a role in the disaster. However, most meteorologists refute this, attributing it instead to natural fluctuations in atmospheric circulation. A tornado results from complex processes within a storm cloud interacting with the Earth's surface. For a tornado to form, specific conditions, such as temperature and humidity, must be met. Therefore, it is impossible to determine with certainty whether climate change will lead to more frequent tornadoes or an increase in their intensity.

Another major threat to agriculture is heavy rainfall, which poses challenges for agricultural landscapes, primarily by accelerating runoff and increasing sediment transport. Additionally, agricultural soil in Czechia suffers from compaction-affecting up to 45% of soils—due to heavy machinery use, insufficient organic matter, and acidification. As a result, the soil loses its ability to retain and infiltrate water (AS ČR 2019). It is expected that heavy rainfall will become increasingly frequent [39]. The greatest risks of surface runoff and erosion occur with wide-row crops such as corn and beet. However, the consequences of heavy rainfall can be mitigated through practices like using intercrops or undersowing.

While these primary threats stem largely from direct weather extremes, their effects are significantly shaped by the physical and ecological characteristics of the landscape. The structure, management, and resilience of agricultural landscapes determine whether these initial threats are absorbed, amplified, or transformed into additional hazards. This interplay gives rise to what we identify as secondary threats—cascading consequences such as drought, wildfires, floods, and landslides, which can result when vulnerable landscapes fail to buffer or mitigate primary climate-induced impacts.

10.1.4 Secondary Threats

Secondary threats such as aforementioned drought, wildfires, floods and landslides are closely linked to how human activities, in particular land use and landscape planning, strengthen or weaken the natural resilience of landscape. Floods are a significant challenge for farmers in many countries, not just in Europe. Farmers report that floods most often occur during the germination phase of crops, leading to substantial damage and reduced yields. According to a study by [32], floods in Croatia caused yield losses of up to 50% between 2015 and 2020. The most commonly grown crops-winter wheat, winter barley, soybeans, and corn—are particularly affected, with local farmers identifying corn as one of the most vulnerable. Over the past five years, up to 80% of Croatian farmers have faced short-term flooding, where soil remains submerged for up to seven days. In Czechia, 60 to 100 flash floods occur annually.

Drought is an increasingly frequent challenge for many European farmers. It arises due to high temperatures, increased evapotranspiration, and changes in precipitation and water flow, causing disruptions in various economic activities. Water consumption for agricultural irrigation is expected to rise in the coming years. In response, the European Council has adopted a regulation allowing the reuse of treated urban wastewater for agricultural irrigation. With climate change intensifying, the reuse of treated water is expected to become an increasingly popular strategy, particularly in southern Europe [15]. Significant drought episodes in 2000, 2003, 2012, and 2015 severely impacted Central Europe, causing estimated damages exceeding 500 million euros in Czechia [43]. Between August 2011 and May 2012, Czechia experienced an extreme drought, with its most important agricultural regions being the hardest hit. The country endured what was classified as a 100-year drought, and according to the Palmer Drought Severity Index, the 2012 drought was the worst in 130 years. This event led to an unprecedented decline in cereal yields over the past 52 years, with winter crops being more severely affected than spring crops. Additionally, signs of hydrological drought were observed in rivers. Similar drought conditions were also recorded in Austria and Slovakia. Sensitivity experiments using global circulation models suggest that droughts of similar intensity could occur at 20-year intervals by 2050 [49]. With climate change leading to rising temperatures and longer, more intense droughts in Czechia, especially in its arid areas investigated by [38] the risk of wildfires is increasing.

Drier landscapes, higher temperatures, and windy weather all contribute to a greater number of fires. For most of the year, vegetation cover on agricultural land serves as a natural barrier to wildfire spread. However, the highest risk occurs during physiological maturity, harvest periods, and prolonged dry spells. In such conditions, agricultural crops can become a wildfire hazard rather than a deterrent. Crop selection plays a crucial role in fire risk. Most root crops do not pose a significant risk at any stage of their development, whereas cereals and rapeseed, which have nearly identical harvest times, present a much higher risk. Additionally, the size of crop plots, whether high-risk or low-risk, affects fire spread. Other important factors include local conditions such as moisture levels, land slope, exposure, and the broader water regime of the landscape. Adaptation measures can help mitigate the risk of natural fires. These include crop diversification, dividing large fields into smaller land blocks, implementing direct sowing (no-till) methods, controlling land drainage, increasing water retention capacity, constructing water reservoirs, and introducing wildfire forecasting systems. While wildfires have historically been more frequent in lowland areas, by 2050, their occurrence is expected to reach unprecedented levels at altitudes above 500 m, particularly in forested regions, due to a combination of drought and adverse wind conditions. Currently, wildfires destroy less than 5% of the carbon stored annually in biomass, but this figure is expected to rise in the near future. Therefore, wildfire risk should be integrated into national climate change mitigation and adaptation strategies. Although Central Europe, including Czechia, is not currently considered a wildfire hotspot compared to the Mediterranean region, this is likely to change. As a result, comprehensive reforms in forest management and fire preparedness will be essential [44].

Building agro-sector resilience to the impacts of climate change and, more broadly, the transformation of whole food systems has been a topic of debate for several decades [6, 8, 9, 22, 47]. However, in recent years, concerns about food security have been growing [45] and amplified by the COVID-19 pandemic [4, 35], as well as the military conflict between Russia and Ukraine. In order to combat such challenges agricultural innovations go beyond climate change mitigation and adaptation but seek for latest technological advances.

10.1.5 Regional Perspective on Agro-Environmental Risks: The Case of Czechia

While agro-environmental security is a global concern, its specific challenges and risk factors vary across regions, depending on geographical, climatic, and socio-economic conditions. To illustrate how these risks manifest in a particular context, this study focuses on Czechia, a landlocked country in Central Europe with a diverse landscape and a temperate continental climate → Fig. 10.1. Czechia serves as a relevant case study due to its strategic location, well-established risk management framework, and active participation in international disaster risk reduction initiatives. By analyzing agro-environmental threats in Czechia, we aim to provide insights that can be applicable to other regions facing similar environmental and socio-economic challenges. In a nutshell the relevance of a risk analysis conducted in Czechia for a global scale, consider the following points:

Geographical and Climatic Representativeness: Czechia, located in the heart of Europe, experiences a temperate continental climate with distinct seasonal variations, including warm summers and cold, snowy winters. These conditions are representative of many mid-latitude regions worldwide, making Czechia a valuable case study for understanding climate-related risks that also affect other landlocked and temperate-zone countries.

Diverse Landscape and Environmental Conditions: With its mix of hilly terrain, low mountains, and river basins, Czechia presents a variety of environmental conditions that are relevant for analyzing risks such as droughts, floods, and extreme weather events. These conditions are shared by numerous regions globally, enhancing the study's applicability beyond Czechia.

Strong Institutional Framework for Disaster Risk Reduction: Czechia has a well-established disaster risk reduction (DRR) framework, with efforts dating back to 2005 when the Czech Hydrometeorological Institute began coordinating national disaster response efforts. Over time, these efforts expanded to include anthropogenic risks, culminating in the establishment of the National Platform for Disaster Risk Reduction in 2015 under the Ministry of the Environment. The structured and evolving approach to risk management is further reinforced by recent strategic documents, such as the Environmental Security Concept 2021–2030 with a view to 2050 [27], which outlines long-term priorities for strengthening national resilience and integrating DRR within broader environmental security policies. Such an attitude makes Czechia a model for other nations seeking to develop or enhance their own DRR strategies.

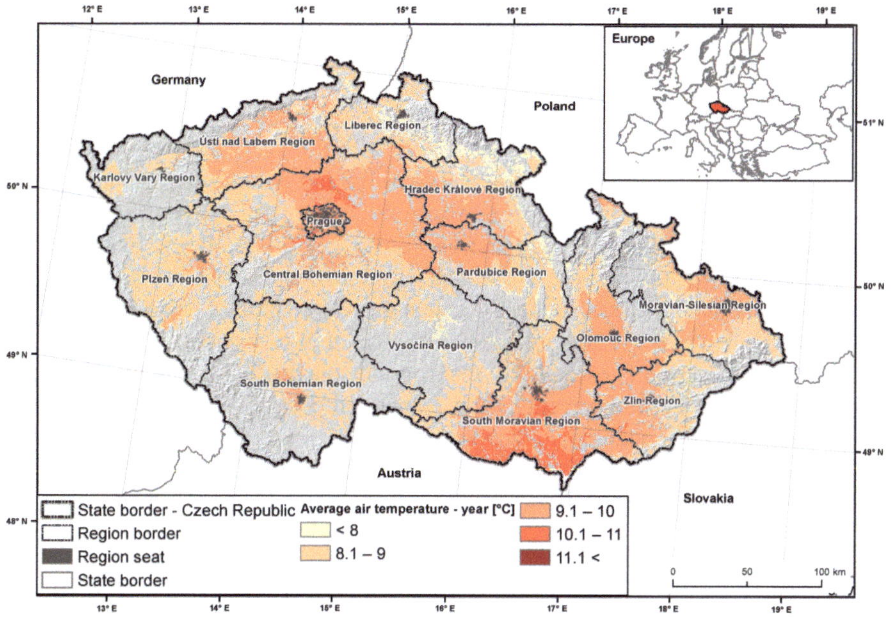

Fig. 10.1 Czechia (also referred as the Czech Republic) located in central Europe (12°–19° E, 48°–51° N), in colors there is an agricultural land up to an altitude to 500 masl according to long-term annual air temperature; adapted from [37]

International Cooperation and Policy Alignment: Czechia actively participates in international disaster risk reduction initiatives, aligning with global frameworks such as the Sendai Framework for Disaster Risk Reduction. This means that insights from Czechia's risk assessments are relevant not only locally but also contribute to broader global discussions on risk mitigation and climate adaptation.

Scalability and Transferability of Risk Analysis Methods: The methodologies used in Czechia's risk assessment, such as Preliminary Risk Analysis (PRA) and Multi-Criteria Risk Analysis (MCRA), are not country-specific but rather universally applicable. The structured approach to evaluating climate-related threats, economic impacts, and social vulnerabilities can be adapted to various regions facing similar environmental and socio-economic challenges.

A Microcosm of Broader Climate and Socioeconomic Risks: While Czechia is a relatively small country, it faces many of the same challenges as larger nations, including extreme weather events, environmental degradation, and the need for sustainable agricultural practices. The findings from Czechia's risk analysis can serve as a reference for other countries, particularly those in Europe and beyond with similar environmental conditions and governance structures.

By considering these factors, Czechia's risk analysis offers insights that are not only locally relevant but also globally transferable, contributing to a broader understanding of agro-environmental security and climate risk mitigation.

10.2 Materials and Methods

10.2.1 Risk Evaluation Process

To maintain focus on agro-environmental security and climate change nexus, only climate-related threats were analyzed. These threats not only have the potential to negatively impact agricultural production but also pose significant risks to other critical systems, including the environment, economy, and social structures.

A comprehensive analysis of these dimensions was conducted using relevant data, which, by its nature, is not always highly technical or purely scientific. However, for this type of research, it is essential to incorporate sources that document the occurrence and impact of specific events, such as media reports. These sources have been particularly useful in quantifying economic damage, loss of human life, and other key consequences. To gain a comprehensive understanding of agro-environmental threats, we followed these steps:

1. Collected data on identified threats.
2. Conducted both a Preliminary Risk Analysis (PRA) and a Multi-Criteria Risk Analysis (MCRA).
3. Compared our PRA and MCRA with the official national risk analysis from 2015 to observe changes over time. The national risk analysis of agro-environmental relevant threats was derived from the comprehensive "Threat Analysis for the Czech Republic", conducted [29] governed by the Ministry of the Interior of the Czech Republic. A total of 72 hazards of both anthropogenic and naturogenic origin were identified and evaluated. A more up-to-date hazard analysis has not been officially prepared to date.
4. Evaluated the role of expertise in risk assessment by comparing our findings with those of Agroecology students in the first year of a master's (non-doctoral) program, who conducted their own PRA and MCRA between 2022 and 2024 (hereafter referred to as "students"). Agroecology is Master's program at Mendel University in Brno, focuses on developing expertise in sustainable agricultural landscapes, emphasizing the understanding of agroecosystems and the fundamental relationships and processes within agricultural environments.
5. Concluded with a preliminary proposal for risk treatment and mitigation action.

All PRA and MCRA assessments followed the same methodology by [29] to ensure consistency and comparability across analyses. Figure 10.2 describes the whole process of risk evaluation.

10.2.1.1 Preliminary Risk Analysis (PRA)

It is a stage where a simple consequence-probability matrix is used to classify risks. The probability of occurrence (P) is categorized from 1 to 3 as Unlikely, Likely, Very likely and the consequences (C) are rated from 1 to 3 as Low, Significant,

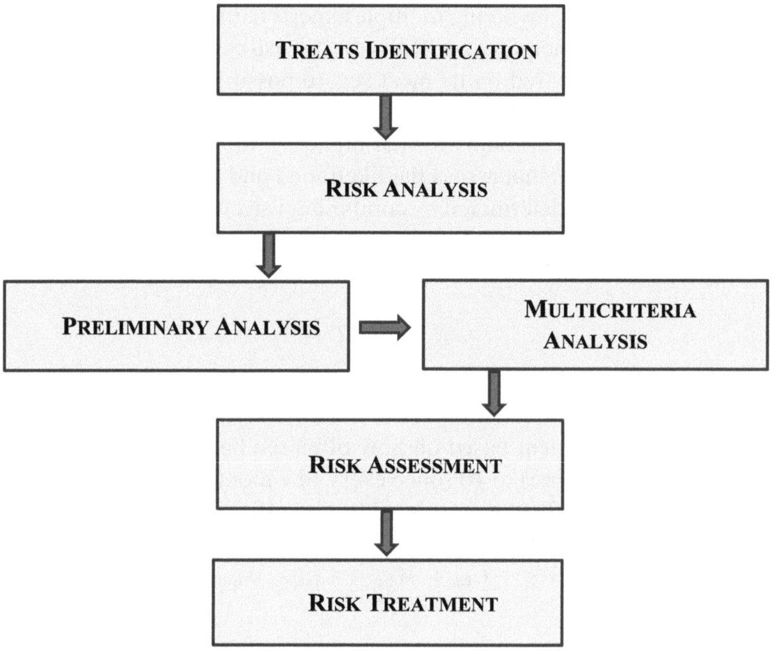

Fig. 10.2 Risk evaluation process

Catastrophic. The risk level (R) is then calculated using the formula Eq. (10.1):

$$R = P \times C \tag{10.1}$$

where:

P = probability
C = consequences.

If the resulting risk score exceeds 3, a multi-criteria analysis is conducted to further assess the degree of risk.

10.2.1.2 Multi-criteria Risk Analysis (MCRA)

MCRA can be conducted either by an individual or a working group. Individual assessments are often influenced by personal biases and subjective opinions, whereas group assessments help minimize individual errors through collective expertise. In many cases, decision-making is limited by the availability of data, meaning the accuracy of the results depends heavily on the expertise and experience of the evaluators. When data is insufficient, expert judgment is used to determine risk levels, introducing the potential for subjective errors. To mitigate this, group decision-making methods can be employed, involving multiple experts to improve objectivity and reliability. A fundamental principle of MCRA is the "worst-case scenario" approach each hazard type is analyzed based on its most severe possible impact within the given country's conditions. The coefficients for each criterion are assigned accordingly, ensuring that the analysis accounts for the highest level of risk. MCRA quantifies risk using multiple factors that assess the likelihood and consequences of a hazard. The final risk score (R) is determined by combining various coefficients that evaluate the frequency, human impact, environmental impact, economic consequences, and social disruption. The general formula Eq. (10.2) for calculating the risk rate is:

$$R = F \times C \tag{10.2}$$

where:

F = Frequency coefficient based on how often the hazard occurs: from 1 (once every 1000 years or more) to 10 (once every few months).
C = Consequence coefficient calculated by Eq. (10.3):

$$C = C_{\mathrm{H}} \times W_{\mathrm{H}} + C_{\mathrm{En}} \times W_{\mathrm{En}} + C_{\mathrm{Ec}} \times W_{\mathrm{Ec}} + C_{\mathrm{S}} \times W_{\mathrm{S}} \tag{10.3}$$

where:

C_{H} = Humam life and health impact coefficient calculated by Eq. (10.4):

$$C_{\mathrm{H}} = \frac{(C_{\mathrm{H1}} + C_{\mathrm{H2}})}{2} \tag{10.4}$$

where:

C_{H1} = fatal impact coefficient from 0 (no human fatalities) to 10 (more than 1000 human fatalities).
C_{H2} = endangerment coefficient from 0 (no people endangered) to 10 (more than one million people endangered).
C_{En} = Environmental impact coefficient measuring damage to environmental components such as water sources, protected areas, and natural habitats; from 0 (no damage) to 10 (very high damage and threat).
C_{Ec} = Economic impact coefficient assessing financial damage, including losses to livestock and restoration costs; from 1 (damage between €42,000 and €4.2 million) to 10 (damage exceeding €208 billion).
C_{S} = Social impact coefficient calculated by Eq. (10.5):

$$C_S = \frac{(C_{\mathrm{S1}} + C_{\mathrm{S2}} + C_{\mathrm{S3}})}{3} \tag{10.5}$$

where:

C_{S1} = constrained persons coefficient, measuring how many people are affected; from 0 (none) to 10 (more than 5 million affected).

C_{S2} = coefficient of expected duration of restrictive conditions measuring how long rescue/recovery efforts last; from 0 (no restrictions) to 10 (more than 25 years).
C_{S3} = society's constraint coefficient assessing societal disruption; from 0 (no constraints) to 10 (extreme societal constraints.
W_H, W_{En}, W_{Ec}, W_S = Weights assigned to each coefficient (W_H = 0.4; W_{En}, W_{Ec}, W_S = 0.2).

Resulting comprehensive MCRA formula is Eq. (10.6):

$$R = F \times \left(\left(\frac{(C_{H1} + C_{H2})}{2} \times 0.4\right) + (C_{En} \times 0.2) + (C_{Ec} \times 0.2) + \left(\frac{(C_{S1} + C_{S2} + C_{S3})}{3} \times 0.2\right)\right) \tag{10.6}$$

10.2.1.3 Risk Assessment and Risk Treatment

In the MCRA framework, risk values (R) theoretically range from 0 to 100. Once the risk level is assessed, a threshold for acceptable risk must be established. Following the recommendation in [29], we have set this threshold at R = 30. If a risk exceeds this level, appropriate mitigation measures or actions must be implemented to reduce it.

The determination of this threshold takes into account realistic constraints, primarily financial resources. Rather than being derived from a fixed formula or statistical rule, the threshold is set deliberately through agreement, balancing risk reduction with practical feasibility.

10.3 Results and Discussion

10.3.1 Comprehensive Risk Analysis

Using PRA (Figs. 10.3, 10.4 and 10.5) and MCRA (Figs. 10.6 and 10.7), we created a comprehensive comparison of the results of all three analyses, where: group **A** stand for our analysis; group **B** stand for official national risk analysis conducted [29], and group **C** stand for analysis of Agroecology students.

It is apparent that our group (A) generally assigned higher risk scores compared to the official national assessment from 2015 (B) and lower scores compared the agroecology students' assessment from 2022 to 2024 (C).

The students' scores are the highest in most cases, probably reflecting both their limited expertise and deep concern about climate change and environmental issues

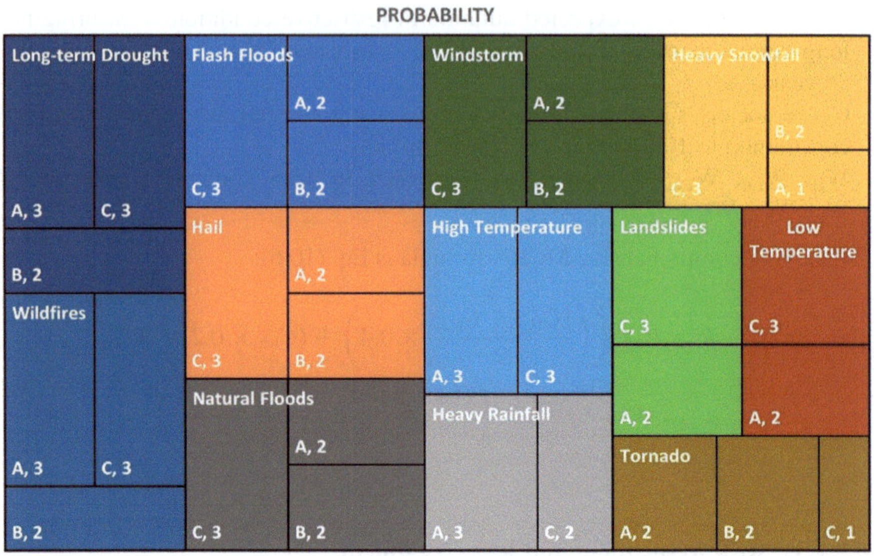

Fig. 10.3 Comparation of detailed results of PRA probability conducted by three groups (A, B, C)

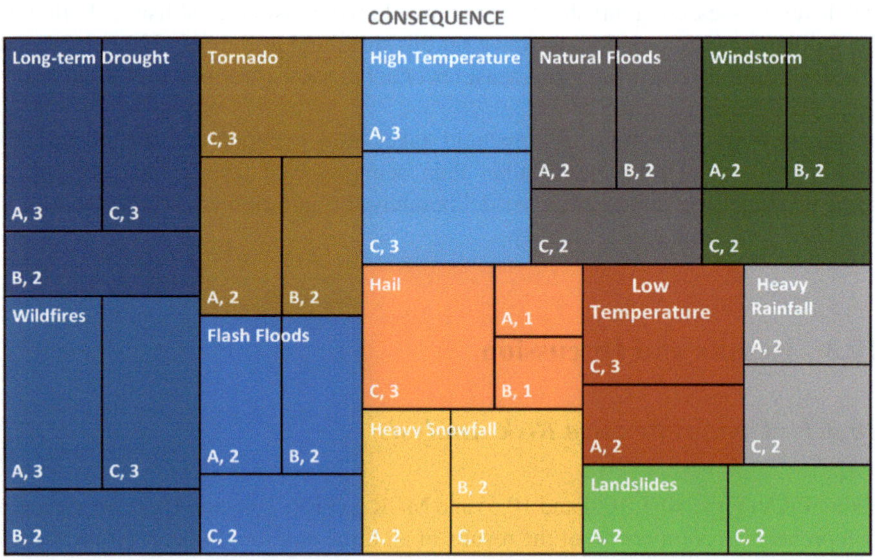

Fig. 10.4 Comparation of detailed results of PRA consequence by three groups (A, B, C)

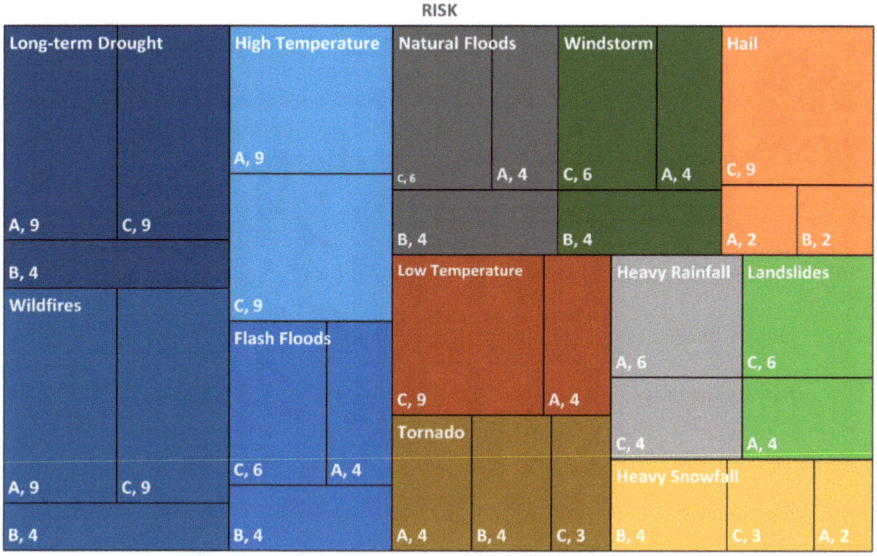

Fig. 10.5 Comparation of detailed results of PRA conducted by three groups (A, B, C)

in general. The three risk analyses reveal notable differences in risk perception and assessment, shaped by both methodological approaches and the evolving understanding of climate risks. Our 2024 analysis reflects more recent data and climate trends, which is why risks such as flash floods, heavy rainfall, and wildfires are assessed as more significant compared to the 2015 official national analysis. The older assessment assigns very low or even zero risk to hazards like high temperatures, heavy rainfall, and tornadoes, likely underestimating their growing impact due to climate change. The differences suggest that extreme weather events have become more frequent and severe over the past decade. The higher scores in Group C may reflect a precautionary bias, which is common in novice assessors who lack formal risk training but are highly sensitized to environmental issues through current academic exposure. While we do not have direct quantitative evidence on their motivation, this interpretation is supported by heuristic reasoning: individuals with limited domain experience often compensate by erring on the side of caution, especially when facing complex and uncertain threats like climate change [33, 46]. Their scoring reflects a high sensitivity to perceived vulnerability and a lower threshold for unacceptable risk. This aligns with findings from [50], whose survey of agricultural students in Central European universities showed that while students are well-informed about environmental risks—particularly those related to climate change—their perceptions are shaped more by academic instruction and media narratives than by practical experience. As such, their judgments may reflect a heightened awareness of vulnerability rather than expert-level calibration of risk trade-offs. Notably, Group C's assessments for hazards such as hail (score: 55) and windstorms (score: 60) were markedly higher than in Groups A and B, possibly indicating a more conservative stance or differing

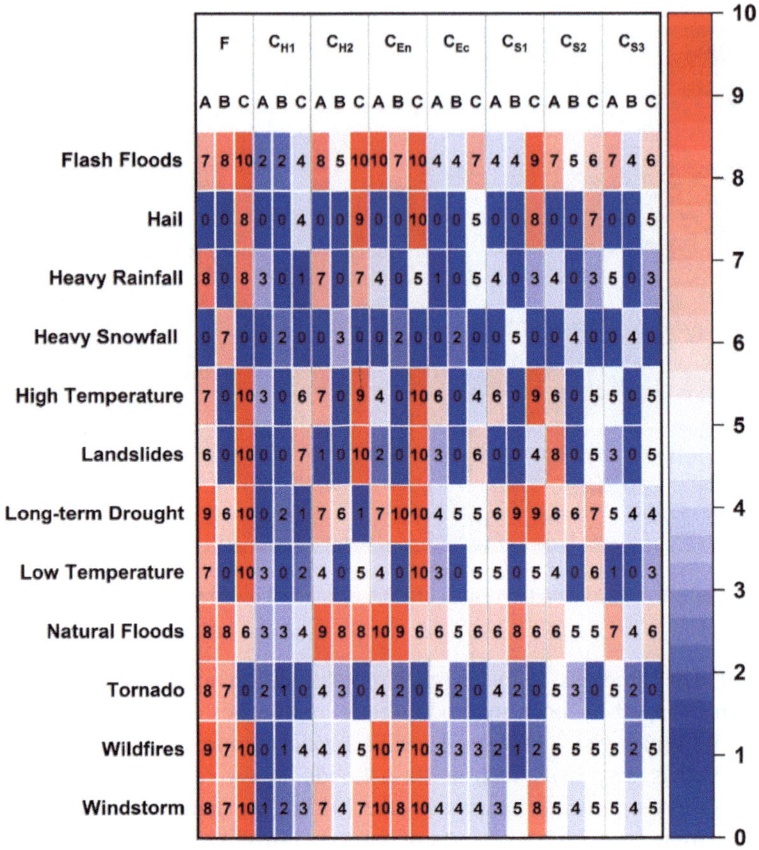

Fig. 10.6 Comparison of MCRA coefficients assigned by three groups (A, B, C)

assumptions about the likelihood and severity of future climate scenarios. Despite these differences, all three analyses highlight long-term drought and natural floods as major risks, indicating a shared recognition of these threats, even as risk perception evolves with newer data and a greater awareness of climate change. These differences in addition highlight the importance of expert judgment, up-to-date data, and comparative analysis in risk assessment, reinforcing the need for continuous review and adaptation in policy and planning.

Once the threshold for an unacceptable level of risk was set at 30, our analysis identified the following threats as meeting or exceeding this level: flash floods, heavy rainfall, high temperature, long-term drought, natural floods, tornado, wildfires and windstorm (see Fig. 10.7). According to the methodology, these threats should be addressed in some way to mitigate their impact. In general risk management, various tools can be employed, including risk and failure plans, as well as specific mitigation and adaptation measures that range from strategic and conceptual to practical and operational, for instance.

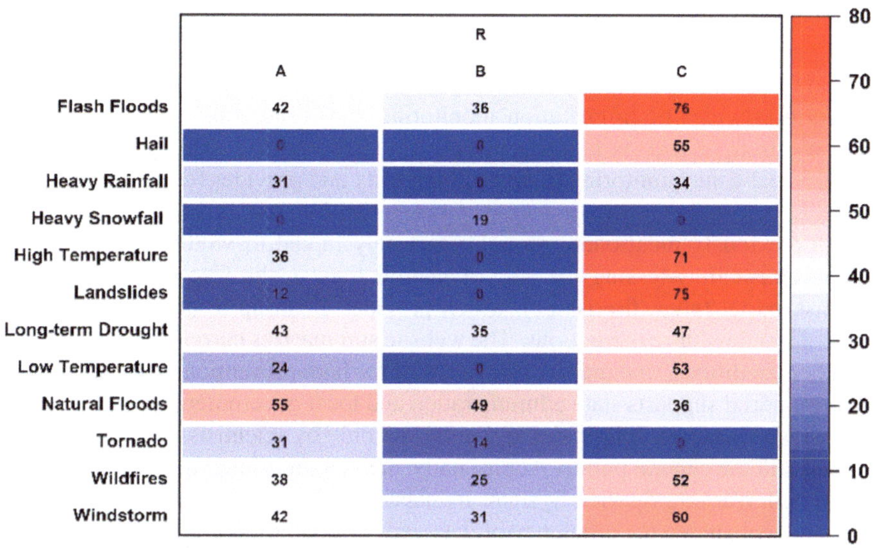

Fig. 10.7 Comparation of final level of risk by MCRA assigned by three groups (A, B, C)

Floods have been widely recognized at all levels of decision-making in Czechia, especially since the severe flooding in 1997. As a response, flood risks were incorporated into the Act No. 254/2001 Coll., on Water and on Amendments to Certain Acts (the Water Act) and monitoring systems for flood risk and profound warning system were deployed. Despite the existence of an integrated warning service system in the country, it is still challenging to predict flash floods in a timely manner due to their small geographic extent. Local warning systems, however, can continuously monitor the current state of watercourses, and if any changes occur, they can immediately send alerts to firefighters and municipal mayors. Recently, the ESRI "Map of Extreme Floods" (MEF) application has been developed, focusing on European flood events. This application is designed to provide geographic characterizations and mappings of hydrological extremes. The primary objectives of the MEF application are to (i) archive, (ii) verify, (iii) correct, (iv) add further data and information, (v) facilitate data exchange, and (vi) provide valuable information to scientists and the public. Currently, the database includes 24 major floods from 1432 to 2002, with an additional 20 floods in preparation. Each flood site is associated with estimated extreme values, economic losses, infrastructure damage, water levels, flood marks, and other relevant data [10].

The opposite extreme of floods, which represent one form of water-related threat, is drought. Regarding drought management at the legal level, a substantial Amendment the Water Act was introduced in 2021, adding a completely new section on drought and its mitigation. Additionally, there are relevant concepts, strategies, and policies addressing drought. To accurately assess the extent of drought, regular monitoring and evaluation are essential. One of the key tools used in agricultural practice in Czechia since 2012 is the web portal Intersucho (www.intersucho.cz). This platform provides high-resolution monitoring and forecasting of agricultural and forestry drought and its impacts in both Czechia and Slovakia using a tool that enables real-time monitoring of drought intensity and provides forecasts for the next 10 days.

As previously mentioned, drought is closely linked to wildfires. A significant advancement in assessing the risk of natural fires in the Czech Republic is the Firerisk portal (www.firerisk.cz), which provides clear, up-to-date information on fire risks and wildfire predictions. The website summarizes current and expected fire weather conditions, making it a valuable tool for both prevention and response. The Firerisk portal supports state administration and local governments in implementing preventive measures. It is designed for use not only by scientists but also by professionals and the general public. Additionally, the system is integrated with the central Integrated Warning Service System, managed by the Czech Hydrometeorological Institute, enhancing the efficiency of data and information flows.

Fires are often intensified by strong winds. In agricultural landscapes, wind speed can be effectively reduced by planting windbreaks, which offer multiple benefits. These include reducing evaporation, increasing dew formation, raising air temperature, enhancing soil and air humidity, trapping snow (and accumulating nutrients), and, most importantly, preventing wind erosion. Windbreaks can also improve crop quality and yield while providing habitats for wildlife, offering shelter for nesting and reproduction. In addition to windbreaks, maintaining vegetation cover helps mitigate the impact of strong winds on bare soil. Wind plays a crucial role in the spread and intensity of wildfires, as strong winds can accelerate the movement of flames, causing fires to spread rapidly and with greater intensity.

In agricultural landscapes, this effect can be especially devastating, but the risk can be mitigated by strategically planting windbreaks, which provide a lot of benefits. They can reduce evaporation, increase dew formation, raise air temperature, and enhance soil and air humidity. These advantages contribute to a more stable environment that supports crop growth, especially in regions where winds can cause extreme fluctuations in temperature and moisture levels. Moreover, windbreaks help trap snow, accumulating nutrients, and provide habitats for wildlife, offering shelter for nesting and reproduction. This makes them a sustainable agricultural practice [20]. Windbreaks also play a crucial role in controlling wind erosion. Strong winds can cause significant soil loss by carrying away the topsoil, leading to reduced soil fertility and crop yield. To counter this, planting windbreaks is an effective method for preventing wind erosion, a concern outlined in certified methodologies for soil and wind protection [7, 30]. By reducing wind speed and providing a protective barrier, windbreaks minimize the risk of soil degradation, which is vital for maintaining agricultural productivity in the long term. By integrating windbreaks with sustainable land management practices, it is possible to address the interconnected risks of wildfires, wind erosion, and soil degradation, ultimately ensuring a more resilient agricultural landscape. Khel et al. [18] discuss how windbreaks function as tools for soil protection, offering a detailed evaluation of their effectiveness and implementation across the Czech landscape. These guidelines help land managers evaluate and optimize the use of windbreaks.

Our research, however, is specifically focused on compiling a list of practical measures directly applicable to agricultural landscapes, primarily at the farm level. For each measure, we provide a rough estimation of its cost and efficiency, and also identify potential trade-offs, including any unintended threats it may exacerbate (Table 10.1). imply that a balanced approach to the measures is needed. Instead of prioritizing one risk over another, cost-effective and efficient integrated solutions should be adopted. Measures should be selected not only for their ability to mitigate multiple threats but also for their financial feasibility and practical impact, ensuring that resources are allocated where they provide the greatest resilience per investment.

10.3.2 Emerging Challenges in Agro-environmental Security

As agricultural systems evolve and climate conditions shift, new challenges in agro-environmental security continue to emerge. Addressing these effectively requires continuous adaptation and improvement of risk assessment methodologies.

One key challenge is the regular updating of risk analyses. Given the increasing impacts of climate change, the introduction of new agricultural practices, and shifts in land use, risk assessments must be frequently revised to remain relevant. Without this, outdated models could lead to ineffective mitigation strategies and unpreparedness for emerging threats.

Another important aspect is the adoption of a more focused methodological approach in risk assessment. A refined methodology could provide a more precise understanding of the damages and vulnerabilities affecting agriculture. However, this level of detail comes with a trade-off—it requires extensive data collection and processing, which can be resource-intensive and demanding. Striking a balance between precision and feasibility will be crucial.

Another challenge that may arise is the increasing interdependence of global food supply chains. As agricultural production becomes more interconnected, localized disruption, whether climate-related, technological failures, or cyberattacks, could have cascading effects on international markets and food security. Factoring in these systemic risks could enhance the robustness of future risk assessments.

Furthermore, the integration of advanced technologies into agricultural practices introduces new dimensions of risk. The adoption of autonomous systems, AI-driven algorithms, and automation is revolutionizing the sector by enhancing efficiency and productivity. Yet these advancements also create cybersecurity vulnerabilities that must be factored into agro-environmental risk assessments. Ensuring the resilience of digital infrastructure and protecting automated systems from cyber threats will be essential components of future risk management.

Despite the promise of Agriculture 4.0 and 5.0, their widespread adoption is often constrained by persistent barriers. Research on this integration remains limited, with

Table 10.1 Impact of agricultural measures: threats mitigated versus potential trade-offs

Measure	Threats Mitigated	Potential Trade-Offs	Cost	Efficiency
Increase soil infiltration (cover crops, reduced tillage, organic matter)	Floods Heavy rainfall Drought	Wildfire	2	3
Controlled drainage & retention systems (ponds, buffer strips)	Floods Heavy rainfall Drought High temperature	No major negatives	3	3
Flood-resistant crops (waterlogging-tolerant varieties)	Floods Heavy rainfall	Drought	2	2
Field design improvements (contour plowing, raised beds)	Floods Heavy rainfall Drought Windstorms	No major negatives	2	3
Drought-resistant crops (heat- and drought-tolerant varieties)	Drought High temperature Wildfire	Floods Heavy rainfall	2	2
Soil moisture conservation (mulching, organic matter increase)	Drought Windstorms	Wildfire	1	3
Efficient irrigation (drip irrigation, rainwater collecting)	Drought High temperature Wildfire	No major negatives	3	3
Crop rotation & diversification (resilient species, maintaining soil health)	Drought Floods Windstorms	No major negatives	1	3
Windbreaks & shelterbelts (tree planting)	Windstorms Drought High temperature	Wildfire	3	3
Soil protection (strip cropping, cover crops)	Floods Heavy rainfall Drought Windstorms	No major negatives	1	2
Fire-resistant crops & buffer zones (clover, alfalfa firebreaks)	Wildfire	No major negatives	2	2
Soil & residue management (avoiding excess dry plant material)	Wildfire	Floods Drought	1	3
Water access & emergency planning	Wildfire Drought	No major negatives	3	3

Footnote Rating Scale (Cost & Efficiency): For cost, 1 = Low (cheap), 2 = Medium, 3 = High (expensive); for efficiency, 1 = Low efficiency, 2 = Medium efficiency, 3 = High efficiency.

only a few countries actively addressing the deployment of disruptive technologies in agriculture [24]. In Central Europe, for instance, IoT-based solutions have struggled with poor data coverage, limited access to affordable computing power, and insufficient data processing services—challenges that disproportionately affect small and medium-sized farms. Social acceptance also poses a significant hurdle, as many farmers, grounded in hands-on experience, may perceive these technologies as unnecessary or overly complex. Therefore, education, training, and targeted investment are critical to enabling a successful transition to a more technology-driven agricultural sector.

While the benefits of Agriculture 4.0 and 5.0 have been widely discussed [14, 23, 31, 34, 48], the cybersecurity risks associated with these advancements are an increasingly urgent concern. The reliance on interconnected digital systems makes agriculture vulnerable to cyber threats, such as attacks on drones used for crop spraying or automated climate control systems in farms [2]. Alarmingly, agricultural businesses are often underinvesting in cybersecurity measures, making them attractive targets for cybercriminals. A recent study by [25] highlights that poor cybersecurity practices, a lack of regulatory frameworks, and outdated IT infrastructure are among the top risks increasing agricultural cyber vulnerabilities. Cybersecurity breaches in agriculture can lead to data loss, system failures, and disruptions in food production, ultimately affecting food security. Addressing these risks requires not only technological advancements in cybersecurity but also policy development, investment, and training to ensure that modern agricultural systems remain resilient against emerging digital threats.

10.4 Conclusion

Our research underscores that building a resilient society capable of facing both existing and emerging threats requires their timely analysis. In the context of agro-environmental risks, regularly updating risk assessments is crucial, and the revision cycle may need to be shorter than the current 10-year interval. The accelerating pace of climate change, evolving agricultural practices, and the emergence of new threats necessitate more frequent reassessments to ensure effective risk management. Without timely updates, outdated assessments may fail to capture critical risks, leaving agricultural systems vulnerable to unforeseen challenges.

An important finding of our study is the notable difference in risk perception across different groups. Students consistently assigned higher risk scores, which may reflect their heightened environmental concerns or possibly a different interpretation of risk factors. Their evaluations could also be influenced by limited experience, as the applied risk assessment method relies heavily on expert judgment. While their perspective highlights growing awareness of environmental issues, expert-driven refinements remain essential for ensuring accuracy in risk evaluations.

Additionally, our study highlights two significant gaps in current risk assessments: the lack of a cybersecurity dimension and the absence of considerations related to food self-sufficiency at the national or regional level. As agriculture becomes increasingly digitalized, cyber threats pose growing risks to food production and supply chains. Integrating these aspects into future risk assessments is crucial for ensuring agro-environmental security in an era of rapid technological advancements.

With the official national risk analysis scheduled for revision in the near future, our findings offer valuable input for its update. We will contribute to its updating by our membership in national disaster risk platform. Given our team's active role in the original national assessment, our updated results will contribute to refining and enhancing the forthcoming revision, ensuring it reflects the latest climate trends, technological risks, and vulnerabilities.

The importance of this topic is further underscored by growing academic and professional interest. Notably, team member Emma Vacková's thesis, *Agro-Environmental Security of Czechia under Global Change*, was awarded the 2024 national prize for the best agro-related thesis. While a student competition, this recognition highlights the increasing relevance of agro-environmental security in academic discourse and its urgency in shaping future research and policy discussions.

Acknowledgements The study was supported by project SS02030040 "PERUN - Prediction, Evaluation and Research for Understanding National sensitivity and impacts of drought and climate change for Czechia", co-financed with state support of the Technology Agency Czech Republic as part of the Program Environment for Life.

References

1. Arsenović D, Lehnert M, Fiedor D et al (2019) Heat-waves and mortality in Czech cities: a case study for the summers of 2015 and 2016. Geogr Pannon 23:162–172. https://doi.org/10.5937/gp23-22853
2. Barreto L, Amaral A (2018) Smart farming: cyber security challenges. In: 2018 International conference on intelligent systems (IS), Funchal, Portugal. IEEE, Piscataway, NJ, pp 870–876. https://doi.org/10.1109/IS.2018.8710531
3. Bastit F, Brunette M, Montagné-Huck C (2023) Pests, wind and fire: a multi-hazard risk review for natural disturbances in forests. Ecol Econ 205:107702. https://doi.org/10.1016/j.ecolecon.2022.107702
4. Blay-Palmer A, Carey R, Valette E et al (2020) Post COVID 19 and food pathways to sustainable transformation. Agr Hum Values 37:517–519. https://doi.org/10.1007/s10460-020-10051-7
5. Da Silveira F, Lermen FH, Amaral FG (2021) An overview of agriculture 4.0 development: systematic review of descriptions, technologies, barriers, advantages, and disadvantages. Comput Electron Agr 189:106405. https://doi.org/10.1016/j.compag.2021.106405
6. Dentoni D, Waddell S, Waddock S (2017) Pathways of transformation in global food and agricultural systems: implications from a large systems change theory perspective. Curr Opin Env Sust 29:8–13. https://doi.org/10.1016/j.cosust.2017.10.003
7. Doležal P, Podhrázská J, Kučera J et al (2017) Risk management of wind erosion: certified methodology. VÚMOP, Brno, v.v.i. Certification authority: SPU, Certificate No. 2/2017 SPU/O
8. El Bilali H (2019) Research on agro-food sustainability transitions: where are food security and nutrition? Food Secur 11:559–577. https://doi.org/10.1007/s12571-019-00922-1

9. El Bilali H, Callenius C, Strassner C et al (2019) Food and nutrition security and sustainability transitions in food systems. Food Energy Secur 8:e00154. https://doi.org/10.1002/fes3.154
10. Elleder L, Šírová J (2023) MEF application—the extreme floods are already in maps! In: EGU General Assembly 2023, Vienna, Austria, EGU23-3160. https://doi.org/10.5194/egusphere-egu23-3160
11. Gill JC, Malamud BD (2014) Reviewing and visualizing the interactions of natural hazards. Rev Geophys 52:680–722. https://doi.org/10.1002/2013RG000445
12. Gonçalves AC, Sousa AMO (2017) The fire in the Mediterranean Region: a case study of forest fires in Portugal. In: Mediterranean identities—environment, society, culture. IntechOpen. https://doi.org/10.5772/intechopen.69410
13. Haloui D, Oufaska K, Oudani M et al (2024) Bridging industry 5.0 and agriculture 5.0: historical perspectives, opportunities, and future perspectives. Sustainability 16:3507. https://doi.org/10.3390/su16093507
14. Hartanto R, Arkeman Y, Hermadi I et al (2019) Intelligent unmanned aerial vehicle for agriculture and agroindustry. In: IOP conference series: earth and environmental science, vol 335, p 012001. https://doi.org/10.1088/1755-1315/335/1/012001
15. Hristov J, Barreiro-Hurle J, Salputra G et al (2021) Reuse of treated water in European agriculture: potential to address water scarcity under climate change. Agr Water Manage 251:106872. https://doi.org/10.1016/j.agwat.2021.106872
16. Chuchma F, Středová H, Středa T (2016) Bioindication of climate development on the basis of long-term phenological observation. In: Proceedings of international PhD students conference—MendelNet 2016, Brno, Mendel University, pp 380–383
17. Kappes MS, Keiler M, von Elverfeldt K et al (2012) Challenges of analyzing multi-hazard risk: a review. Nat Hazards 64:1925–1958. https://doi.org/10.1007/s11069-012-0294-2
18. Khel T, Řeháček D, Kučera J et al (2017) Methodology for evaluating the effectiveness and implementation of windbreaks in the landscape as a tool for protecting soil from wind erosion. Certified methodology for practice, Research Institute for Soil and Water Conservation, v.v.i., Prague
19. Kučera J, Fukalová P, Středová H et al (2024) Evaluation of the spatial structure of windbreaks from digital photography. J Ecol Eng 25:379–389. https://doi.org/10.12911/22998993/192473
20. Kučera J, Podhrázská J, Středa T et al (2024) New methods of assessing wind erosion risk: Innovative approach to soil protection respecting windbreak effect. Contrib Geophys Geod 54:371–388. https://doi.org/10.31577/congeo.2024.54.4.4
21. Lamichhane JR (2021) Rising risks of late-spring frosts in a changing climate. Nat Clim Change 11:554–555. https://doi.org/10.1038/s41558-021-01090-x
22. Lamine C (2011) Transition pathways towards a robust ecologization of agriculture and the need for system redesign. Cases from organic farming and IPM. J Rural Stud 27:209–219. https://doi.org/10.1016/j.jrurstud.2011.02.001
23. Lu Y, Liu M, Li C et al (2022) Precision fertilization and irrigation: progress and applications. AgriEngineering 4:626–655. https://doi.org/10.3390/agriengineering4030041
24. Lubag M, Bonifacio J, Tan JM et al (2023) Diversified impacts of enabling a technology-intensified agricultural supply chain on the quality of life in hinterland communities. Sustainability 15:12809. https://doi.org/10.3390/su151712809
25. Maraveas C, Rajarajan M, Arvanitis KG et al (2024) Cybersecurity threats and mitigation measures in agriculture 4.0 and 5.0. Smart Agr Technol 9:100616. https://doi.org/10.1016/j.atech.2024.100616
26. Ministry of the Environment (MŽP) and Czech Hydrometeorological Institute (ČHMÚ) (2015) Climate change adaptation strategy in the Czech Republic. Ministry of the Environment [online]. Ministry of the Environment, Prague. Available at: https://mzp.gov.cz/system/files/2025-03/OAZK_Narodni_adaptacni_strategie-aktualizace_20211026.pdf. Accessed 2024-04-03
27. Ministry of the Environment (MŽP) (2020) Environmental security concept 2021–2030 with a view to 2050. Ministry of the Environment [online]. Ministry of the Environment, Prague. Available at: https://mzp.gov.cz/system/files/2025-01/SOBaKR-KoncepceEnvironmentalniBezpecnosti-20250106.pdf. Accessed 2024-04-02

28. Moreira F, Leal M, Bergonse R et al (2023) Recent trends in fire regimes and associated territorial features in a fire-prone Mediterranean Region. Fire 6:60. https://doi.org/10.3390/fire6020060
29. Paulus F, Krömer A, Petr J et al (2015) Threat analysis for the Czech Republic—final report [online]. Fire Rescue Service of the Czech Republic, Prague. Available at: https://www.hzscr.cz/soubor/analyza-hrozeb-zprava-pdf.aspx. Accessed 2024-04-02
30. Podhrázská J, Pejchal M, Kučera J et al (2021) Establishment and maintenance of windbreaks under deteriorated pedoclimatic conditions: certified methodology. VÚMOP, Brno, v.v.i. Certification authority: SPU, Certificate No. 1/2022 SPU/O
31. Rose DC, Chilvers J (2018) Agriculture 4.0: broadening responsible innovation in an era of smart farming. Front Sustain Food Syst 2:87. https://doi.org/10.3389/fsufs.2018.00087
32. Senko H, Pole L, Mešić A et al (2022) Farmers observations on the impact of excessive rain and flooding on agricultural land in Croatia. J Cent Eur Agric 23:125–137. https://doi.org/10.5513/JCEA01/23.1.3292
33. Slovic P, Finucane ML, Peters E et al (2005) Affect, risk, and decision making. Health Psychol 24:4 (Suppl S35–S40). https://doi.org/10.1037/0278-6133.24.4.S35
34. Soussi A, Zero E, Sacile R et al (2024) Smart sensors and smart data for precision agriculture: a review. Sensors 24:2647. https://doi.org/10.3390/s24082647
35. Stephens EC, Martin G, van Wijk M et al (2020) Editorial: Impacts of COVID-19 on agricultural and food systems worldwide and on progress to the sustainable development goals. Agric Syst 183:102873. https://doi.org/10.1016/j.agsy.2020.102873
36. Středová H, Fukalová P, Chuchma F et al (2020) A complex method for estimation of multiple abiotic hazards in forest ecosystems. Water 12:2872. https://doi.org/10.3390/w12102872
37. Středová H, Podhrázská J, Chuchma F et al (2021) The road map to classify the potential risk of wind erosion. ISPRS Int J Geo-Inf 10:269. https://doi.org/10.3390/ijgi10040269
38. Středová H, Rožnovský J, Středa T (2013) Predisposition of drought occurrence in selected arid areas of the Czech Republic. Contrib Geophys Geod 43:237–252. https://doi.org/10.2478/congeo-2013-0015
39. Středová H, Štěpánek P, Středa T et al (2023) Effect of erosive efficient rains specification on rainfall erosivity factor. Contrib Geophys Geod 53:301–318. https://doi.org/10.31577/congeo.2023.53.3.7
40. Středová H, Toman F (2012) Erosion potential of snow cover in the Czech Republic. Acta Univ Agric Silvic Mendel Brun 60:117–124
41. Sulik S (2022) Convective environment and development of a tornadic supercell in the Czech Republic on 24 June 2021. Meteorol Hydrol Water Manage 10:60–72. https://doi.org/10.26491/mhwm/150023
42. Tilloy A, Malamud BD, Winter H et al (2019) A review of quantification methodologies for multi-hazard interrelationships. Earth-Sci Rev 196:102881. https://doi.org/10.1016/j.earscirev.2019.102881
43. Trnka M, Hlavinka P, Možný M et al (2020) Czech drought monitor system for monitoring and forecasting agricultural drought and drought impacts. Int J Climatol 40:5941–5958. https://doi.org/10.1002/joc.6557
44. Trnka M, Možný M, Jurečka F et al (2021) Observed and estimated consequences of climate change for the fire weather regime in the moist-temperate climate of the Czech Republic. Agr Forest Meteorol 310:108583. https://doi.org/10.1016/j.agrformet.2021.108583
45. Vågsholm I, Arzoomand NS, Boqvist S (2020) Food security, safety, and sustainability—getting the trade-offs right. Front Sustain Food Syst 4. https://doi.org/10.3389/fsufs.2020.00016
46. Weber EU (2006) Experience-based and description-based perceptions of long-term risk: why global warming does not scare us (yet). Clim Change 77:103–120. https://doi.org/10.1007/s10584-006-9060-3
47. Weber H, Poeggel K, Eakin H et al (2020) What are the ingredients for food systems change towards sustainability? Environ Res Lett 15:113001. https://doi.org/10.1088/1748-9326/ab99fd

48. Yépez-Ponce DF, Salcedo JV, Rosero-Montalvo PD et al (2023) Mobile robotics in smart farming: current trends and applications. Front Artif Intell 6:1213330. https://doi.org/10.3389/frai.2023.1213330
49. Zahradníček P, Trnka M, Brázdil R et al (2015) The extreme drought episode of August 2011–May 2012 in the Czech Republic. Int J Climatol 35:3335–3352. https://doi.org/10.1002/joc.4211
50. Zsembeli J, Kovács E, Harsányi E et al (2023) Effectiveness of higher education in Central European universities towards climate change adaptation in the agricultural sector: a comprehensive survey. Prog Agric Eng Sci 20:271–306. https://doi.org/10.1556/446.2024.00141

Chapter 11
Beyond the Trend: Climate Variability as a Threat Multiplier for Critical Infrastructure and Global Cyber-Physical-Social Systems

Christina Karamperidou

Abstract Large-scale interannual and decadal variability of the coupled ocean-atmosphere system modulates the location, frequency, and duration of extreme weather events, such as floods, heatwaves, and droughts. These fluctuations can amplify demands on critical cyber-physical infrastructure to manage such extremes, thus increasing society's vulnerability to attacks on these systems. When climate variability compounds underlying long-term climate trends, these threats intensify further. Although outside of NATO's area of responsibility, the tropical Pacific is a prime example of these complex interactions, exerting a disproportionately large influence on global weather and climate patterns. This chapter begins by examining the complexity of tropical Pacific large-scale climate, including uncertainties in the region's mean climate shifts and how such shifts are related to its prominent year-to-year variability phenomenon, the El Niño-Southern Oscillation. Following this, the chapter highlights how the superposed impacts of tropical Pacific climate variability and change could create strategic windows of opportunity for cyber-physical attacks worldwide, focusing on the heightened vulnerability of data center operation to heat extremes, which cyber attackers have weaponized to disrupt data center operation by targeting the centers' cooling systems.

11.1 Introduction

The Earth system consists of interconnected processes across the ocean, atmosphere, and land. These processes exhibit fluctuations at multiple timescales, from short-term weather to seasonal climate variations, to longer interannual variability and multidecadal and centennial shifts.

C. Karamperidou (✉)
Department of Atmospheric Sciences, University of Hawai'i at Mānoa, Honolulu, HI, USA
e-mail: ckaramp@hawaii.edu

M. Huber et al. (eds.), *Critical Infrastructure Systems and Extreme Threat Nexus: Preparing for Resilience and Security*, NATO Science for Peace and Security Series C: Environmental Security, https://doi.org/10.1007/978-3-032-13170-6_11

This constant variability creates a *weather-to-climate continuum* that has superposed impacts on various sectors, requiring careful decision-making spanning multiple timescales. The tropics exert a major influence on this continuum because of the key role that tropical temperature and precipitation patterns–especially those in the Pacific Ocean–play in distributing energy and moisture worldwide.

Thus, although outside of NATO's area of responsibility, the tropical Pacific and its complex, multiscale climate are crucial for evaluating weather-to-climate threats to cyber-physical-social systems around the globe. Its significance stems from its disproportionately large influence on global weather and climate patterns [1], its well-documented role as a source of weather and climate predictability [2], and its usefulness in constraining climate projection uncertainties at interannual to longer time scales [3], where such uncertainties grow more pronounced and can shape strategic decisions.

This chapter aims to highlight the challenges and opportunities associated with the role of tropical Pacific climate variability in assessing and anticipating threats to global cyber-physical-social systems. It will first examine the complexity of tropical Pacific climate, including uncertainties in the region's mean climate shifts and their impacts on its interannual variability. Following this, the chapter will demonstrate how the superposed impacts of tropical Pacific climate variability and change could impact climate conditions in ways that can create strategic windows of opportunity for cyber-physical attacks that exploit the heightened vulnerability of cyberinfrastructure and its cooling management systems to heat extremes. Drawing from this example, the chapter will conclude with a summary and recommendations towards advancing climate-resilient and cyber-secure cyberinfrastructure.

11.2 Background and Motivation

The weather-to-climate continuum has always influenced strategic decision-making. A key element of this decision-making is the ability to reduce the uncertainties tied to forecasting weather and climate, and anticipating their impacts.

11.2.1 Strategic Decision-Making in the Weather-to-Climate Continuum

Since the beginning of modern warfare and governance, weather and climate have played a central role in shaping outcomes that have determined the course of history. In the Battle of Salamis (480 BC) during the Greco-Persian Wars, Themistocles' knowledge of prevailing winds and forecasting of local weather conditions dictated the Greek naval strategy, leading to a decisive victory against a numerically superior enemy [4]. Centuries later, during D-Day, the Allied Forces' weather forecasts proved crucial for the success of the amphibious invasion that decided the outcome of the war. Weather reconnaissance flights and the interception of meteorological observations and analyses were an essential aspect of World War II warfare for both sides [5]. On one hand, the decryption of the Enigma machine allowed the Allies to integrate meteorological observations from the Nazi-occupied regions in their weather analyses. On the other hand, the Germans routinely intercepted American coded surface and upper-air analyses, which were transmitted under weaker security to expedite critical tactical decisions. These examples illustrate that weather and climate information, along with the associated risk analysis, have always been a critical factor in decision-making at all strategic levels.

This historical reliance on weather and climate insights finds a modern parallel in how strategic decisions are made today. For example, the U.S. Department of Defense integrates weather and climate risk into operational and infrastructure planning across its decision time continuum, from 1–10 day missions to long-term strategic planning [6]. The decision support tools used to quantify such risks are subject to varying levels of confidence. For assault planning, staging, etc., at the weather scales (1–10 days), forecasts are based on remote sensing, satellite data, and weather forecast models that provide high confidence. For exercise planning at seasonal time scales, it is important to assess the probability of mission-limiting weather, the needs for cold/warm weather gear, and to consider projections of snow or rainfall amounts. The available tools for seasonal climate predictions, such as climate model simulations and statistical analyses based on historical climate data, carry medium confidence. Low confidence and significant science gaps confound interannual (1–4 years), decadal (10–20+ years) and multidecadal (60+ years) climate projections. This in turn results in low confidence when assessing risk at interannual to longer timescales; for example, wildfire and hurricane risk, heat risk, and future wargames conditions (e.g., ice-free Arctic, sea-level driven coastal migration, water-energy security scenarios) [6].

However, modern warfare is not solely defined by operational readiness and battlefield victories. Cyberwarfare has become integral to modern conflict [7, 8], while the protection and resilience of critical cyber-physical-social infrastructure underpins societal functioning both during times of conflict and at peace. From manufacturing to water and energy systems to public health services, transportation networks, data centers, and military installations, critical infrastructure is also vulnerable to extreme weather and climate events. This vulnerability is compounded by shifts in background climate variability at interannual-to-multidecadal timescales, and by long-term climate trends.

Much like with tactical operations and exercise planning, safeguarding critical infrastructure against weather and climate extremes hinges upon identifying, quantifying, and reducing the sources of the associated forecasting and projection uncertainties.

11.2.2 Sources of Climate Uncertainty and Predictability

Uncertainty in climate projections stems from our incomplete knowledge of climate, insufficient representation of climate system, and limitations in computing resources; the sources of uncertainty operate at different, but often overlapping, time scales [9].

First, there are inherent uncertainties across timescales due to the nature of the climate system and its internal variability. For example, it has been shown that even in the absence of any external climate forcings[1], the frequency and amplitude characteristics [10] as well as the predictability [11] of the prominent phenomenon of tropical Pacific interannual climate variability—the El Niño-Southern Oscillation (ENSO)–vary at multidecadal-to-centennial time scales. A second source of uncertainty, referred to as model or structural uncertainty, is due to the difference across climate models in the implementations of numerical methods, varied mathematical expressions, and diverse parameterizations of small-scale processes [12]. Lastly, the lack of confidence in the socioeconomic pathway that the global community will adopt in the following decades leads to the so-called scenario uncertainty [13].

While reducing scenario uncertainty is a function of complex socioeconomic conditions, assessing and reducing uncertainties associated with internal variability, model skill in simulating critical climate phenomena, and model disagreements in projecting shifts in these phenomena is a subject of intense scientific focus. The motivation for this focus stems from climate variability's dual function:

> As much as the various phenomena of interannual and decadal climate variability are a source of uncertainty in weather and climate projections, they are also a source of predictability.

Exemplifying this, the advances in tropical oceanography and meteorology in the1960s[14, 15] and1970s[16–18] led to the realization in the1980s [19, 20] that ENSO is the largest coherent single course of global climate variability on interannual-to-decadal timescales and thus the largest source of global climate predictability [21–25]. Since the first successful prediction of El Niño using a dynamical model by Cane et al. [26, 27], advances in modeling and predicting ENSO have shaped seasonal predictions of a wide array of climate variables and phenomena around the globe. This includes rainfall and atmospheric circulation extremes [23, 28], tropical cyclones [29], streamflow extremes and drought [23], heat extremes [30], human health [31], trade and the economy [32], and conflict [33]. In addition, ENSO's multidecadal "cousin", tropical Pacific decadal climate variability, has emerged as the most significant source of modulation of global climate [34–37] at

[1] The term external climate forcings refers to the impacts on Earth's radiation balance by factors that are considered external to the climate's internally generated variability. These include changes in the concentration of greenhouse gases, changes in solar activity, tectonic activity, volcanic eruptions, etc.

longer time scales, as well as the most significant source of model uncertainty in projections of future climate around the globe [3, 38–40].

11.3 South of the Tropic of Cancer: Equatorial Climate Variability and Change

The equatorial Pacific climate is shaped by complex, interacting processes that operate across a wide range of timescales. This section examines the dynamics of this region, focusing on the El Niño-Southern Oscillation (ENSO) and how its variability relates to broader, long-term shifts in the region's climate.

11.3.1 The ENSO Phenomenon, Its Predictability, and Uncertainties

Under normal conditions, surface ocean temperatures in the eastern and central equatorial Pacific are approximately 4°C cooler than those in the western tropical Pacific (Fig. 11.1a). This pronounced east-west temperature gradient is maintained by easterly trade winds along the equator. Every 2 to 8 years, however, these trade winds weaken, allowing warmer water from the western Pacific to shift eastward via oceanic waves, resulting in anomalously warm sea surface temperatures (SST) in the central and east Pacific (Fig. 11.1c,d). These warm conditions are collectively known as *El Niño* events. When the peak temperature anomalies are concentrated in the far eastern Pacific (Fig. 11.1c), the phenomenon is called an *Eastern Pacific (EP) El Niño* event, whereas if they are concentrated in the central Pacific (Fig. 11.1d) an event is referred to as a *Central Pacific (CP) El Niño*. Conversely, when trade winds are stronger than normal, the eastern and central Pacific are colder than usual (Fig. 11.1b)–a phase known as *La Niña*.

Together, El Niño (warm phase) and La Niña (cold phase) along with their atmospheric pressure variations compose the *El Niño-Southern Oscillation (ENSO)* phenomenon. The strength of individual ENSO events is typically quantified by the magnitude of the sea surface temperature anomalies in the equatorial Pacific, often averaged over the so-called NINO3.4 region (5°S–5°N, 170°W–120°W; Fig. 11.1b,e).

Research on the dynamics, impacts, and prediction of ENSO intensified after the devastating impacts of the 1982 and 1997 El Niño events. These strong events exemplified ENSO's severe global impacts that include drought and wildfires [42], ecological disasters [43, 44], marine heatwaves [45, 46], and coral bleaching [47],

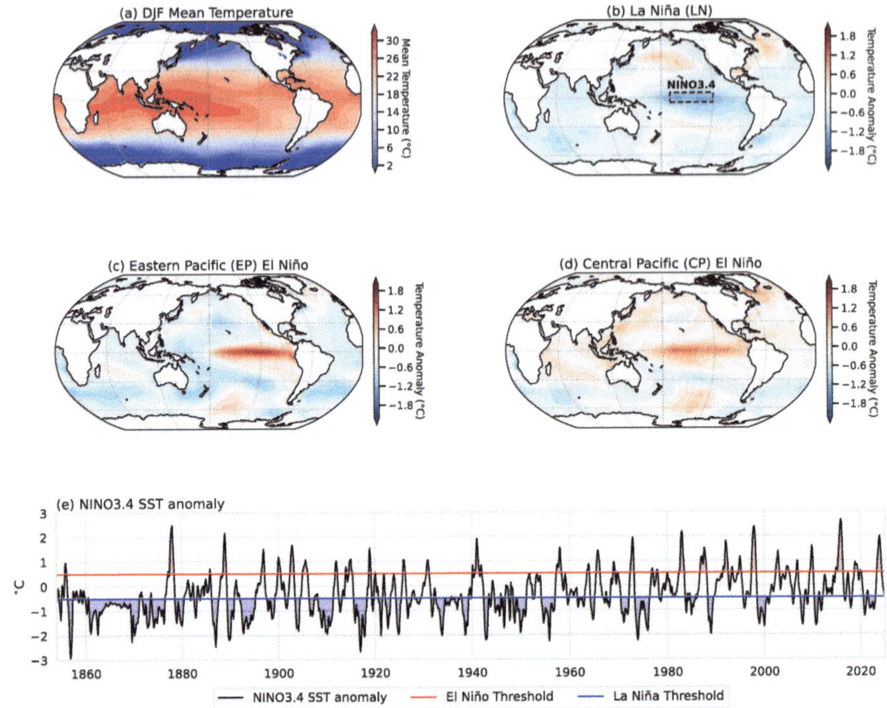

Fig. 11.1 Mean and anomalous sea surface temperatures associated with the El Niño-Southern Oscillation (ENSO) **a** During normal conditions in December-January-February (DJF), the eastern Pacific is colder than the western Pacific. **b** During La Niña events, temperatures in the central and eastern equatorial Pacific are colder than normal (blue shading). During **c** Central (CP) and **d** Eastern (EP) Pacific El Niño events, warm temperature anomalies (red shading) peak in the eastern and central Pacific, respectively. **e** The strength of ENSO events is defined by the average temperature anomaly in the NINO3.4 region (5°S–5°N, 170°W–120°W), with El Niño and La Niña events occurring when the warm or cold anomaly exceeds 0.5°C and −0.5°C, respectively. Anomalies are defined as the departure from the average monthly temperatures in the climate normal period 1981–2010. ENSO patterns and indices are calculated using the NOAA Extended Reconstructed SST V5 (ERSSTv5) dataset [41]

among others. Following the large 1997 El Niño event, a strong La Nina in 1998 led to river floods in China that resulted in thousands of deaths and the displacement of more than 200 million people [48], while flooding in Bangladesh led to food shortages and spread of waterborne diseases that killed thousands and affected millions [49, 50]. These severe, far-reaching impacts created much impetus to improve ENSO prediction using a hierarchy of models, from statistical to dynamical to machine learning models.

Coupled ocean-atmosphere general circulation models (GCMs)[2] initialized with observations can achieve reasonable predictive skill for certain ENSO indices, such as the NINO3.4 region sea surface temperature anomalies (Fig. 11.1e), even at lead times of up to nine months [51]. Machine learning models that were recently developed for NINO3.4 index prediction can achieve good skill at even longer lead times [52, 53]. Yet, translating these predictions into impacts remains challenging, especially because ENSO's "flavors"—eastern Pacific (EP) or central Pacific (CP)–each result in distinct regional climate responses [28, 54–56].

Accurately forecasting EP versus CP El Niño events using state-of-the-art dynamical or statistical models remains a major obstacle, while few studies using machine learning models have focused specifically on predicting these distinct ENSO flavors [57]. Fewer than half of the operational models can distinguish between EP and CP events at seasonal lead times [58], and difficulties persist even for strong events, as evidenced by underestimations of the 1997–98 and 2015–16 El Niños [59, 60]. Systematic biases, such as overly warm or wet conditions near the eastern Pacific and the cold SST bias along the equator, shift the effective location of ENSO variability within models and complicate the representation of its nonlinearities [3, 61, 63, 64]. Indeed, many state-of-the-art climate models fail to capture the full range of ENSO nonlinearity and diversity [3, 65], which casts doubt on their utility for predicting both the intensity and the spatial pattern of warming–particularly in the far eastern Pacific [66]. Although multi-model ensembles can mitigate some errors and quantify uncertainties [67], models and their initialization systems are not independent [67, 68]. This can result in misplaced forecast consensus, as illustrated by the failed predictions of a large El Niño event in 2014 [69]. To further complicate things, ENSO predictability varies naturally at multidecadal timescales [11], which implies that a skillful prediction system today may be unreliable tomorrow [70].

These persistent forecasting uncertainties have cascading implications for cyber-physical-social systems that depend on reliable climate information, such as infrastructure management, energy grids, disaster relief and community resilience planning, because each ENSO flavor can lead to drastically different regional risks, thereby complicating strategic decisions and response actions.

11.3.2 Multi-scale Interactions in Tropical Pacific Climate

To improve our understanding and modeling of ENSO properties (frequency, strength, predictability) and impacts, it is essential to decipher their interplay with

[2] General Circulation Models use mathematical equations to characterize how energy and matter interact in different parts of the ocean, atmosphere, and land (source: climate.gov).

changes in the tropical Pacific mean climate state at (multi)decadal and centennial time scales [3, 39, 71, 72].

Tropical Pacific decadal variability encompasses phenomena like the Interdecadal Pacific Oscillation (IPO) [73] and Pacific Decadal Oscillation (PDO) [74–76] (Fig. 11.2). Although the precise mechanisms driving these multidecadal shifts remain a subject of active research, it is widely acknowledged that multidecadal variations can modulate the frequency and intensity of El Niño and La Niña events and the relative occurrence of Eastern vs. Central Pacific events, thereby influencing the magnitude and spatial footprint of ENSO-driven teleconnections [37, 77, 78]. For example, during the positive phase of the PDO (Fig. 11.2a), whereby the SST in the tropical Pacific resembles persistent El Niño-like conditions, the response to the interannual ENSO events of the frequency of persistent mid-latitude weather systems can change sign [78, 79].

Tropical Pacific decadal-scale variability can also affect longer-term climate trends by enhancing or masking underlying warming signals, leading to periods of slower surface temperature increases followed by accelerated warming phases [34, 80]. As a result, understanding and predicting tropical Pacific decadal-to-centennial variability is critical for understanding and anticipating changes in ENSO characteristics and improving seasonal-to-decadal climate forecasts, which in turn inform risk assessment and adaptation strategies across multiple sectors.

At multidecadal-to-centennial timescales, most climate models project a warming of the eastern tropical Pacific under future emissions scenarios (Fig. 11.3c,d), which is reminiscent of the warming anomalies during El Niño events (Fig. 11.1c,d). Under such an El Niño-like change in the surface ocean temperatures of the tropical Pacific, one may anticipate corresponding El Niño-like changes in global weather and climate as a response. For example, the projected decrease in the frequency of atmospheric blocking events in the Northern Hemisphere has been attributed to the projected strong warming of the tropics, which strengthens the jet stream and consequently hinders block formation [81–85]. Atmospheric blocking events are persistent, high-impact weather patterns in the middle and high latitudes that divert the jet stream and storm tracks for days to weeks, and can be associated with long-lived, severe heatwaves [86–89], cold spells, droughts, and flooding [90]. Projections of regional blocking frequency in response to future climate scenarios have been shown to be sensitive to the pattern of sea surface temperature anomalies in the tropical Pacific, and in particular, whether the projected warming is more similar to Eastern Pacific or Central Pacific El Niño [91].

However, there is considerable debate surrounding the robustness of this model consensus on the mean tropical Pacific climate change. This debate stems from the fact that in the past 40 years, there has been a cooling trend in the eastern tropical Pacific (Fig. 11.3a), manifesting as a negative trend of the PDO Index (Fig. 11.2b). Climate models fail to simulate this trend as they show an El Niño-like warming trend (Fig. 11.3b) which they carry forward in their projections for the end of the 21st century (Fig. 11.3c,d) [3, 38, 39].

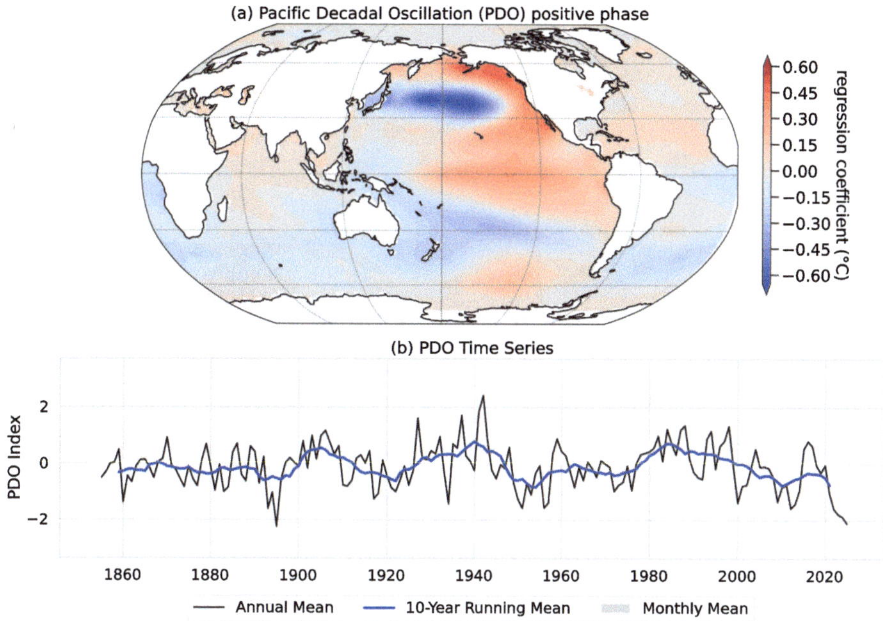

Fig. 11.2 Pacific decadal oscillation (PDO) pattern and time series. a Global sea surface temperature pattern during the positive phase of the PDO, shown by the regression coefficient between the PDO index (subfigure b) and sea surface temperatures. The positive PDO phase is characterized by a warm wedge of higher than normal sea surface temperatures in the eastern equatorial Pacific and a cold horseshoe pattern of lower than normal sea surface temperatures connecting the north, west and southern Pacific. **b** The PDO Index is defined as the principal component of the leading pattern (EOF) of sea surface temperature anomalies in the North Pacific basin polewards of 20°N. Grey bars in subfigure **b** show the monthly mean PDO Index values, while the black and blue lines show the annual mean and 10-year running mean values of the index, respectively. PDO patterns and indices are calculated using the NOAA Extended Reconstructed SST V5 (ERSSTv5) dataset [41]

Therefore, despite the critical importance of both the multidecadal mean state of the tropical Pacific and its interannual variability (i.e., ENSO), our understanding of their present-day relationship and how it may evolve in the future remains limited. This uncertainty carries significant implications for efforts to anticipate and mitigate the risks that weather and climate extremes pose to the operation and security of critical cyber-physical-social systems.

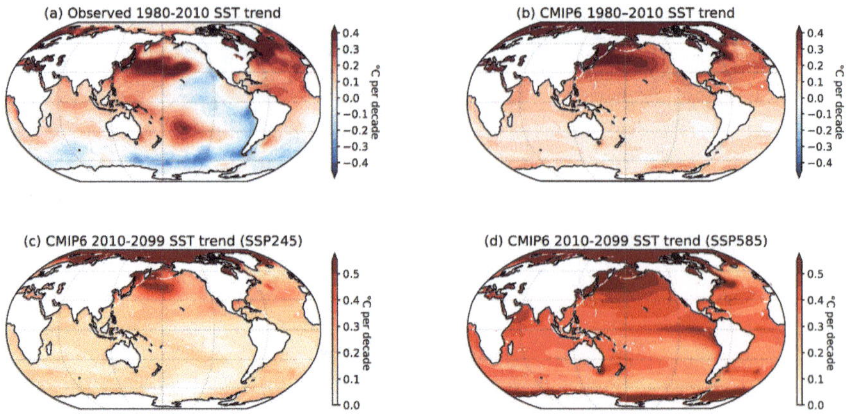

Fig. 11.3 Observed and simulated sea surface temperature trends. a Observed sea surface temperature (SST) trends calculated over 1980–2010 using the ERSSTv5 dataset [41], showing a distinctive cooling pattern in the central equatorial Pacific and a horseshoe warming pattern connecting the north, west, and southern Pacific. **b** Average SST trends from climate model simulations calculated over 1980–2010, lacking the observed cooling pattern; models are forced by historical forcing. **c** Average SST trends from climate model simulations calculated over 2010–2099; models are forced by the middle-of-the-road Shared Socioeconomic Pathway 2–4.5 (SSP245) and **d** the high-emissions SSP 5-8.5 (SSP585). The number of simulations used to calculate the climate model trends is 63, 43, and 40 for the three scenarios (historical, SSP245, and SSP585), respectively, and is based on the availability of output. Climate model output is from models participating in the Coupled Model Intercomparison Project Phase 6 (CMIP6). Only one simulation per model was used to ensure equal weighting of their sea surface temperature trends when computing their multi-model average

11.4 Tropical Pacific Climate Variability as a Threat Multiplier for Cyber-Physical Systems

To illustrate the implications of the uncertainties in understanding and projecting climate variability, this section examines how heat extremes, amplified by tropical Pacific climate variability, can threaten the operation of data centers, underscoring ENSO's role as a threat multiplier for vital infrastructure.

11.4.1 Extreme Heat Threats to Data Center Operation and Security

Data centers and their operation exemplify how critical cyber-physical systems, essential to all sectors of modern society, from banking and power grid operation to military training and operations and disaster relief to space exploration, medical research and academia, are vulnerable to weather and climate extremes, as they rank among the most energy and water-intensive infrastructure classes. The recent surge in artificial intelligence and machine learning technology has exponentially increased computational demand and electricity consumption by data centers. According to the International Energy Agency, the electricity consumption by data centers is estimated to rise from 460 TWh in 2022 to a range between 620–1,050 TWh by 2026, which is equivalent to adding at least one Sweden or at most one Germany to global electricity demand [92].

Approximately 40% of a data center's power is dedicated to cooling; this percentage is lower for centers using economizers that leverage outside air or water to reduce energy costs, and higher for centers that employ mechanical cooling via compressors and refrigeration. The effectiveness and sustainability of these cooling systems hinge on ambient air temperature and humidity, with humidity being a determining factor for the energy required for cooling, the reliability of operation, and the lifetime of the equipment. The recommended inlet air temperature and humidity for typical information technology equipment is 18°C to 27°C at 10–70% relative humidity and dew point below 15°C [93]. Such recommended and allowable envelopes depend on equipment density and ambient air pollution levels; for instance, certain equipment classes allow dew points up to 24°C. When the outdoor temperature and humidity exceed these ranges, free cooling, whether by direct outside air or evaporative methods, becomes inefficient or even unfeasible, and full-capacity mechanical cooling is required.

Weather extremes pose growing threats to data center operations as they can disrupt power supply, impair cooling capacity, force shutdowns, and constrain optimal data center locations.

Recent heatwaves caused major outages at significant data centers, including those operated by Oracle and Google in London and Twitter in Sacramento in 2022. Over recent decades, exposure to humid and hot conditions has increased across populated regions [94], and is projected to further increase under warmer mean climate states [95]. This will exacerbate the threat to data center operation, with the projected number of existing data centers exposed to at least one day per year of dangerous wet-bulb temperatures[3] projected to be multiple times larger as the background climate gets warmer (Fig. 11.6). Additionally, compound weather extremes, such as tropical cyclones and flooding followed by heatwaves [97], exemplified by Puerto Rico post-hurricane Maria, intensify challenges in maintaining data center operations and complicate disaster response efforts that often depend on reliable information networks. Recent studies project a rise in coinciding river flooding and extreme heat,

[3] Wet-bulb temperature is the lowest temperature to which air can be cooled by the evaporation of water into it, at a constant pressure. At 100% relative humidity, the wet-bulb temperature is equal to the air temperature (dry-bulb temperature). Wet-bulb temperatures that exceed 31°C are considered dangerous and pose risks of hyperthermia, heat stroke, or death, even for young, healthy individuals [96], as under these conditions the human body cannot dissipate metabolic heat through sensible or latent cooling.

particularly east of the Rockies [98, 99]. Such conditions strain data center operation by compromising water supply essential for their cooling systems, increasing energy demand to manage humidity, and disrupting power supply for prolonged periods.

At the same time, data center cooling management systems have increasingly been used as a backdoor for cyberattacks (e.g., Iran, 2019; Singapore, 2022). Cybersecurity measures for cooling management systems are often inadequate or not aligned with advanced cybersecurity protocols, making them vulnerable to breaches. This vulnerability has been highlighted in real-life incidents, such as the 2013 Target Data Breach and a 2009 cyberattack on a Texas hospital's HVAC system. Research environments have also demonstrated these risks; for instance, in 2011, researchers showed that a device plugged into an external power outlet can jam the signals that control the air-conditioning systems [100]. Leaked documents over the years have revealed adversarial actor intent to map and/or target Building Management Systems (BMS), emphasizing the urgent need to enhance cyber defenses in data centers and their components, especially as smart BMS technologies become more widespread.

The documented weaponization of heat extremes by cyber-attackers underscores the need for data center cooling innovations to be both **cyber-secure** and **climate-resilient**.

11.4.2 Compounded Threats from Climate Variability and Change

The heatwaves of 2022 and consequent disruptions in state-of-the-art data centers highlight the role of tropical Pacific climate variability and change in modulating risk to the operation and security of data centers. Both the dry-bulb and wet-bulb temperatures in June-September 2022 (Fig. 11.4a,b) exceeded high percentiles (Fig. 11.4c,d), straining operations of data centers that are generally designed based on mean climatological conditions. Somewhat counterintuitively, the year 2022 was marked by colder than normal conditions in the tropical Pacific, that is, La Niña conditions. In fact, the 2022 La Niña conditions were part of a persistent event that lasted for three years (2021–2023; Fig. 11.1e), dubbed a "triple-dip" La Niña.

The precise manifestation of ENSO's impact on humid heat extremes varies by region and by the phase of ENSO (Fig. 11.1), which can exacerbate or mitigate heat and humidity risk (Fig. 11.5). For example, during Central Pacific El Niño years, it is approximately 50% more likely that wet-bulb temperatures in Europe exceed their 95th percentile compared to the baseline risk, while during La Niña events this

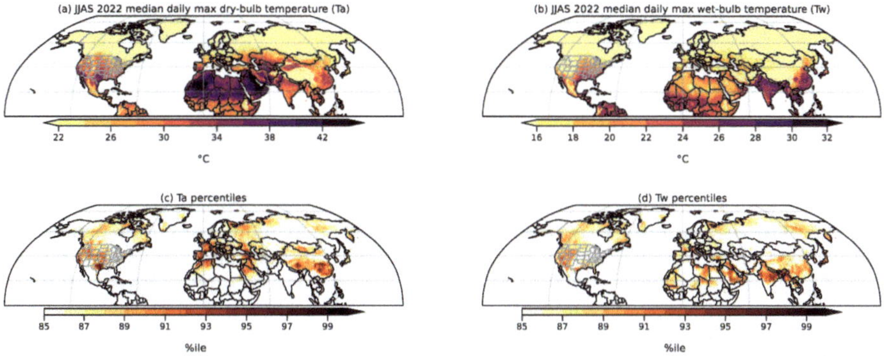

Fig. 11.4 Heat conditions in June-September (JJAS) 2022 that resulted in data center operation disruptions. **a** Median daily maximum dry-bulb (air) temperature (Ta); **b** Median daily maximum wet-bulb temperature (Tw); **c** The percentile corresponding to Ta values shown in subfigure a; **d** The percentile corresponding to Tw values shown in subfigure **b**. Percentiles are based on the 1981–2010 period. Daily wet-bulb temperatures are estimated from maximum daily temperature, dewpoint temperature, and pressure from the Centre for Medium-Range Weather Forecasts ERA5 reanalysis dataset [101]

risk is lower than the baseline (Fig. 11.5a,b)[4]. Such varied responses highlight the importance of examining ENSO in the context of its different phases and flavors rather than assuming a uniform effect on temperature extremes.

The ENSO impacts on the relative risk of extreme wet-bulb temperatures influence data center overheating risk, cooling system efficiency, the duration of power outages that affect data center operations, as well as their vulnerability to cyber attacks that weaponize heat, and the ability and speed of recovery are two separate objects.

Further, these relative risks may change as climate shifts at multidecadal and centennial timescales. It is worth noting that the La Niña-related heat extremes in the summer of 2022 happened within the context of the 40-year cooling trend in the eastern tropical Pacific or, alternatively, the persistently negative phase of the Pacific Decadal Oscillation (Fig. 11.2b). These underlying La Niña-like climate conditions are associated with increased blocking events [78] during the recent decades.

[4] Here, the relative risk of extreme humid-heat events under different ENSO conditions is calculated based on the relative risk ratio of wet-bulb temperature (Tw) exceeding the 95th percentile of daily maximum wet bulb temperatures in the period 1981–2010. The condition-specific risk for La Niña, EP El Niño and CP El Niño conditions was computed as the ratio of extreme days to total days for each ENSO condition. The relative risk ratio for each condition was calculated by normalizing the condition-specific risk against the overall risk. The overall risk is calculated as the ratio of extreme days over the total number of days in the dataset.

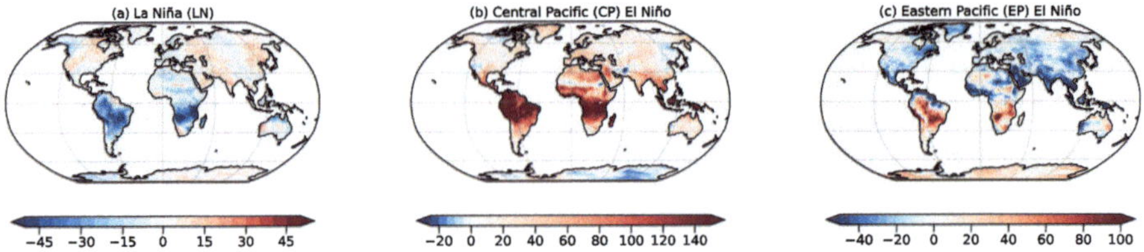

Fig. 11.5 The percent difference in risk of extreme daily maximum wet-bulb temperatures under ENSO conditions compared to the overall risk. The daily maximum wet-bulb temperature is considered extreme if it exceeds its 95th percentile, calculated based on the period 1981–2010. The year is defined from April to the following March to align with the typical development, peak, and decay seasons of ENSO events. When the difference is positive, the relative risk ratio is above 1, and the risk of exceeding the threshold is higher under the specific condition compared to the overall risk. For example, a value of 50% on these maps indicates that the risk is 50% higher under this condition than the overall risk. The patterns of relative risk are largely of opposite sign in La Niña vs. Central Pacific El Niño, but with asymmetric magnitudes and regional deviations. Under Eastern Pacific El Niño conditions, the impact on relative risk can be of opposite sign to that during Central Pacific El Niño, particularly in the tropics and the Northern Hemisphere. Daily wet-bulb temperatures are estimated from maximum daily temperature, dewpoint temperature, and pressure from the Centre for Medium-Range Weather Forecasts ERA5 reanalysis dataset [101]

Blocking events can lead to long-lived severe heatwaves, such as the Pacific Northwest heatwave of June 2021 [86], which happened during the triple-dip La Nina of 2021–2023 and led to hundreds of attributable deaths, mass mortalities of marine life, and severe agricultural yield decreases, and wildfires. Such ENSO influences have changed in the past [78] and are expected to change in the future as well [21], regardless of whether ENSO characteristics (frequency, strength, duration of events) themselves change.

Indeed, even without changes in event strength, frequency, and patterns, when ENSO events are superposed on warmer background conditions, they increase the area and population exposure to humid heat extremes [30]. All ENSO phases and flavors lead to a multifold increase in the number of data centers that are exposed to dangerous wet-bulb temperatures (maximum daily Tw > 31°C) for at least one day per year as compared to the number exposed under climatological conditions across all future background warming levels (Fig. 11.6). This multiplying effect of ENSO events on a global scale regardless of the phase and flavor is largely due to the fact that the impacts of El Niño vs La Niña are not necessarily of opposite sign everywhere (also see Fig. 11.5), while the asymmetry in the maximum strength of the two opposite phases also plays a role.

In addition, at each background warming level, ENSO events result in data center exposure that is comparable to or exceeds that anticipated climatologically under much warmer background climates (Fig. 11.6). This holds true even for low background warming levels (e.g., 1.5°C or 2°C), where the number of affected data centers is very small under climatological conditions but multiplies during ENSO events. The exponential increase in data center exposure during El Niño events at progressively warmer background climates can have significant implications for preparedness and

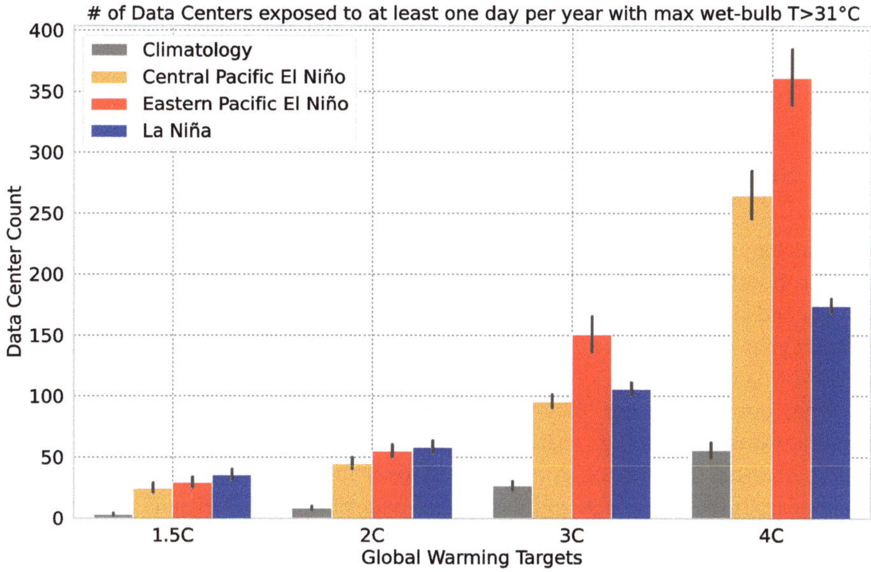

Fig. 11.6 Data center exposure to extreme humid heat under different background climate warming levels and ENSO events. Number of existing data centers exposed to at least one day with dangerous wet-bulb temperature (Tw>31°C) under four global warming targets representing warmer background climate conditions by 1.5°C, 2°C, 3°C, and 4°C. Shades of gray, blue, orange, and red represent climatological, La Niña, Central Pacific El Niño, and Eastern Pacific El Niño conditions (June through following May), respectively. Error bars for each ENSO flavor represent one standard deviation determined from the multi-model ensemble. These calculations follow the methods of Menzo et al. 2025 [30] and use the data of Vecellio et al. [95] and Kong and Huber [102]. Data center locations are based on the Data Center Map Database and are used here with permission by the database administrators for academic research purposes

mitigation of cyber-physical threats that may seek to weaponize this hazard at large scales. Here, a wet-bulb threshold of 31°C illustrates a complex cyber-physical-social scenario in which temperature and humidity are extreme enough to trigger a public health emergency [96] requiring significant emergency management efforts for human safety that depend on critical infrastructure, while also forcing data centers to run mechanical cooling at full capacity, leading to a 30–40% surge in electricity demand and elevated overheating risk.

The disproportionate impact of ENSO on the exposure of not only humans [30] but also the built environment and the infrastructure that enables today's way of life, such as its impact on data center exposure to humid heat extremes presented here (Fig. 11.6), underscores the complexity of the climate-cyber-physical-social system nexus in present-day climate and more so in an ever-changing climate.

This chapter will conclude with a summary and some final thoughts on the challenges and opportunities lying ahead.

11.5 Summary and Discussion

Tropical Pacific climate variability, dominated by its prominent phenomenon–the El Niño Southern Oscillation–exerts a profound influence on global and regional weather and climate extremes through complex teleconnections that modulate risks to critical cyber-physical-social systems. The exact regional manifestation of ENSO's impact on such extremes depends not only on the phase of ENSO (El Niño vs. La Niña) but also on the "flavor", which shifts the location of peak sea surface temperature anomalies, e.g., eastern vs. central Pacific El Niño. Complicating matters further, changes in the background climate over multidecadal or longer timescales can alter both the frequency and the spatial pattern of ENSO events, reshaping the spatiotemporal distribution of ENSO-related extremes.

Despite notable progress, significant disagreements remain in how climate models represent tropical Pacific variability across interannual (ENSO), multidecadal, and centennial scales, resulting in diverging projections for weather and hydroclimate extremes. Indeed, the magnitude and pattern of tropical Pacific SST changes at multidecadal scales are still a subject of considerable inter-model disagreement and bias [3, 38, 103]. These uncertainties translate into large risk envelopes for critical infrastructure, since changes in extremes reflect both the changing mean state of the tropical Pacific and the relative occurrence of different ENSO phases and flavors [72].

To illustrate the multiplicative role of tropical Pacific climate variability in creating compound operational risks, this chapter examined the issue of data center vulnerability to humid heat. Humid hot conditions can be conducive to unintentional disruptions to data center operation due to power outages and straining of their cooling systems. At the same time, cyber attackers can exploit elevated heat conditions to target weaknesses in cooling management systems, aiming to prolong outages and hinder recovery.

On the one hand, the increasing frequency, intensity, and duration of heat extremes threaten the power and water supply for data center cooling, which is generally designed based on historical climate information. On the other hand, ENSO events heighten the probability of record-breaking temperatures in many regions and serve as analogs for warmer background climates and associated shifts in the probability of humid heat extremes [30]. Indeed, data centers' global exposure to humid heat during ENSO events may exceed that expected under certain warmer baseline conditions, as illustrated in an example of extreme cyber-physical-social system exposure whereby wet-bulb temperatures exceed a threshold of 31°C. Meanwhile, the exponential increase in computational demand due to the expansion and rapid integration of artificial intelligence and machine learning further amplifies cooling needs and strains existing infrastructures.

Such vulnerabilities can be weaponized when attackers leverage cybersecurity weaknesses within building management and thermal control systems. Technological progress has partially mitigated these challenges by allowing data centers to operate at higher temperatures and by introducing innovations like submerged cooling and on-chip cooling. However, holistic solutions necessitate additional progress in improving weather-to-climate modeling and risk assessments and integration of the physical infrastructure with cybersecurity measures to deter, detect, and recover from cyber-attacks on cooling management systems.

11.6 Recommendations and Conclusion

This chapter highlighted the role of tropical Pacific climate variability as a modulator of weather-to-climate threats to cyber-physical-social systems using the example of ENSO-related heat extremes and their impact on the operation and security of data centers worldwide. The analysis presented here underscores several priority areas:

- **Develop frameworks for threat assurance and mitigation of compound weather and cyber threats to cyber-physical-social systems**: This includes developing scenario-building and stress-testing capabilities for existing and new critical infrastructure systems.
- **Improve weather-to-climate models:** Promote the development of numerical and ML/AI models of the weather-to-climate continuum to provide industry-relevant information and reduce uncertainties in weather predictions and climate projections. Additional special focus on capturing the feedbacks that drive heatwave severity, for example by refining land-atmosphere coupling in climate models or improving their spatial resolutions, is warranted.
- **Unravel the interplay between the time scales of climate variability in the tropical Pacific and its impacts on compound extreme events:** Given its impact on global climate and weather extremes, a focus on improved understanding, reduction of simulation biases, and enhanced predictability of tropical Pacific climate is imperative for developing strategic interventions to protect critical infrastructure globally.
- **Invest in technological innovations with integrated enhanced cybersecurity:** Innovations in cooling technology and building retrofits are essential to deter and counter the impacts of the compound weather and cyber

threats. Their integration with secure thermal management systems is critical for reducing the likelihood and severity of cyberattacks that exploit weather-driven vulnerabilities.

Ultimately, anticipating and mitigating the compounding impacts of tropical Pacific climate variability on global weather-to-climate extremes remains crucial for protecting infrastructure far beyond the Pacific Rim. Adding another layer of complexity, recent studies showed that, while the impact of large-scale geoengineering interventions on tropical Pacific climate variability remains unclear [104, 105], there is potential for the application of targeted, regional geoengineering solutions, such as marine cloud brightening, to dampen developing El Niño events [106, 107]. Yet, questions of ethics, governance, and trade-offs abound, given ENSO's geographically diverse impacts on phenomena like hurricane formation, agriculture, trade, and heat extremes, as illustrated in this chapter. Even within individual nations, such interventions could pit one region's welfare against another's, revealing deep uncertainties about who should decide whether or how to alter a naturally evolving El Niño event. In this light, continued interdisciplinary research, robust stakeholder engagement, and thorough risk assessments will be paramount to guide any future discussions or pursuit of large-scale climate interventions.

Acknowledgements CK is partially supported by National Science Foundation grants #AGS-2202663 and #EEC-2330175. Computing resources provided by NCAR's Computational and Information Systems Laboratory, sponsored by the National Science Foundation [108], were used to perform the reported analysis.

Competing Interests The author has no conflicts of interest to declare that are relevant to the content of this chapter.

References

1. McPhaden MJ, Zebiak SE, Glantz MH (2006) ENSO as an integrating concept in earth science. Science 314:1740–1745. https://www.science.org/doi/abs/10.1126/science.1132588
2. Hoerling MP, Kumar A, Zhong M. El (1997) Niño, la Niña, and the nonlinearity of their teleconnections. J Climate 10:1769–1786
3. Karamperidou C, Jin F-F, Conroy JL (2017) The importance of ENSO nonlinearities in tropical Pacific response to external forcing. Climate Dynam 49:2695–2704
4. Herodotus (2018) 8:70–95
5. Persson A (2020) Right for the wrong reason?: A new look at the 6 June 1944 D-Day forecast by a neutral Swede. Bullet Amer Meteorol Soc 101:E993–E1006. https://journals.ametsoc.org/view/journals/bams/101/7/bamsD180311.xml
6. Department of Defense, Office of the Undersecretary of Defense (Acquisition and Sustainment) (2024) Department of defense 2024–2027 climate adaptation plan. Report submitted to national climate task force and federal chief sustainability officer. 5 Sept 2024
7. Warner M (2012) Cybersecurity: a pre-history. Intell National Secur 27:781–799. https://doi.org/10.1080/02684527.2012.708530

8. Liff AP (2012) Cyberwar: a new 'absolute weapon'? the proliferation of cyberwarfare capabilities and interstate war. J Strategic Stud 35:401–428. https://doi.org/10.1080/01402390.2012.663252
9. Hawkins E, Sutton R (2009) The potential to narrow uncertainty in regional climate predictions. Bullet Amer Meteorol Soc 90:1095–1108
10. Wittenberg AT (2009) Are historical records sufficient to constrain ENSO simulations?. Geophys Res Lett 36. https://agupubs.onlinelibrary.wiley.com/doi/abs/10.1029/2009GL038710
11. Karamperidou C, Cane MA, Lall U, Wittenberg AT (2014) Intrinsic modulation of ENSO predictability viewed through a local Lyapunov lens. Climate Dynam 42:253–270. https://doi.org/10.1007/s00382-013-1759-z
12. IPCC (2023) In: Climate change 2021—The physical science basis: working group I contribution to the sixth assessment report of the intergovernmental panel on climate change. Cambridge University Press
13. Wu Y, Miao C, Fan X, Gou J, Zhang Q, Zheng H (2022) Quantifying the uncertainty sources of future climate projections and narrowing uncertainties with bias correction techniques. Earth's Futur 10:e2022EF002963. https://doi.org/10.1029/2022EF002963
14. Bjerknes J (1966) A possible response of the atmospheric Hadley circulation to equatorial anomalies of ocean temperature. Tellus 18:820–829
15. Bjerknes J (1969) Atmospheric teleconnections from the equatorial Pacific. Monthly Weather Rev 97:163–172
16. Wyrtki K (1975) El Niño—The dynamic response of the equatorial pacific Ocean to atmospheric forcing. J Phys Oceanogr 5:572–584. https://doi.org/10.1175/1520-0485(1975)005<0572:ENTDRO>2.0.CO;2
17. McCreary J (1976) Eastern tropical ocean response to changing wind systems, with application to El Niño. J Phys Oceanography 6:632–645
18. Hurlburt HE, Kindle JC, O'Brien JJ (1976) A numerical simulation of the onset of El Niño. J Phys Oceanography 6:621–631
19. Philander SGH, Yamagata T, Pacanowski RC (1984) Unstable air-sea interactions in the tropics. J Atmospheric Sci 41:604–613
20. Ropelewski CF, Halpert MS (1987) Global and regional scale precipitation patterns associated with the el niño/southern oscillation. Monthly Weather Rev 115:1606–1626
21. Taschetto AS, Ummenhofer CC, Stuecker MF, Dommenget D, Ashok K, Rodrigues RR, Yeh S-W (2020) ENSO atmospheric teleconnections. In: McPhaden MJ, Santoso A, Cai W (eds) El Niño Southern Oscillation in a changing climate, pp 209–335. https://doi.org/10.1002/9781119548164.ch14
22. Sprintall J, Cravatte S, Dewitte B, Du Y, Gupta AS (2020) ENSO oceanic teleconnections. In: McPhaden MJ, Santoso A, Cai W (eds) El Niño Southern Oscillation in a changing climate, pp 337–359. https://doi.org/10.1002/9781119548164.ch15
23. Goddard L, Gershunov A (2020) Impact of El Niño on weather and climate extremes. In: McPhaden MJ, Santoso A, Cai W (eds) El Niño Southern oscillation in a changing climate, pp 361–375. https://doi.org/10.1002/9781119548164.ch16
24. Cane MA, Evans M (2000) Do the tropics rule? Science 290:1107–1108. https://www.science.org/doi/abs/10.1126/science.290.5494.1107
25. Seager R, Chiang JCH, Shaman J (2015) Do the tropics rule? assessing the state of tropical climate science. Bullet Amer Meteorol Soc 96:ES211–ES214. https://journals.ametsoc.org/view/journals/bams/96/12/bams-d-15-00043.1.xml
26. Cane MA, Zebiak SE, Dolan SC (1986) Experimental forecasts of El Niño. Nature 321:827–832
27. Zebiak SE, Cane MA (1987) A model El Niño-Southern oscillation. Monthly Weather Rev 115:2262–2278
28. McKenna M, Karamperidou C (2023) The impacts of El Niño diversity on Northern hemisphere atmospheric blocking. Geophys Res Lett 50:e2023GL104284. https://agupubs.onlinelibrary.wiley.com/doi/abs/10.1029/2023GL104284. E2023GL104284 2023GL104284

29. Lin I-I, Camargo SJ, Patricola CM, Boucharel J, Chand S, Klotzbach P, Chan JCL, Wang B, Chang P, Li T, Jin F-F (2020) ENSO and tropical cyclones. In: McPhaden MJ, Santoso A, Cai W (eds) El Niño Southern Oscillation in a changing climate, pp 377–408. https://doi.org/10.1002/9781119548164.ch17
30. Menzo ZM, Karamperidou C, Kong Q, Huber M (2025) El Niño enhances exposure to humid heat extremes with regionally varying impacts during Eastern versus Central Pacific events. Geophys Res Lett 52:e2024GL112387. https://agupubs.onlinelibrary.wiley.com/doi/abs/10.1029/2024GL112387
31. Kovats RS, Bouma M, Hajat S, Worrall E, Haines A (2003) El Niño and health. The Lancet 362:1481–1489. https://api.semanticscholar.org/CorpusID:6854919
32. Zebiak SE et al (2015) Investigating El Niño–Southern oscillation and society relationships. Wiley Interdisciplinary Rev: Climate Change 6. https://api.semanticscholar.org/CorpusID:41382380
33. Hsiang SM, Meng KC, Cane MA (2011) Civil conflicts are associated with the global climate. Nature 476:438–441
34. Kosaka XSY (2013) Recent global-warming hiatus tied to equatorial pacific surface cooling. Nature 501:403–407
35. Meehl GA, Teng H (2014) Regional precipitation simulations for the mid-1970s shift and early-2000s hiatus. Geophys Res Lett 41:7658–7665. https://agupubs.onlinelibrary.wiley.com/doi/abs/10.1002/2014GL061778
36. Henley BJ (2017) Pacific decadal climate variability: indices, patterns and tropical-extratropical interactions. Global Planetary Change 155:42–55. https://www.sciencedirect.com/science/article/pii/S0921818116304398
37. Liguori G, Di Lorenzo E (2018) Meridional modes and increasing Pacific decadal variability under anthropogenic forcing. Geophys Res Lett 45:983–991. https://agupubs.onlinelibrary.wiley.com/doi/abs/10.1002/2017GL076548
38. Seager R et al (2019) Strengthening tropical Pacific zonal sea surface temperature gradient consistent with rising greenhouse gases. Nat Climate Change 9:517–522
39. Cai W et al (2021) Changing El Niño–Southern Oscillation in a warming climate. Nature Rev Earth Environ 2:628–644. https://www.nature.com/articles/s43017-021-00199-z
40. Jiang F, Seager R, Cane MA (2024) A climate change signal in the tropical Pacific emerges from decadal variability. Nature Commun 15:8291. Nature Publishing Group. https://www.nature.com/articles/s41467-024-52731-6
41. Huang B et al (2017) Extended reconstructed sea surface temperature, version 5 (ersstv5): upgrades, validations, and intercomparisons. J Climate 30:8179–8205
42. King AD, Karoly DJ, van Oldenborgh GJ (2016) Climate change and El Niño increase likelihood of Indonesian heat and drought. Bullet Amer Meteorol Soc 97:S113–S117. https://journals.ametsoc.org/view/journals/bams/97/12/bams-d-16-0164.1.xml
43. Valle CA, Cruz F, Cruz JB, Merlen G, Coulter MC (1987) The impact of the 1982–1983 El Niño-Southern Oscillation on seabirds in the Galapagos Islands, Ecuador. J Geophys Res: Oceans 92:14437–14444. https://agupubs.onlinelibrary.wiley.com/doi/abs/10.1029/JC092iC13p14437
44. McPhaden MJ, Karamperidou C (2022) La Niña came to Eden. Bullet Amer Meteorol Soc 103:E2862–E2877. https://journals.ametsoc.org/view/journals/bams/103/12/BAMS-D-21-0343.1.xml
45. Sen Gupta TMBJA (2020) Drivers and impacts of the most extreme marine heatwave events. Scient Reports 10:19359
46. Holbrook SGAOE NJ (2020) Keeping pace with marine heatwaves. Nat Rev Earth Environ 1:482–493
47. Glynn JLBACCMO, Peter W, Maté (2001) Coral bleaching and mortality in Panama and Ecuador during the 1997–1998 El Niño–Southern Oscillation event: spatial/temporal patterns and comparisons with the 1982–1983 event. Bullet Marine Sci 69:79–109
48. Jonkman S (2001) Global perspectives on loss of human life caused by floods. Nat Hazards 34:151–175

49. Kunii O, Nakamura S, Abdur R, Wakai S (2002) The impact on health and risk factors of the Diarrhoea epidemics in the 1998 Bangladesh floods. Public Health 116:68–74
50. del Ninno C, Dorosh P, Smith LC, Roy DK (2001) The 1998 floods in Bangladesh: disaster impacts, household coping strategies, and responses. In: Research reports 122, International Food Policy Research Institute (IFPRI). https://EconPapers.repec.org/RePEc:fpr:resrep:122
51. L'Heureux ML, Levine AFZ, Newman M, Ganter C, Luo J-J, Tippett MK, Stockdale TN (2020) ENSO prediction. In: McPhaden MJ, Santoso A, Cai W (eds) El Niño Southern Oscillation in a changing climate, pp 227–246. https://doi.org/10.1002/9781119548164.ch10
52. Ham YG, Kim JH, Luo JJ (2019) Deep learning for multi-year ENSO forecasts. Nature 573:568–572. https://doi.org/10.1038/s41586-019-1559-7
53. Ham Y-G, Kim J-H, Kim E-S, On K-W (2021) Unified deep learning model for El Niño/Southern oscillation forecasts by incorporating seasonality in climate data. Sci Bullet 66:1358–1366. https://linkinghub.elsevier.com/retrieve/pii/S2095927321002243
54. Ashok K, Behera SK, Rao SA, Weng H, Yamagata T (2007) El Niño Modoki and its possible teleconnection. J Geophys Res: Oceans 112. https://agupubs.onlinelibrary.wiley.com/doi/abs/10.1029/2006JC003798
55. Kiefer J, Karamperidou C (2019) High-resolution modeling of ENSO-induced precipitation in the tropical Andes: implications for proxy interpretation. Paleoceanography Paleoclimatol 34:217–236. https://agupubs.onlinelibrary.wiley.com/doi/abs/10.1029/2018PA003423
56. Lu B-Y, Chu P-S, Kim S-H, Karamperidou C (2020) Hawaiian regional climate variability during two types of El Niño. J Climate 33:9929–9943. https://journals.ametsoc.org/view/journals/clim/33/22/jcliD190985.xml
57. Rivera Tello GA, Takahashi K, Karamperidou C (2023) Explained predictions of strong Eastern Pacific El Niño events using deep learning. Scient Reports 13:21150. Nature Publishing Group. https://www.nature.com/articles/s41598-023-45739-3
58. Ren H-L et al (2019) Seasonal predictability of winter ENSO types in operational dynamical model predictions. Climate Dynam 52:3869–3890. https://doi.org/10.1007/s00382-018-4366-1
59. Barnston AG, He Y, Glantz MH (1999) Predictive skill of statistical and dynamical climate models in SST forecasts during the 1997–98 El Niño episode and the 1998 La Niña onset. Bullet Amer Meteorol Soc 80:217–243. https://doi.org/10.1175/1520-0477(1999)080<0217:PSOSAD>2.0.CO;2
60. L'Heureux ML et al (2016) Observing and predicting the 2015-16 El Niño. Bullet Amer Meteorol Soc. http://doi.org/10.1175/BAMS-D-16-0009.1
61. Takahashi K, Montecinos A, Goubanova K, Dewitte B (2011) ENSO regimes: reinterpreting the canonical and Modoki El Niño. Geophys Res Lett 38. Wiley & Sons, Ltd. https://doi.org/10.1029/2011GL047364
62. Karamperidou C, Jin F-F, Conroy JL (2017) The importance of ENSO nonlinearities in tropical Pacific response to external forcing. Climate Dynam 49:2695–2704. https://doi.org/10.1007/s00382-016-3475-y
63. Cai W et al (2021) Changing El Niño–Southern oscillation in a warming climate. Nature Rev Earth Environ 2:628–644. https://doi.org/10.1038/s43017-021-00199-z
64. Takahashi K, Dewitte B (2016) Strong and moderate nonlinear El Niño regimes. Climate Dynam 46, 1627–1645. https://doi.org/10.1007/s00382-015-2665-3
65. Karamperidou C, DiNezio PN (2022) Holocene hydroclimatic variability in the tropical Pacific explained by changing ENSO diversity. Nature Commun 13, 7244. https://doi.org/10.1038/s41467-022-34880-8
66. Ding H, Newman M, Alexander MA, Wittenberg AT (2018) Skillful climate forecasts of the tropical Indo-Pacific ocean using model-analogs. J Climate 31:5437–5459. http://doi.org/10.1175/JCLI-D-17-0661.1
67. Becker EJ, Kirtman BP, L'Heureux M, Muñoz AG, Pegion K (2022) A decade of the North American multimodel ensemble (NMME): research, application, and future directions. Bullet Amer Meteorol Soc E973–E995. https://www.doi.org/10.1175/BAMS-D-20-0327.1

68. Abramowitz G et al (2019) ESD reviews: model dependence in multi-model climate ensembles: weighting, sub-selection and out-of-sample testing. Earth Syst Dynam 10:91–105. https://doi.org/10.5194/esd-10-91-2019
69. McPhaden MJ (2015) Playing hide and seek with El Niño. Nature Climate Change 5:791–795. https://doi.org/10.1038/nclimate2775
70. Weisheimer A et al (2022) Variability of ENSO forecast skill in 2-year global reforecasts over the 20th century. Geophys Res Lett 49:e2022GL097885. https://agupubs.onlinelibrary.wiley.com/doi/abs/10.1029/2022GL097885. E2022GL097885 2022GL097885
71. Wyman, D. A., Conroy, J. L. & Karamperidou, C. The Tropical Pacific ENSO–Mean State Relationship in Climate Models over the Last Millennium. *Journal of Climate***33**, 7539–7551 (2020). https://journals.ametsoc.org/view/journals/clim/33/17/jcliD190673.xml
72. Karamperidou C, DiNezio PN (2022) Holocene hydroclimatic variability in the tropical Pacific explained by changing ENSO diversity. Nature Commun 13:7244; Number: 1 Publisher: Nature Publishing Group. https://www.nature.com/articles/s41467-022-34880-8
73. Henley BJ, Gergis J, Karoly DJ, Power S, Kennedy J, Folland CK (2015) A tripole index for the interdecadal Pacific oscillation. Climate Dynam 45:3077–3090
74. Mantua NJ, Hare SR, Zhang Y, Wallace JM, Francis RC (1997) A Pacific interdecadal climate oscillation with impacts on salmon production. Bull Amer Meteor Soc 78:1069–1079
75. Mantua HSNJ (2002) The Pacific decadal oscillation. J Oceanography 58:35–44
76. Newman M et al (2016) The Pacific Decadal Oscillation, revisited. J Climate 29:4399–4427. https://journals.ametsoc.org/view/journals/clim/29/12/jcli-d-15-0508.1.xml
77. Power S, Casey T, Folland C, Colman A, Mehta V (1999) Inter-decadal modulation of the impact of ENSO on Australia. Climate Dynamics 15:319–324
78. Karamperidou C (2024) Extracting paleoweather from paleoclimate through a deep learning reconstruction of last millennium atmospheric blocking. Commun Earth Environ 5:535. https://doi.org/10.1038/s43247-024-01687-y
79. Lupo AR (2021) Atmospheric blocking events: a review. Annals of the New York Acad Sci 1504:5–24. https://onlinelibrary.wiley.com/doi/abs/10.1111/nyas.14557
80. England M, McGregor S, Spence P et al (2014) Recent intensification of wind-driven circulation in the Pacific and the ongoing warming hiatus. Nature Clim Change 4:222–227. https://doi.org/10.1038/nclimate2106
81. Barnes EA, Polvani LM (2015) CMIP5 projections of Arctic amplification, of the North American/North Atlantic circulation, and of their relationship. J Climate 28:5254–5271. https://journals.ametsoc.org/view/journals/clim/28/13/jcli-d-14-00589.1.xml
82. Kennedy D, Parker T, Woollings T, Harvey B, Shaffrey L (2016) The response of high-impact blocking weather systems to climate change. Geophys Res Lett 43:7250–7258. https://onlinelibrary.wiley.com/doi/abs/10.1002/2016GL069725
83. Woollings T et al (2018) Blocking and its response to climate change. Current Climate Change Reports 4:287–300. https://doi.org/10.1007/s40641-018-0108-z
84. Patterson M, Bracegirdle T, Woollings T (2019) Southern hemisphere atmospheric blocking in CMIP5 and future changes in the Australia–New Zealand sector. Geophys Res Lett 46:9281–9290. https://onlinelibrary.wiley.com/doi/abs/10.1029/2019GL083264
85. Davini P, D'Andrea F (2020) From CMIP3 to CMIP6: Northern hemisphere atmospheric blocking simulation in present and future climate. J Climate 33:10021–10038. https://journals.ametsoc.org/doi/10.1175/JCLI-D-19-0862.1
86. White RH, Anderson S, Booth JF et al (2023) The unprecedented Pacific Northwest heatwave of June 2021. Nat Commun 14:727. https://doi.org/10.1038/s41467-023-36289-3
87. Chan PW, Catto JL, Collins M (2022) Heatwave–blocking relation change likely dominates over decrease in blocking frequency under global warming. npj Climate Atmospheric Sci 5:1–8. Nature Publishing Group. https://www.nature.com/articles/s41612-022-00290-2
88. Kautz L-A et al (2022) Atmospheric blocking and weather extremes over the euro-Atlantic sector—a review. Weather Climate Dynam 3:305–336. https://wcd.copernicus.org/articles/3/305/2022/

89. Narinesingh V, Booth JF, Ming Y (2023) Northern hemisphere heat extremes in a warmer climate: more probable but less colocated with blocking. Geophys Res Lett 50. e2022GL101211 (2023). https://agupubs.onlinelibrary.wiley.com/doi/abs/10.1029/2022GL101211
90. Booth JF, Narinesingh V, Towey KL, Jeyaratnam J (2021) Storm surge, blocking, and cyclones: a compound hazards analysis for the northeast United States. J Appl Meteorol Climatol 60:1531–1544. https://journals.ametsoc.org/view/journals/apme/60/11/JAMC-D-21-0062.1.xml
91. Matsueda M, Endo H (2017) The robustness of future changes in Northern Hemisphere blocking: a large ensemble projection with multiple sea surface temperature patterns. Geophys Res Lett 44:5158–5166. https://agupubs.onlinelibrary.wiley.com/doi/abs/10.1002/2017GL073336. _eprint: https://agupubs.onlinelibrary.wiley.com/doi/pdf/10.1002/2017GL073336
92. IEA (International Energy Agency) (2001) Electricity 2024. Research reports, IEA (International Energy Agency). https://www.iea.org/reports/electricity-2024
93. ASHRAE (American Society of Heating, Refrigerating and Air-Conditioning Engineers, Inc.) (2021) Equipment Thermal Guidelines for Data Processing Environments. Technical Report, 5th edn, Revised and Expanded
94. Rogers CDW et al (2021) Recent increases in exposure to extreme humid-heat events disproportionately affect populated regions. Geophys Res Lett 48:e2021GL094183. https://agupubs.onlinelibrary.wiley.com/doi/abs/10.1029/2021GL094183. E2021GL094183 2021GL094183
95. Vecellio DJ, Kong Q, Kenney WL, Huber M (2023) Greatly enhanced risk to humans as a consequence of empirically determined lower moist heat stress tolerance. Proc National Acad Sci 120: e2305427120. https://www.pnas.org/doi/abs/10.1073/pnas.2305427120
96. Vecellio DJ, Wolf ST, Cottle RM, Kenney WL (2022) Evaluating the 35°c wet-bulb temperature adaptability threshold for young, healthy subjects (PSU Heat Project). J Appl Physiol 132:340–345
97. Simpkins G (2023) Compound tropical cyclone–heatwaves on the rise. Nat Rev Earth Environ 4:3. https://doi.org/10.1038/s43017-022-00386-6
98. Gu L et al (2022) Global increases in compound flood-hot extreme hazards under climate warming. Geophys Res Lett 49:e2022GL097726. https://agupubs.onlinelibrary.wiley.com/doi/abs/10.1029/2022GL097726
99. You J, Wang S, Zhang B, Raymond C, Matthews T (2023) Growing threats from swings between hot and wet extremes in a warmer world. Geophys Res Lett 50:e2023GL104075. https://agupubs.onlinelibrary.wiley.com/doi/abs/10.1029/2023GL104075
100. Kennedy D, Simon R (2011) Pentesting over Power Lines. In: DEF CON 19. https://doi.org/10.5446/40555
101. Hersbach H et al (2020) The ERA5 global reanalysis. Quarter J Royal Meteorol Soc 146:1999–2049
102. Kong Q, Huber M (2025) A global high-resolution and bias-corrected dataset of CMIP6 projected heat stress metrics. Sci Data 12(1):246. https://doi.org/10.1038/s41597-025-04527-6. PMID: 39939321; PMCID: PMC11821900
103. Karamperidou C, Stuecker MF, Timmermann A, Yun K-S, Lee S-S, Jin F-F, Santoso A, McPhaden MJ, Cai W (2020) ENSO in a changing climate. In: McPhaden MJ, Santoso A, Cai W (eds) El Niño Southern oscillation in a changing climate, pp 471–484. https://doi.org/10.1002/9781119548164.ch21
104. Gabriel CJ, Robock A (2015) Stratospheric geoengineering impacts on El Niño/Southern Oscillation. Atmospheric Chem Phys 15:11949–11966. https://acp.copernicus.org/articles/15/11949/2015/
105. Guo A, Moore JC, Ji D (2018) Tropical atmospheric circulation response to the G1 sunshade geoengineering radiative forcing experiment. Atmospheric Chem Phys 18:8689–8706. https://acp.copernicus.org/articles/18/8689/2018/
106. Haywood JM, Jones A, Jones AC, Halloran P, Rasch PJ (2023) Climate intervention using marine cloud brightening (MCB) compared with stratospheric aerosol injection (SAI) in the UKESM1 climate model. Atmospheric Chem Phys 23:15305–15324. https://acp.copernicus.org/articles/23/15305/2023/

107. Wan JS, Fasullo JT, Rosenbloom N, Chen C-CJ, Ricke K (2024) Targeted marine cloud brightening can dampen El Niño. https://arxiv.org/abs/2406.07853
108. Computational and Information Systems Laboratory (2019) Cheyenne: HPE/SGI ICE XA system. National Center for Atmospheric Research, Boulder, CO

Part V
Social and Economic Considerations

Chapter 12
Wiring Resilience: Metrics for a Weather-Adapted Power Grid

Thushara Gunda, Raquel L. Valdez, Zachary Kilwein, Rachael Alfant, J. Kyle Skolfield, Ryan Adams, and Nicole D. Jackson

Abstract The power grid is a critical infrastructure that supports various societal activities. While numerous investments have aimed to improve reliability of the power grid, the system is still highly vulnerable to extreme weather impacts. This vulnerability is driven by the complicated nature of extreme weather events and their concurrent impacts to multiple systems components. Although a need for resilience approaches has been identified, no clear measures exist to guide resilience analyses. To address this gap, we leverage multiple disciplines to identify quantitative resilience measures that can be used to support characterization of extreme weather impacts on the grid. We then demonstrate the value of these multi-disciplinary measures using a case study for a synthetic grid subject to a heatwave and a cyber attack involving disconnection of services. The implemented metrics demonstrate that the compounded load shedding from cyber attacks during a heatwave are likely due to the decreased redundancy of assets. Future explorations of these metrics under multiple types of threats and active integration into modeling assessments could help validate outcomes with real-world behaviors. Such continued explorations will advance current understanding and resilience assessments of the power grid as it continues to evolve and adapt to a range of extreme weather conditions.

Highlights:

- Summarize the complicated nature of extreme weather events influence on grid operations
- Describe resilience theory and current practices (including limitations) for incorporating resilience in current grid assessments
- Demonstrate the value of leveraging multi-disciplinary metrics to advance resilience insights to the power grid

T. Gunda (✉) · R. L. Valdez · Z. Kilwein · R. Alfant · J. Kyle Skolfield · R. Adams · N. D. Jackson
Sandia National Laboratories, Albuquerque, NM, USA
e-mail: tgunda@sandia.gov

M. Huber et al. (eds.), *Critical Infrastructure Systems and Extreme Threat Nexus: Preparing for Resilience and Security*, NATO Science for Peace and Security Series C: Environmental Security, https://doi.org/10.1007/978-3-032-13170-6_12

12.1 Introduction

The power grid can be described as the largest engineered machine that humans have ever created [1]. In many countries, the power grid provides a backbone for many aspects of modern society, with uninterrupted services contributing to the safety, security, and prosperity of the communities it serves. For example, in the United States (U.S.), the power grid has been built-out since the 20th century for increased commercial, industrial, and residential use. Today, the U.S. grid is a complex system consisting of interconnecting physical, cyber, and social components, spanning power assets, communications, and decision activities (Fig. 12.1). Physical components of the power grid connect the various generation and storage assets (e.g., natural gas power plants, hydropower, and batteries) to end users through transmission and distribution lines [2]. In recent decades, cyber components have also become crucial in enhancing efficiency, security, and reliability of the overall power grid [3], with sensors and communication devices providing real-time information to support system operations. Social components (e.g., regulations, ownership structures) also influence decision-making associated with grid operations and planning (e.g., [4]). These physical, cyber, and social components rarely work in isolation, but rather are interconnected and influence each other to support the overall power grid.[1]

The primary objective of the power grid is to provide reliable energy with minimal service disturbances or interruptions. Accordingly, grid operators have historically focused on metrics and regulations that emphasize reliability, often measured using system uptime or load shed (i.e., the amount of power that was not delivered) [8]. Regulations also govern consistency of operations (e.g., [9]) and depending on the region, may dictate a certain level of generation required to be held in reserve ready to connect to the grid in the event of significant voltage drops [10]. System conditions are monitored through supervisory control and data acquisition (SCADA) systems or energy management systems (EMS), which also enable control (e.g., changing set-points) to maintain stability in the grid. Entities are also complementing engineering-based strategies with market-based techniques (for both capacity and reserves) to better value resources in times of need and leverage the flexibility of storage devices (which can serve as both supply or load) to minimize cost for grid operations [11].

While these measures have increased overall reliability, grid operations during major events continue to remain a challenge [12]. Namely, extreme conditions require operations that can respond to or anticipate low-probability, high-risk disruptions using simulation, forecasting, and control. However, grid operators tend to prioritize reliability (which emphasize continued service during standard operating conditions) rather than resilience (which emphasizes withstanding, operating through, and recovering from major events) [8, 13]. This is especially the case for extreme weather events (e.g., hurricanes, floods, droughts, and heatwaves), which pose substantial risks to normal infrastructure operations resulting in loss of life and substantial economic costs [14, 15]. In addition, the increasing severity and frequency

[1] The interested reader should consult extended resources on the history of the U.S. grid (e.g., [5]), including associated mathematical modeling (e.g., [6]) for more information.

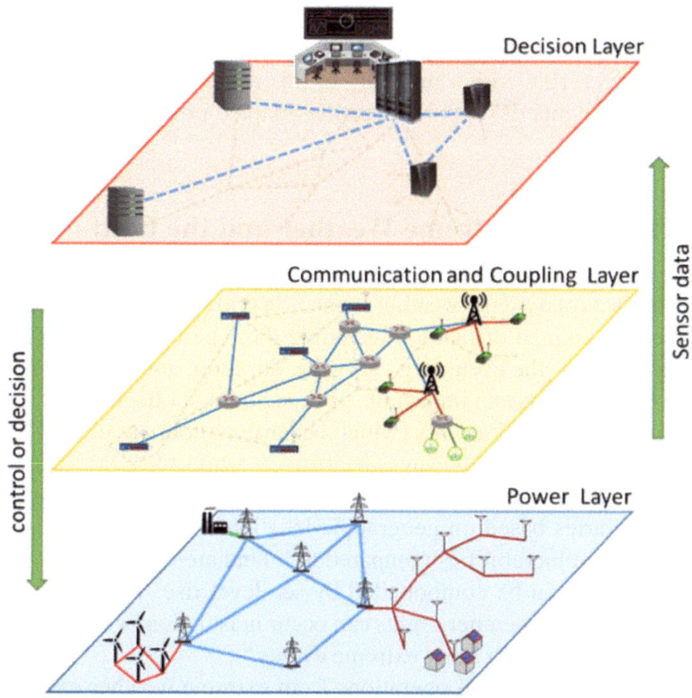

Fig. 12.1 Power grid operations are influenced by various interacting physical (e.g., power assets), cyber (e.g., communication), and social (e.g., decision-making) components [7]

of extreme weather events [16] has introduced considerable uncertainty into energy systems globally. For example, the South Australian blackout of 2016 disrupted power for over 1.7 million people due to a severe 50-year storm (i.e., expected to occur once every 50 years on average). This storm, which included at least two tornadoes and approximately 80,000 lightning strikes, resulted in devastating, cascading impacts spanning disruption of essential services, transportation issues, and emergency response access [17].

Unfortunately, current capabilities to measure the impact of extreme weather events on resilient performance of power systems are underdeveloped [18]. More attention is especially needed to understand power grid performance during extreme events (including compounded events with adversarial attacks) that better accounts for network topologies. This opportunity motivates the focus of this chapter, which has three primary objectives: (1) summarize the complicated nature of extreme weather events influence on grid operations (Sect. 12.2), (2) describe resilience theory and current practices (including limitations) for incorporating resilience in current grid assessments (Sect. 12.3), and (3) demonstrate the value of leveraging multi-disciplinary metrics to advance resilience insights to the power grid (Sect. 12.4). The latter is done through the use of a case study for a synthetic power grid that is exposed to heatwave and cyber attack scenarios. Then in Sect. 12.5, we reflect on the case study implementation, highlighting gaps and opportunities for future work. Finally, we conclude in Sect. 12.6 with a summary of this work's contributions and need for continued advancements to support a weather-adapted power grid.

12.2 Background: Extreme Weather and the Grid

The specific impacts of extreme weather events[2] on grid operations are highly variable and influenced by a range of factors, including the type of event, geographic location, and magnitude of the event (e.g., severity, duration, and scale). Multiple types of extreme weather events can impact the grid, including both chronic events (which often last months) and acute events (which occur over hours to days) [20]. Chronic events include droughts and coldwaves/heatwaves, while acute events include wildfires, hurricanes, floods, ice, and wind conditions (Table 12.1). Exposure to these events typically varies based on geographic location. For instance, coastal regions may face different vulnerabilities compared to inland areas due to their exposure to storm surges, which can be compounded by sea level rise[3] [22]. It should also be noted that these extreme weather events can occur in isolation or can be compounded (i.e., co-occur together) with other extreme events.[4]

The specific impacts to grid operations from extreme weather events is heavily influenced by the nature and scale of the weather event (e.g., localized or widespread). For instance, a polar vortex (i.e., a large area of low pressure and cold air, characterized by strong winds that usually surrounds Earth's poles [23]) in January 2014 brought extreme cold temperatures to the Midwest and Northeast U.S., leading to a significant surge in electricity demand for heating. This increased demand, coupled with operational challenges at power plants due to frozen equipment and natural gas supply constraints (with more than 10,000 megawatts (MW) of unintended losses in generation), resulted in widespread power outages and strain on the interconnected electrical grid across multiple states [24]. Concurrent impacts to both generators and loads extend to other types of extreme weather events as well (Fig. 12.2). For instance, during a heatwave, the demand for electricity elevates load on the grid due to increased use of cooling. Simultaneously, heatwaves can affect water availability for cooling processes in power plants, thereby limiting their operational capacity

---

[2] Extreme weather events are defined as significant and unusual climate or weather phenomena that rank above a threshold value near the upper or lower ends of the range of historical measurements in a given region. For example, classifications of extreme temperature events are location-specific, both in terms of the thresholds for temperatures as well as duration [19].

[3] Sea level rise can lead to gradual inundation of coastal infrastructure (e.g., substations, power plants), which can be heightened during storm surges and lead to flooding. Increased exposure to systems with high salinity water can also reduce infrastructure integrity [21].

[4] Co-occurring extreme events could both be weather-related (such as the case of droughts often preceding wildfires or ice events occurring during a coldwave) or a combination of a weather event and an adversarial attack (such as a cyber attack during an extreme weather event).

Table 12.1 Extreme chronic (A) and acute (B) weather-related threats and their associated physical (p) and cyber (c) impacts to power grid operations

Weather threat	Threats	Sources
A. Chronic		
Drought	Reduced water supply for cooling and hydropower generation (p). Increased risk of cyber attacks on water management systems (c)	[25–29]
Heatwaves & Coldwaves	Increased demand for cooling/heating, leading to grid stress (p). Thermal inefficiencies (p). Cyber vulnerabilities in demand response systems (c). Data overload affecting grid management systems (c)	[29–31]
B. Acute		
Flood	Damage to infrastructure (e.g., power lines, substations) (p). Increased potential for cyber attacks during recovery efforts (c). Compromised communication systems affecting emergency response (c)	[29, 32]
Wind	Damage to transmission lines and generation facilities (p). Increased risk of cyber vulnerabilities during high winds (c). Disruption of control systems for wind energy generation (c)	[29, 33, 34]
Hurricane	Severe damage to infrastructure (e.g., power lines, substations, generation facilities) (p). Flooding and wind damage leading to prolonged outages (p). Increased risk of cyber attacks during chaotic conditions (c). Disruption of communication and control systems affecting emergency response and recovery (c)	[29, 35]
Wildfire	Destruction of physical infrastructure (e.g., power poles) (p). Cybersecurity risks during emergency response (c). Disruption of communication networks for emergency services (c)	[29, 36, 37]
Ice	Ice accumulation leading to downed power lines (p). Vulnerability of grid control systems to cyber attacks during ice storms (c). Failure of monitoring systems due to physical damage (p,c)	[29, 38, 39]

and exacerbating the strain on available resources [25]. Similar issues are observed during wildfires, hurricanes, and other extreme events (see Table 12.1).

In addition to physical impacts, grid operations can also be disrupted due to lack of site access by operators [40] and due to failures in cyber components. Cyber failures can occur in multiple ways, including data overload on control and communication systems as well as damages to associated physical assets (e.g., communication lines) [41]. Additionally, compromised infrastructure is also vulnerable to malicious activity as evidenced by the preponderance of cyber attacks on the grid [29, 42, 43]. Malicious actors can attack remote monitoring systems through false data injection,

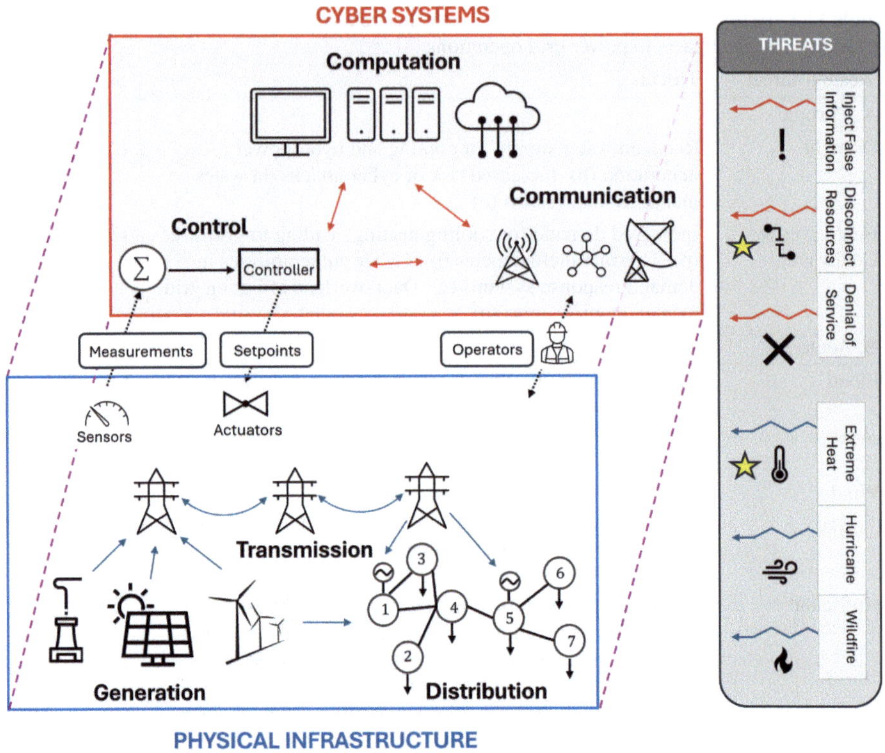

Fig. 12.2 Illustration of principal components of the power grid, including threat vectors from extreme weather events and cyber attacks. Starred items reflect the threats evaluated in the Texas case study

disconnecting resources,[5] and denial of services[6] [42] (Fig. 12.2). Concerns about cyber attacks are especially pronounced during extreme weather conditions, when adversaries could exploit the chaos and resource strain caused by extreme weather to launch attacks on critical infrastructure [46]. However, limited data on cyber attacks challenge our abilities to effectively understand current vulnerabilities and risks from these compounding threats [47, 48]. For example, false data injection is considered a relatively simple attack that can potentially degrade system performance or even cause cascading failures [49–51]. However, detection of false data injections (even large ones) remains a significant challenge [52, 53]. Thus, more attention is needed to understand and characterize resilience of the power grid during possible co-occurring extreme weather events and cyber attacks [54]. In the following section, we describe

[5] This occurs when a cyber attacker gains access to a process control network and can send out control actions that switch off lines, substations, or generators. This can cause massive outages and slow response time, as was seen in the 2015 cyber attack in Ukraine [44].

[6] This occurs when a cyber attacker locks out users from access to information, devices, and other network resources [45].

one opportunity we are pursuing to leverage multi-disciplinary metrics for resilience analyses for power grids facing both singular and compounding extreme weather threats.

12.3 Resilience Theory and Metrics

Resilience as a concept has gained prominence[7] for informing risk mitigation activities associated with the inevitable disruption of systems [59]. Derived from the Latin word “resilire" (which means “to spring back"), resilience of engineered systems (e.g., power grid) focuses on the ability of a system to return to an equilibrium or steady state after a disturbance [55, 60]. Three fundamental resilience characteristics have been identified that enable system stability and recovery: (1) robustness, (2) redundancy, and (3) resourcefulness (hereafter referred to as the “3 Rs of resilience”) [61, 62]. The “3Rs” of resilience focus on different aspects of the system to support its ability to withstand, operate through, or recover from disruptions (Table 12.2). For example, robustness typically incorporates use of continuous monitoring, modularity, and adaptive decision-making to ensure early detection and response to potential issues (continuous monitoring), allow for interchangeable units that can be independently modified or replaced in system design (modularity), and enable informed and flexible responses to changing conditions and emerging challenges (adaptive decision-making). In contrast, redundancy focuses on the depth (i.e., multiple backups that can take over if primary functions fail) and breadth (i.e., variations in responses to address different disturbances and challenges) of system components. Finally, resourcefulness emphasizes the capacity for self-organization and the attributes of creativity and innovation, which are linked to the development of novel solutions and adaptive strategies.

While research into the “3Rs” of resilience spans numerous disciplines, applying these principles to complex systems (i.e., systems of interconnected and interdependent components with emergent properties not easily predicted by analyzing individual components [63]), such as the power grid, can be quite challenging. These issues are further exacerbated in the context of extreme weather events that can concurrently stress multiple components of a complex system (Table 12.1). As noted previously, while empirical metrics (e.g., load shed) are relevant for understanding grid performance, we posit that expansion of resilience metrics is needed to more effectively characterize *both* outcomes of interest as well as contributing features to better support multiple resilience purposes, spanning evaluation to learning and improvement [64]. Since there is no standard approach for resilience applications in energy systems [65], we draw on various disciplines, which have advanced the

[7] The theory of resilience has been adapted by multiple research disciplines (including organizational behavior, social sciences, economics, and engineering) [55]. In engineering, resilience is commonly applied in fields such as supply chain management [56], infrastructure design [57], and disaster recovery [58], where quick recovery to maintain functionality is crucial.

Table 12.2 Resilience characteristics and associated resilience metrics

Characteristic	Description	Metric	Grid application
Robustness	Capacity of a system to absorb and withstand disturbances. Typically incorporates continuous monitoring, modularity, and adaptive decision-making	Entropy	Evaluate relative diversity of power grid components (e.g., generators, buses)
Redundancy	Availability of alternate components or pathways to maintain system functionality. Enhanced by the depth (quantity of system functions) and breadth (diversity of solutions and strategies)	Functional redundancy	Measure interchangeability through similarity of components of the power grid (e.g., connections between generators and lines)
Resourcefulness	Ability of a system to effectively respond to and recover from disturbances. Enhanced by incorporating the capacity for self-organization and the attributes of creativity and innovation	Global reaching centrality	Characterize relative importance of assets across the power grid network

application of resilience metrics in their domains, including information theory [66], biosciences [67], and complex systems sciences [68].

Drawing on this scholarship, we downselected three metrics that align with the 3Rs of resilience to support power grid analyses: entropy, functional redundancy (FR), and global reaching centrality (GRC) [69]. Specifically, entropy, which indicates the order within a system, can be used to quantify the relative diversity of components within the overall power grid (e.g., generators, buses), with higher entropy values indicating more diversity within a system [70]. In contrast, FR, which measures the interchangeability of components within a population, can be used to understand the similarity between two nodes on the grid (e.g., buses) with respect to their outgoing edges of the nodes (e.g., transmission lines), with higher values of FR indicating more redundancy within a system [71]. Finally, GRC, which measures the relative importance of different nodes within a network, can be used to characterize the capacity of a network to manage possible disruptions, wherein low values of GRC could indicate that all nodes are equally important (i.e., "central") of a graph [72]. A summary of these possible applications to the grid are captured in Table 12.2. Given the various ways that threats can manifest on a system, it should be noted that these metrics are not objective measures.[8] But rather, their utility lies in relative comparisons of the values across different scenarios and also enables an evaluation of potential tradeoffs between the resilience components (e.g., redundancy and diversity) [73]. To help demonstrate the utility of these metrics, we conduct a case study evaluating a power grid system performance under physical and cyber threats. More information about this case study and associated findings are captured in the following section.

12.4 Case Study

A common workflow for resilience assessments consists of identifying a system of interest, introducing a disruption, and then evaluating system outcomes to inform opportunities for improvement [74]. For this case study, the system of interest is a synthetic power grid that is loosely modeled after the power system in Texas, U.S. [75]. The Texas synthetic power grid (hereafter referred to as TX-SYN; available at [76]) contains thousands of assets (including 1,500 substations; 500 generators, 2,100 transmission lines; and 1,725 buses) with a total maximum generation capacity of 96,291 MW and total load of 80,396 MW [75].

Egret,[9] an optimal power flow model, was used to simulate grid operations during different conditions. The disruptions of interest are an extreme weather event (heatwave, which often affect central Texas) and a cyber attack (disconnection of services). These disruptions are evaluated both in isolation and as a compounding event (i.e., co-occurring heatwave and cyber attack). So, Egret executions were conducted for four scenarios: (1) a baseline scenario (i.e., no disruptions); (2) heatwave; (3) cyber attack; and (4) compounded threats (i.e., co-occurring heatwave and cyber attack).

System outcomes from the four scenarios were evaluated using both empirical metrics and multi-disciplinary, resilience metrics. The two empirical metrics involve load shed (both total and mean values per bus) in MW. The three multi-disciplinary metrics (entropy, FR, and GRC noted above) were implemented[10] on different parts of the grid system (Table 12.2). Translation of the multi-disciplinary metrics to the power grid requires re-envisioning of the structure of the TX-SYN as a graph network. In this context, buses can be thought of as nodes of the graph network while generators

[8] In other words, there is no fixed value of entropy, FR, or GRC at which a system is deemed resilient.

[9] Egret is a Python-based package for solving unit commitment and economic dispatch (i.e., optimal power flow) problems that can be downloaded from GitHub [77].

[10] These three metrics were tabulated using open-source Python package, pyRoCS, which can downloaded from GitHub [69].

and loads serve as attributes of the nodes, and transmission lines and transformers are attributes of the edges of the graph.

12.4.1 Scenario Details

For the baseline, the network was executed using default settings within the TX-SYN. In contrast, the heatwave scenario assumes a 1-hour heatwave that reaches 33° C,[11] which reflects the mean temperatures observed during four extreme heat events in central Texas [78, 79]. For this scenario, various physical assets are derated (based on values drawn from the literature), which demonstrates deficiencies in operations under this extreme event. The cyber attack scenario simulates an adversarial attack that deliberately disables parts of the grid. In this scenario, 10% of generators in central Texas (comprising about 3% of all generators on the grid) are randomly disabled. Finally, the compounded scenario considers both threats simultaneously: an intentional threat actor disabling physical grid components during a heatwave. In this compounded scenario, grid components are derated (as in the heatwave scenario) in conjunction with generators being randomly disabled in central Texas (as in the cyber attack) (Table 12.3).

12.4.2 Results and Implications

From an empirical perspective, the four scenarios highlight notable differences in the consequences associated with the disruptions. In particular, neither the baseline or cyber attack demonstrated any issues with load shed. The lack of load shed with the cyber attack could reflect high resource adequacy built into the TX-SYN. Significantly more load shed is observed, however, in the heatwave scenario than the baseline scenario. This indicates that the derating of components resulting from extreme heat created a generation-load imbalance, and nearly 13,000 MW of load was shed as a consequence. While an isolated cyber attack did not result in load shed, the presence of a cyber attack during the heatwave did result in additional load shed (3,042 MW greater than heatwave alone) (Table 12.4). These findings are similar to others who've identified the escalated impacts associated with co-occurring extreme weather events and cyber attacks [54].

Evaluation of the multidisciplinary metrics provides insights into some of the contributing features. For example, GRC, which measures the relative importance of nodes on the networks, remained unchanged across scenarios. This is an expected outcome, since the edges of the TX-SYN (i.e., lines and transformers) were not disconnected or disabled in any of the scenarios. However, changes were observed

[11] While it does not reflect the hottest conditions or total duration observed, this elevated temperature reflects repeat exposures that could stress system performance.

Table 12.3 Summary of assets impacted for the scenarios in the synthetic Texas grid

Scenario	Asset	Adjustment [Source]
Baseline	All	None, represents standard grid operations. The lines are loaded to no more than 85% capacity, with voltages between the 0.97 to 1.05 per-unit range [75].
Heatwave	Generators	1% decrease in generation for each °C above 16 °C [80]
	Transmission lines	For lines above 66 kV, 1.3% derate for each °C above 25 °C [81]
	Loads	Increase by 1.8% per °C above ambient temperature of 22 °C, using the formulation provided by [82]
	Transformers	0.7% derate for each °C above 30 °C [39]
	Buses	Same as baseline
Cyber attack	Generators	Randomly disable 10% of Zone 7 generators
	All except generators	Same as baseline
Compounded Threat	All	Heatwave and cyber attack adjustments combined

in the other metrics. For example, entropy, which was used to measure the relative diversity of grid components, varied across the scenarios. The baseline scenario had the highest entropy value (i.e., order), while the cyber attack scenario had the second most entropy, and the heatwave and compounded threats had the least entropy values (Table 12.4). This indicates that the extreme weather event had a much larger affect on the overall order of the system state than the cyber attack. In contrast, the cyber attack had a larger impact on the FR values than the heatwave (relative to the baseline). Both the entropy and FR values where lowest, however, under compounded threats, the scenario that demonstrated the highest load shed (Table 12.4).

These outcomes underscore that the multi-disciplinary metrics may be capturing important resilience features of the TX-SYN operations under different threats. Namely, load shed may be more correlated to reductions in robustness (i.e., entropy) than redundancy (i.e., FR). However, during compounded threats, both robustness and redundancy may need to be considered together to understand overall power grid performance. This indicates significant promise for continued evaluation of these multi-disciplinary metrics to serve as leading indicators of system resilience.

Table 12.4 Summary of load shed and multidisciplinary resilience metrics

Scenario	Total load shed (MW)	Mean load shed (per Bus)	Entropy	FR	GRC
Baseline	0.00	0.00	8.41	7.84e-4	4.63e-2
Heatwave	12,999	6.50	7.47	3.95e-3	4.63e-2
Cyber attack	0.00	0.00	8.05	3.91e-3	4.63e-2
Compounded threat	16,041	8.02	7.46	1.77e-3	4.63e-2

Entropy, FR, and GRC are constant for lines, loads, and transformers, and were thus omitted from the table

12.5 Discussion

The power grid is an important critical infrastructure that is increasingly subjected to extreme weather events. Evaluations of historical events noted that extreme weather events (e.g., storms) contribute to more than 50% of the large black-outs around the world [83]. Although various efforts[12] are currently underway to understand these interactions, significant gaps still persist in extreme weather assessments on the power grid. This chapter describes some of the challenges associated with effectively assessing resilience and presents one opportunity involving use of multi-disciplinary metrics to better characterize resilience features (i.e., "3Rs of resilience") that could be impacting system performance outcomes.

While this work introduced a couple of novel concepts, centered around use of multi-disciplinary resilience metrics to characterize compounding threats, there is much that still needs to be resolved. This includes further exploration of (1) multi-disciplinary metrics across different compounding events, (2) model representations of power grid operations, and (3) validation of simulation behaviors against real-world outcomes. The case study only evaluated a single type of extreme event at a relatively short timescale. However, further exploration is needed to evaluate the utility of the metrics under different types of extreme events (e.g., hurricanes or chronic threats, such as droughts) and compounded cyber attacks (e.g., denial of service or injection of false information) (Fig. 12.2). Longer event timelines also require exploration, especially since operators have a number of actions they can take to manage optimal power flow in the system [90], including connecting or disconnecting generation, loads [91], and lines [92, 93]. Evaluation of existing plans in place in the event of significant disruptions [94], such as local outages and even complete black-outs [95, 96], could also be evaluated under these longer time horizons, which may warrant additional metrics (e.g., characterizing causal complexity of the system) to be incorporated into analyses [69]. As the grid system representation and analysis

[12] For example, several models linking grid information and weather conditions have been developed to predict power outages and grid resilience under extreme conditions (e.g., [84–86] for hurricanes, [87] for tropical storms, and [88, 89] for high winds).

evolves, a metrics framework could also be introduced to facilitate implementation and analysis of multi-disciplinary quantitative measures.

Another area that requires further attention is better accounting for these different types of responses (across scales, actors, and processes), which may also require leveraging alternate modeling approaches [97]. Although optimal power flow models have been used to simulate and study the impact of extreme events on grid behaviors, such as graph analyses [98–100], hetero-functional graphs may provide a more comprehensive representation of a system's functionality by mapping individual processes into logically-feasible sequences [101]. Similarly, agent-based modeling (ABM) frameworks can facilitate emulation of real-time decision-making by allowing agents to respond to disruptions using the same monitoring tools and system actions as human operators [102]. For instance, during a physical disruption like a heatwave, agents can take action at various critical moments, such as when a weather report is issued or when there is an increase in system load. By integrating these feedback loops, these alternate approaches could better capture the system architecture across the built, social, and nature layers of the power grid. This enables us to analyze not only the "what" (e.g., load shedding, re-closing breakers, or false data injection) but also the "how" and "when" of actions that impact the performance of the power grid (Fig. 12.1).

Finally, it should be noted that the power grid is rapidly evolving. In particular, the speed at which system architectures are connecting components as well as strategies employed by operators is influencing our ability to understand the response of the overall system to these disruptions. Thus, new techniques (e.g., artificial intelligence (AI) and machine learning (ML))[13] are needed to support rapid data collection and validation of simulations. The application possibilities for these cutting edge tools span multiple aspects of power grids, including causal identification of power grid failures to generation of load forecasts and smart grid operations [104]. They can also be used to identify trends in system evolution through use of generative AI models that can synthesize information across unstructured datasets (e.g., media content). Such insights (e.g., increasing size and complexity of transmission and distribution systems due to rising electrification and demand) must be captured more comprehensively to ensure that the evaluated systems (in modeling and simulations) accurately reflect the size and complexity of the dynamic power grid across both space and time [2].

The increasing frequency of record-breaking numbers of extreme weather events motivates continued attention towards understanding how energy systems can withstand and recover from these events [105]. For example, the U.S. experienced 27 climate and weather-related events exceeding billion dollars in damages in 2024, which exceeded the annual average of 9 events observed from 1980 to 2024 [15]. This chapter demonstrates how resilience theory can be translated into practical measures (by leveraging multiple disciplines) to develop insights that can augment cur-

[13] In addition to evaluating large, interconnected grids, AI/ML tools can also be used to evaluate microgrids and storage solutions, which are increasingly being considered as an active solution for rapid response during extreme weather conditions [28, 103].

rent approaches for power grid assessments. Further maturation of these approaches and integration with other cutting-edge techniques (e.g., ABMs, AI/ML) can enable us to explore more complex interactions and assess the sensitivities of current systems (and potential adaptation strategies) under a range of realistic stressors. While the focus of this chapter is on extreme weather impact evaluations of power grids, implementation of multi-disciplinary resilience approaches can be expanded to other resources, infrastructures, and emerging technologies that are critical to the stability and security of our nations.

12.6 Conclusion

This chapter explores the relationships between the power grid and extreme weather events, including the potential for compounded risks from cyber attacks. It also underlines the importance of translating resilience theory into quantitative resilience metrics that can be used to improve characterization of compounding vulnerabilities within the electricity system. Continued attention for addressing resilience gaps, however, is needed to further refine the metrics used, capture relevant feedbacks in grid models, and validate implemented techniques. These advancements will be critical for enabling the power grid to be weather-adapted and resilient as it evolves to meet increasing demands of our societies.

Acknowledgements This work was supported by the Laboratory Directed Research and Development program at Sandia National Laboratories, a multimission laboratory managed and operated by National Technology and Engineering Solutions of Sandia LLC, a wholly owned subsidiary of Honeywell International Inc. for the U.S. Department of Energy's National Nuclear Security Administration under contract DE-NA0003525.

References

1. Bakke G (2016) The grid: the fraying wires between Americans and our energy future. Bloomsbury Publishing USA
2. National Academies of Sciences and Division on Engineering and Physical Sciences and Board on Energy and Environmental Systems and Committee on Enhancing the Resilience of the Nation's Electric Power Transmission and Distribution System (2017) Enhancing the resilience of the nation's electricity system. National Academies Press
3. Arghandeh R, von Meier A, Mehrmanesh L, Mili L (2016) On the definition of cyber-physical resilience in power systems. Renew Sustain Energy Rev 58:1060–1069
4. Federal Energy Regulatory Commission (FERC) (2025) Western energy markets explainer. https://www.ferc.gov/OPP/western-markets-explainer
5. Cohn JA (2017) The grid: biography of an American technology. MIT Press
6. Zhu J (2015) Optimization of power system operation. IEEE Press Series on Power Engineering, 2nd edn. Wiley
7. Aravinthan V, Balachandran T, Ben-Idris M, Fei W, Heidari-Kapourchali M, Hettiarachchige-Don A, Jiang JN, Lei H, Liu C-C, Mitra J, Ni M, Papic M, Parvania M, Sephary M, Singh

C, Srivastava A, Stefanov A, Sun H, Tindemans S (2018) Reliability modeling considerations for emerging cyber-physical power systems. In: 2018 IEEE international conference on probabilistic methods applied to power systems (PMAPS), pp 1–7
8. Bhusal N, Abdelmalak M, Kamruzzaman M, Benidris M (2020) Power system resilience: current practices, challenges, and future directions. IEEE Access 8:18064–18086
9. Watson WF (2017) NERC mandatory reliability standards: a 10-year assessment. Electricity J 30(2):9–14
10. Castillo A, Neill RPO (2013) Survey of approaches to solving ACOPF-optimal power flow paper 4, Tech rep. Federal Energy Regulatory Commission
11. Chen H, Pilong C, Rocha-Garrido P, Frogg D, Jayachandran M, Manno D, Sexauer J, Callaghan C, Dropkin R, Mulhern J et al (2022) Grid resilience with high renewable penetration: a PJM approach. IEEE Trans Sustain Energy 14(2):1169–1177
12. Denholm P (2024) Top 10 things to know about power grid reliability. https://www2.nrel.gov/news/program/2024/top-10-things-to-know-about-power-grid-reliability
13. Underwood A (2024) Reliability, resilience, and the power grid: what they mean, how they're different, and why communities need both. https://www.psehealthyenergy.org/reliability-resilience-and-the-power-grid/
14. U.S. Environmental Protection Agency (US EPA) (2022) Climate change impacts on energy. https://www.epa.gov/climateimpacts/climate-change-impacts-energy
15. Smith AB (2024) 2024: an active year of U.S. billion-dollar weather and climate disasters. http://www.climate.gov/news-features/blogs/beyond-data/2024-active-year-us-billion-dollar-weather-and-climate-disasters
16. Intergovernmental Panel On Climate Change IPCC (2022) Global warming of 1.5°C: IPCC special report on impacts of global warming of 1.5°C above pre-industrial levels in context of strengthening response to climate change, sustainable development, and efforts to eradicate poverty, 1st edn. Cambridge University Press
17. Owens C (2021) South Australia 2016 blackout. https://www.theblackoutreport.co.uk/2021/09/28/south-australia-blackout-2016/
18. Collins MT, Whiting M, Schellenberg JA, Schwartz LC (2025) Bridging the gap on data, metrics, and analyses for grid resilience to weather events: information that utilities can provide regulators, state energy offices, and other stakeholders, Tech rep. Lawrence Berkeley National Laboratory
19. Herring D (2025) What is an "extreme event"? is there evidence that global warming has caused or contributed to any particular extreme event?. http://www.climate.gov/news-features/climate-qa/what-extreme-event-there-evidence-global-warming-has-caused-or-contributed
20. Stephenson DB, Diaz H, Murnane R (2008) Definition, diagnosis, and origin of extreme weather and climate events. Climate Extremes Soc 340:11–23
21. Azevedo de Almeida B, Mostafavi A (2016) Resilience of infrastructure systems to sea-level rise in coastal areas: impacts, adaptation measures, and implementation challenges. Sustainability 8(11):1115
22. Intergovernmental Panel On Climate Change (IPCC). The Ocean and cryosphere in a changing climate: special report of the Intergovernmental Panel on Climate Change, 1st edn. Cambridge University Press
23. National Oceanic and Administrative Administration (NOAA) (2025) What is the polar vortex?. https://www.weather.gov/safety/cold-polar-vortex
24. North American Electric Reliability Corporation (2014) Polar vortex review, Tech rep. NERC
25. Voisin N, Dyreson A, Fu T, O'Connell M, Turner SWD, Zhou T, Macknick J (2020) Impact of climate change on water availability and its propagation through the western U.S. power grid. Appl Energy 276:115467
26. Kimmell TA, Veil JA. Impact of drought on U.S. steam electric power plant cooling water intakes and related water resource management issues, Tech Rep DOE/NETL-2009/1364, Argonne National Lab. (ANL), Argonne, IL
27. CYFIRMA (2024) U.S. water structure's vulnerability to cyber attacks. https://www.cyfirma.com/research/u-s-water-structures-vulnerability-to-cyber-attacks/

28. Thomson J, Hardin K, Sanborn S (2024) As extreme heat disrupts grids, renewables, distributed energy can help. https://deloitte.wsj.com/sustainable-business/as-extreme-heat-disrupts-grids-renewables-distributed-energy-can-help-d6d692b8
29. Paul B, Sarker A, Abhi SH, Das SK, Ali MF, Islam MM, Islam MR, Moyeen SI, Rahman Badal MF, Ahamed MH, Sarker SK, Das P, Hasan MM, Saqib N (2024) Potential smart grid vulnerabilities to cyber attacks: current threats and existing mitigation strategies. Heliyon 10(19):e37980
30. Mohd Zainuddin N, Abd Rahman MS, Ab Kadir MZA, Nik Ali NH, Ali Z, Osman M, Mansor M, Mohd Ariffin A, Syahmi Abd Rahman M, Nor SFM, Mohamad Nasir NAF (2020) Review of thermal stress and condition monitoring technologies for overhead transmission lines: issues and challenges. IEEE Access 8:120053–120081
31. Zuo J, Pullen S, Palmer J, Bennetts H, Chileshe N, Ma T (2025) Impacts of heat waves and corresponding measures: a review. J Cleaner Prod 92:1–12
32. Souto L, Yip J, Wu W-Y, Austgen B, Kutanoglu E, Hasenbein J, Yang Z-L, King CW, Santoso S (2022) Power system resilience to floods: modeling, impact assessment, and mid-term mitigation strategies. Int J Electrical Power Energy Syst 135:107545
33. Mohammadi Darestani Y, Shafieezadeh A, Cha K (2020) Effect of modelling complexities on extreme wind hazard performance of steel lattice transmission towers. Structure Infrastructure Eng 16(6):898–915
34. Li SH (2023) Impact of climate change on wind energy across North America under climate change scenario RCP8.5. Atmos Res 288:106722
35. Guikema SD, Nateghi R, Quiring SM, Staid A, Reilly AC, Gao M (2014) Predicting hurricane power outages to support storm response planning. IEEE Access 2:1364–1373
36. Choobineh M, Mohagheghi S (2016) Power grid vulnerability assessment against wildfires using probabilistic progression estimation model. In: 2016 IEEE power and energy society general meeting (PESGM), pp 1–5. ISSN: 1944-9933
37. Umunnakwe A, Parvania M, Nguyen H, Horel JD, Davis KR (2022) Data-driven spatio-temporal analysis of wildfire risk to power systems operation. IET Generation Transmission Distrib 16(13):2531–2546
38. Aquino J, Prado JCD, Nazaripouya H, Bertoletti AZ (2023) Enhancing power grid resilience against ice storms: State-of-the-art, challenges, needs, and opportunities. IEEE Access 11:60792–60806
39. Dumas M, Kc B, Cunliff CI (2019) Extreme weather and climate vulnerabilities of the electric grid: a summary of environmental sensitivity quantification methods, Tech Rep. ORNL/TM-2019/1252, 1558514, Oak Ridge National Laboratories
40. Tidwell VC, Gunda T, Gayoso N (2021) Plant-level characteristics could aid in the assessment of water-related threats to the electric power sector. Appl Energy 282:116161
41. Department of Homeland Security (DHS) (2024) Implications of extreme weather events on U.S. telecommunications infrastructure, Tech Rep. Analytic Exchange Program
42. Krause T, Ernst R, Klaer B, Hacker I, Henze M (2021) Cybersecurity in power grids: challenges and opportunities. Sensors 21(18):6225
43. Harris R, Molzahn DK (2024) Detecting and mitigating data integrity attacks on distributed algorithms for optimal power flow using machine learning. HICSS, pp 3170–3181
44. Lee RM, Assante MJ, Conway T (2016) Analysis of the cyber attack on the Ukrainian power grid | defense use case, Protocol TLP: White 1-29, Electricity Information Sharing and Analysis Center
45. Cybersecurity & Infrastructure Security Agency (CISA) (2021) Understanding denial-of-service attacks. https://www.cisa.gov/news-events/news/understanding-denial-service-attacks
46. Galliher C (2023) Power struggles: grid attacks and extremism in an age of climate volatility. https://www.lawfaremedia.org/article/power-struggle-grid-attacks-and-extremism-in-an-age-of-climate-volatility
47. Cremer F, Sheehan B, Fortmann M, Kia AN, Mullins M, Murphy F, Materne S (2022) Cyber risk and cybersecurity: a systematic review of data availability. The Geneva papers on risk and insurance. Issues Pract 47(3):698

48. National Academies of Sciences and Division on Engineering and Physical Sciences and Board on Energy and Environmental Systems and Committee on Enhancing the Resilience of the Nation's Electric Power Transmission and Distribution System (2025) Cybercrime classification and measurement. National Academies Press
49. Ouyang M, Fang Y (2017) A mathematical framework to optimize critical infrastructure resilience against intentional attacks. Comput-Aided Civil Infrastructure Eng 32(11):909–929
50. Zhang J, Sankar L (2016) Physical system consequences of unobservable state-and-topology cyber-physical attacks. IEEE Trans Smart Grid 7(4):2016–2025
51. Zhang J, Chu Z, Sankar L, Kosut O (2018) Can attackers with limited information exploit historical data to mount successful false data injection attacks on power systems? IEEE Trans Power Syst 33(5):4775–4786
52. Sahu A, Mao Z, Wlazlo P, Huang H, Davis K, Goulart A, Zonouz S (2021) Multi-source multi-domain data fusion for cyberattack detection in power systems. IEEE Access 9:119118–119138
53. Alkhraijah M, Harris R, Litchfield S, Huggins D, Molzahn DK (2023) Detecting shared data manipulation in distributed optimization algorithms, arXiv
54. Avraam C, Ceferino L, Dvorkin Y (2023) Operational and economy-wide impacts of compound cyber-attacks and extreme weather events on electric power networks. Appl Energy 349:121577
55. Hosseini S, Barker K, Ramirez-Marquez JE (2016) A review of definitions and measures of system resilience. Reliab Eng Syst Safety 145:47–61
56. Ivanov D (2018) Revealing interfaces of supply chain resilience and sustainability: a simulation study. Int J Prod Res 56(10):3507–3523
57. Shadabfar M, Mahsuli M, Zhang Y, Xue Y, Ayyub BM, Huang H, Medina RA (2022) Resilience-based design of infrastructure: review of models, methodologies, and computational tools. ASCE-ASME J Risk Uncertainty Eng Syst Part A: Civil Eng 8(1):03121004
58. Mayer B (2019) A review of the literature on community resilience and disaster recovery. Curr Environ Health Rep 6(3):167–173
59. Linkov I, Trump BD (2019) The science and practice of resilience. In: Systems and decisions. Springer International Publishing
60. Holling CS (1973) Resilience and stability of ecological systems. Ann Rev Ecol Syst 4:1–23
61. Wright C, Kiparoglou V, Williams M, Hilton J (2012) A framework for resilience thinking. Proc Comput Sci 8:45–52
62. World Economic Forum (2013) Global risks 2013. Insight report. World Economic Forum, 8th edn
63. Fraccascia L, Giannoccaro I, Albino V (2018) Resilience of complex systems: state of the art and directions for future research. Complexity 2018(1):3421529
64. Behn RD (2003) Why measure performance? Different purposes require different measures. Public Admin Rev 63(5):586–606
65. Jesse B-J, Heinrichs HU, Kuckshinrichs W (2019) Adapting the theory of resilience to energy systems: a review and outlook. Energy Sustain Soc 9:27
66. Shaqlaih A, White L, Zaman M (2013) Resilient modulus modeling with information theory approach. Int J Geomechanics 13(4):384–389
67. Mainali KP, Slud E, Singer MC, Fagan WF (2022) A better index for analysis of co-occurrence and similarity. Sci Adv 8(4):eabj9204
68. Naugle A, Verzi S, Lakkaraju K, Swiler L, Warrender C, Bernard M, Romero V (2023) Feedback density and causal complexity of simulation model structure. J Simul 17(3):229–239
69. Wachtel A, Gunda T, Caskey S, Cooper R, Womack T, Bonney K, Kliesner K (2025) pyRoCS: a python package to evaluate the resilience of complex systems. SoftwareX 29:101977
70. Shannon CE (1948) A mathematical theory of communication. Bell Syst Technical J 27(3):379–423

71. Ricotta C, de Bello F, Moretti M, Caccianiga M, Cerabolini BE, Pavoine S (2016) Measuring the functional redundancy of biological communities: a quantitative guide. Methods Ecol Evol 7(11):1386–1395
72. Lakkaraju K, Naugle AB, Verzi SJ, Swiler LP, Livesay M, Warrender CE, Bernard ML, Romero VJ (2019) Complexity metrics for agent based models of social systems, Tech rep. Sandia National Lab (SNL-NM), Albuquerque, NM (United States)
73. Brand FS, Jax K (2007) Focusing the meaning(s) of resilience: resilience as a descriptive concept and a boundary object. Ecol Soc 12(1)
74. Wachtel A, Jones KA, Baca MJ, O'Neill-Carrillo E, DeMenno MB (2020) Sandia's integrated methodology for energy and infrastructure resilience analysis, Tech rep. Sandia National Lab (SNL-NM), Albuquerque, NM (United States)
75. Birchfield AB, Xu T, Gegner KM, Shetye KS, Overbye TJ (2017) Grid structural characteristics as validation criteria for synthetic networks. IEEE Trans Power Syst 32:3258–3265
76. Texas A&M Engineering (2025) Electric grid test case repository: Texas2k series25 case, Tech rep. Texas A&M University
77. Knueven B, Laird C, Watson J-P, Bynum M, Castillo A (2019) Egret v. 0.1 (beta). https://doi.org/10.11578/dc.20190311.2
78. National Climatic Data Center, NESDIS, NOAA, U.S. Department of Commerce (2024) Storm events database, version 3.1. https://www.ncdc.noaa.gov/stormevents/
79. National Aeronautics and Space Administration (NASA) (2025) Modern-era retrospective analysis for research and applications, version 2. https://gmao.gsfc.nasa.gov/reanalysis/merra-2/
80. Sathaye J, Dale L, Larsen P, Fitts G, Koy K, Lewis S, Lucena A (2011) Estimating risk to California energy infrastructure from projected climate change, Tech rep. Lawrence Berkeley National Lab (LBNL), Berkeley, CA (United States)
81. Bartos M, Chester M, Johnson N, Gorman B, Eisenberg D, Linkov I, Bates M (2016) Impacts of rising air temperatures on electric transmission ampacity and peak electricity load in the United States. Environ Res Lett 11(11):114008
82. Allen MR, Fernandez SJ, Fu JS, Olama MM (2016) Impacts of climate change on sub-regional electricity demand and distribution in the southern United States. Nature Energy 1(8):1–9
83. Rahman KMJ, Munnee MM, Khan S (2016) Largest blackouts around the world: Trends and data analyses. In: 2016 IEEE international WIE conference on electrical and computer engineering (WIECON-ECE), pp 155–159
84. Johnson PM, Jackson ND, Baroud H, Staid A (2024) Can socio-economic indicators of vulnerability help predict spatial variations in the duration and severity of power outages due to tropical cyclones? Environ Res Lett 19(4):044048
85. Alemazkoor N, Rachunok B, Chavas DR, Staid A, Louhghalam A, Nateghi R, Tootkaboni M (2020) Hurricane-induced power outage risk under climate change is primarily driven by the uncertainty in projections of future hurricane frequency. Sci Rep 10(1):15270
86. Nateghi R, Guikema SD, Quiring SM (2014) Forecasting hurricane-induced power outage durations. Natural Hazards 74:1795–1811
87. Watson PL, Spaulding A, Koukoula M, Anagnostou E (2022) Improved quantitative prediction of power outages caused by extreme weather events. Weather Climate Extremes 37:100487
88. Panteli M, Mancarella P (2015) Modeling and evaluating the resilience of critical electrical power infrastructure to extreme weather events. IEEE Syst J 11(3):1733–1742
89. Mukherjee S, Nateghi R, Hastak M (2018) A multi-hazard approach to assess severe weather-induced major power outage risks in the US. Reliab Eng Syst Safety 175:283–305
90. Skolfield JK, Escobedo AR (2021) Operations research in optimal power flow: a guide to recent and emerging methodologies and applications. Eur J Oper Res
91. Hussain T, Alam S, Hansen TM, Suryanarayanan S (2021) A computationally improved heuristic algorithm for transmission switching using line flow thresholds for load shed reduction, Tech rep. Idaho National Lab
92. Brown WE, Moreno-Centeno E (2020) Transmission-line switching for load shed prevention via an accelerated linear programming approximation of AC power flows. IEEE Trans Power Syst 8950

93. Zhang Y, Bansal M, Escobedo AR (2020) Risk-neutral and risk-averse transmission switching for load shed recovery with uncertain renewable generation and demand. IET Generation Transmission Distrib 14(21):4936–4945
94. Xu M, Ouyang M, Hong L, Mao Z, Xu X (2022) Resilience-driven repair sequencing decision under uncertainty for critical infrastructure systems. Reliab Eng Syst Safety 221:108378
95. Patsakis G, Rajan D, Aravena I, Rios J, Oren S (2018) Optimal black start allocation for power system restoration. IEEE Trans Power Syst 33(6):6766–6776
96. Coffrin C, Van Hentenryck P (2015) Transmission system restoration with co-optimization of repairs, load pickups, and generation dispatch. Int J Electrical Power Energy Syst 72:144–154
97. Oliveira BM, Boumans R, Fath BD, Othoniel B, Liu W, Harari J (2022) Prototype of social-ecological system's resilience analysis using a dynamic index. Ecol Indicators 141:109113
98. Raschke M, Bilis E, Kröger W (2011) Vulnerability of the swiss electric power transmission grid against natural hazards. In: 11th international conference on applications of statistics and probability in civil engineering, pp 1407–1414
99. Jones CB, Bresloff CJ, Darbali-Zamora R (2023) Electric grid vulnerability analysis to identify communities prone to wildfires. IEEE Access 11:35630–35638
100. Schweikert AE, Nield L, Otto E, Deinert MR (2019) Resilience and critical power system infrastructure: lessons learned from natural disasters and future research needs. https://papers.ssrn.com/abstract=3430510
101. Thompson DJ, Schoonenberg WC, Farid AM (2020) A hetero-functional graph analysis of electric power system structural resilience. In: 2020 IEEE power & energy society innovative smart grid technologies conference (ISGT), pp 1–5. ISSN: 2472-8152
102. Ghaffari A, Hung F, Yang YCE, Lu J, Li X (2024) Development of a coupled agent-based generation expansion planning tool with a power dispatch model. Energy Climate Change 5:100156
103. U.S. Department of Energy (2025) Grid systems. https://www.energy.gov/oe/grid-systems
104. Khaleel M, Abulifa SA, Abulifa AA (2023) Artificial intelligent techniques for identifying the cause of disturbances in the power grid. Brilliance Res Artif Intell 3(1):19–31
105. Mohanty A, Ramasamy AK, Verayiah R, Bastia S, Dash SS, Cuce E, Khan TMY, Soudagar MEM (2024) Power system resilience and strategies for a sustainable infrastructure: a review. Alexandria Eng J 105:261–279

Chapter 13
From Misinformation to Climate Crisis: Navigating Vulnerabilities in the Cyber-Physical-Social Systems

Tooba Aamir, Marthie Grobler, and Giovanni Russello

Abstract Within the *cyber-physical-social-climate nexus*, all systems are deeply interdependent: cyber infrastructure facilitates communication, data processing, and automation across physical systems (such as power grids and networks), while social infrastructure provides the human capital and societal norms necessary for the system's functionality. Any disruption within any of these components, whether due to human error or system mismanagement, can propagate throughout the network, amplifying vulnerabilities and creating a significantly scaled impact. This chapter explores the critical role of human vulnerabilities within the cyber-physical-social-climate nexus, focusing on the interdependencies across cyber, physical, and social systems and how these interdependencies can scale in a climate context. While cyber and physical vulnerabilities are readily apparent, social vulnerabilities (such as misinformation, resistance to policy change, and lack of public awareness) often go unaddressed despite their profound impact on resilience and climate adaptation. Social infrastructure, including human capital, societal norms, and policy frameworks, shapes community responses and underpins adaptive capacity, yet it is also a significant point of failure when overlooked. This chapter examines how human cognitive biases, risk misperception, and decision-making silos within interconnected systems can lead to resource misallocation and weakened policy effectiveness. These factors are analyzed to demonstrate how inadequate responses across cyber-physical-social layers can cascade, amplifying climate-related risks. By addressing these human factors and aligning decision-making frameworks, we aim to strengthen resilience and foster cohesive adaptation strategies that account for the intricate interrelations of cyber-physical-social-climate systems.

Keywords Cyber risks · Climate threats · Resilience · Interconnectedness · Digital divide · Critical infrastructure · Cascading failures

---

T. Aamir (✉) · M. Grobler
CSIRO's Data61, Clayton, VIC, Australia
e-mail: tooba.aamir@data61.csiro.au

G. Russello
University of Auckland, Auckland, New Zealand

M. Huber et al. (eds.), *Critical Infrastructure Systems and Extreme Threat Nexus: Preparing for Resilience and Security*, NATO Science for Peace and Security Series C: Environmental Security, https://doi.org/10.1007/978-3-032-13170-6_13

13.1 Introduction

The cyber-physical-social-climate nexus underscores the intricate interplay of each domain, each integral to modern society's functionality and resilience. These domains are deeply interconnected, with *cyber* infrastructure enabling communication, automation, and data-driven decision-making across *physical* systems, while *social* infrastructure provides the human capital, societal norms, and regulatory frameworks that sustain their operation. The cyber-physical-social nexus underscores that the performance and resilience of any one system are heavily influenced by its interactions with the others [1]. Adding *climate*-related interactions to this already complex nexus brings about an array of new complications.

As the world faces the escalating impacts of extreme weather events, the importance of understanding the cyber-physical-social-climate nexus becomes increasingly evident. Rising temperatures and other climate-related hazards threaten critical infrastructure, increasing the risk of cascading failures across interconnected systems. In addition, the role humans play is of increasing concern. Although climate and weather are distinct, with climate referring to long-term atmospheric patterns and weather to short-term conditions, climate risks are scientifically linked to shifts in the frequency and intensity of certain weather extremes. Advances in event attribution science have strengthened the connection between human-caused climate risks and specific extreme events, such as the 2020 Australian wildfires and European heatwaves [2]. For instance, misinformation about energy policies or climate adaptation measures can delay necessary upgrades to critical infrastructure or erode public trust in mitigation strategies [3]. These challenges are compounded by the deliberate spread of disinformation by vested interests aiming to undermine climate action. Studies show that such disinformation campaigns not only skew public perception but also hinder effective policy implementation by amplifying doubts about the scientific consensus on climate risks [4]. Addressing these vulnerabilities requires a holistic approach that safeguards individual systems and interdependencies to maintain essential societal functions like energy supply and public safety.

Of particular interest is new risks tied to human factors, such as errors, cognitive biases, and decision-making silos, that are introduced with new technological advancements [5]. We specifically consider the escalating challenge posed by digitally disseminated climate misinformation and disinformation – a crucial vulnerability in today's interconnected world [6]. This information significantly skews public perception and obstructs effective policy implementation, posing widespread challenges that threaten the integrity and adaptability of interconnected systems. Such issues demand heightened attention from governments and international bodies, as they are central to safeguarding national and global security infrastructure against emerging threats. Addressing these vulnerabilities requires a multifaceted approach that targets enhanced digital literacy and situational awareness to improve the public's ability to respond to crises with discernment. For instance, initiatives like the United Nations (UN)' *Verified* campaign aim to combat the 'infodemic' of misinformation by disseminating reliable, fact-based content [7]. Furthermore, improving cross-sectoral coordination and integration of cyber, physical, and social strategies is essential for developing resilience against the compounded risks posed by misinformation in the face of climate risks.

Human involvement in the cyber-physical-social-climate nexus introduces several vulnerabilities, primarily due to limited situational awareness, prevalent cognitive biases, and the widespread amplification of misinformation. These vulnerabilities become particularly acute during natural disasters and crises, where cybercriminals can exploit them. For example, phishing attacks during such times often play on the urgency and fear surrounding an event, tricking the public into divulging sensitive information or donating to fraudulent causes[1] [8]. The aftermath of Hurricane Sandy saw a notable increase in such scams, with criminals taking advantage of the chaos and the outpouring of global support to launch various phishing schemes aimed at siphoning funds intended for disaster relief[2]. Another example was during the 2020 Australian bushfires, when misinformation regarding the cause of the fires spread rapidly through social media, misleading the public and placing additional strains on emergency response efforts. This led to misplaced public anger and hindered the efficient allocation of resources, demonstrating how false information can exacerbate the real-time management of disaster responses [9].

Similarly, during the COVID-19 pandemic, a myriad of conspiracy theories about the virus origins and vaccine safety proliferated, significantly impacting public health responses and vaccine uptake rates worldwide [10, 11]. Sentiment analysis of over 2 million tweets showed a 35% rise in negative sentiment linked to vaccine conspiracy hashtags (e.g., #Plandemic, #DoNotComply), resulting in a decline in public trust correlated with vaccination hesitancy in the US, UK, and Brazil [10, 11]. Overall, recent studies indicate a significant decline in public trust towards institutions handling climate risk initiatives. This erosion hampers effective communication and policy implementation and sentiment shifts after misinformation spikes (e.g., -25% trust drop post-#ArsonEmergency, +22% sentiment gain in UN *Verified* campaign) [7, 10, 12]. Therefore, social systems, including public perceptions, misinformation, and policy resistance, often serve as amplifiers of vulnerabilities across the nexus.

This chapter looks at the vulnerabilities that relate to human behavior in this cyber-physical-social-climate nexus, examining how human actions, decisions, and behaviors can compromise the integrity and adaptability of interconnected systems. It highlights how climate misinformation propagates through digital channels, distorts risk perception, fuels policy resistance, and compromises the effectiveness of critical infrastructure systems. Through empirical case studies and evidence-based analysis, the chapter develops an integrated set of recommendations that combines technological, policy, and community-based interventions to enhance system-wide resilience.

---

[1] https://www.everbridge.com/blog/double-trouble-when-climate-change-and-cyber-crime-collide/.

[2] https://www.forbes.com/sites/causeintegration/2015/10/05/charity-scams-put-the-disaster-in-disaster-relief/.

By tackling the root causes of misinformation and enhancing the capacity of critical infrastructures to withstand and adapt to these challenges, we can significantly improve the resilience of societies against the cascading effects of misinformation in the cyber-physical-social-climate nexus. The primary objectives of this chapter are to:

- Examine how climate misinformation and disinformation affect interdependent cyber, physical, and social systems, contributing to cascading risks and system-wide disruptions (Sects. 13.2 and 13.3).
- Assess the amplification of infrastructure vulnerabilities during natural disasters and emergencies due to communication failures and misleading information (Sect. 13.3).
- Evaluate current and emerging strategies, including governance initiatives, cybersecurity measures, and artificial intelligence (AI) tools, to counter misinformation and protect information integrity (Sects. 13.4 and 13.6).
- Explore actionable, cross-sectoral interventions to enhance public trust, bolster infrastructure resilience, and improve climate adaptation outcomes across the cyber-physical-social-climate nexus (Sects. 13.5 and 13.7).

By addressing these objectives, the chapter provides a comprehensive overview of the multifaceted challenges presented by misinformation within the cyber-physical-social-climate nexus, offering insights into how these challenges can be effectively managed and mitigated through coordinated efforts across multiple domains.

13.2 Misinformation and Disinformation: Definitions, Differences, and Systemic Impacts

Within the cyber-physical-social-climate nexus, social vulnerabilities represent a significant challenge that, when unaddressed, can amplify risks and create substantial barriers to effective adaptation and response. Social systems, encompassing societal norms, human capital, and policy frameworks, are crucial in supporting the functionality of both cyber and physical systems and enabling resilience.

Although misinformation and disinformation are both forms of false information, they differ in intent. *Misinformation* is shared unintentionally, while disinformation is deliberately constructed and disseminated with the intention of misleading. It often arises from misunderstandings or errors and spreads through digital platforms and social networks. Misinformation fosters confusion, scepticism, and resistance among stakeholders, which can hinder the effective operation of cyber and physical systems. This includes resistance to adopting secure systems or implementing critical updates, increasing vulnerability to cyberattacks [13]. *Disinformation*, on the other hand, is deliberately created and disseminated with the intent to deceive and manipulate public opinion or disrupt systems. It is a crafted falsehood designed to manipulate individuals or groups and often involves organized campaigns to destabilize trust in factual information and institutions [14]. This distinction is critical in framing our analysis. While disinformation is a growing threat, this chapter primarily focuses on misinformation and its cascading effects on trust, policy resistance, and system resilience, acknowledging disinformation where relevant.

13.2.1 Systemic Breakdowns: Impact on the Cyber-Physical-Social Nexus

Climate mis- and disinformation have a major impact on the cyber-physical-social nexus. It weakens the effectiveness and resilience of these interconnected systems and erodes societal trust in climate adaptation strategies, further complicating the challenges faced by these integrated networks.

In the *cyber* domain, misinformation about the security of system upgrades or new technologies can cause delays in adopting essential cybersecurity measures[3]. Such hesitation exposes cyber systems to increased risks of attacks, especially during critical periods such as natural disasters or significant public events, when robust defences are most needed. Cybersecurity frameworks such as the EU's Digital Services Act[4] and platforms such as Meta's SimSearchNet++[5] exemplify proactive technical measures that identify, mitigate, and prevent the amplification of false content, including through bot detection, content provenance tools, and systemic risk audits. Instances of misinformation undermine the overall security posture of cyber infrastructure, increasing its vulnerability to cyberattacks.

The impact extends into the *physical* domain, where misinformation about climate adaptation measures can lead to resistance. This resistance can result in project delays or inadequate implementations related to upgrading flood defences or installing renewable energy systems, which leaves communities more vulnerable to the exacerbated effects of climate risks, such as increased frequency and severity of extreme weather events. The consequences of misinformation are most profound in the *social* domain. Misinformation erodes public trust in scientific data and governmental agencies, which is crucial for collective climate action. It polarises public opinion and creates significant barriers to achieving a consensus for implementing large-scale climate resilience and adaptation strategies. False narratives, while sometimes intentionally created, often occupy a grey area between misinformation and disinformation. They may blend unintentional errors with deliberate manipulation, making their classification complex. This is particularly evident in contexts such as the COVID-19 pandemic, where both misinformed sharing and targeted disinformation coexisted [10, 11]. False narratives are particularly destructive in the climate context.

---

[3] https://www.weforum.org/stories/2021/10/the-next-decade-will-be-defined-by-climate-change-and-cyber-risks/.

[4] https://commission.europa.eu/strategy-and-policy/priorities-2019-2024/europe-fit-digital-age/digital-services-act.

[5] https://ai.meta.com/blog/heres-how-were-using-ai-to-help-detect-misinformation/.

Coordinated false narratives campaigns can drastically undermine public support for environmental policies by fostering scepticism about the reality of climate risks.

Beyond its immediate impacts on public perception and infrastructure readiness, climate misinformation also undermines broader systemic responses to climate risks. It contributes to policy resistance by fueling scepticism and division among stakeholders, delaying or obstructing necessary regulatory measures. Furthermore, by undermining scientific consensus, misinformation erodes trust in climate science and institutions, reducing the credibility of evidence-based adaptation and mitigation policies. It also distracts from viable climate solutions, redirecting attention and resources toward debunked theories or unproven alternatives. At an international level, misinformation disrupts global cooperation by fostering distrust among nations and complicating collective climate action. These effects illustrate how misinformation not only compromises public understanding but also weakens the institutional and diplomatic frameworks essential for coordinated climate resilience.

In the physical domain, misinformation can have harmful *short-term effects* during emergencies such as natural disasters [15]. For example, false information about evacuation routes, resource availability, or emergency protocols can delay responses, create panic, and worsen cascading failures in cyber and physical systems. A notable instance occurred during the 2022 floods in Lagos, Nigeria, where the government had to combat false narratives by leveraging social media and reputable media outlets to provide live updates and factual information, hold press conferences, release official statements, and engage with community leaders to ensure accurate information reached a broader audience [16]. Similarly, *in the long run*, misinformation about physical infrastructure projects, such as automated transportation systems or smart cities, can result in delays, resistance, or even rejection of essential upgrades, leaving physical systems susceptible to disruptions [17]. An example of this is the reluctance to adopt innovative grid technologies, making energy infrastructure vulnerable to inefficiencies and failures. In addition, distrust in cyber technologies, fuelled by misinformation, amplifies vulnerabilities across multiple domains, impeding the adoption of advanced cybersecurity measures and innovative infrastructure solutions.

From a social perspective, the effects of misinformation also unfold across temporal dimensions. *In the short term*, i.e., the response phase, spreading false information during emergencies, such as incorrect evacuation routes, erroneous safety guidelines, or fabricated resource shortages, can cause immediate confusion, panic, and misallocation of critical resources. For example, during the 2022 floods in Lagos, Nigeria, inaccurate information about evacuation zones delayed resident responses and diverted emergency services [16]. Evacuation was delayed between three and six hours in certain districts due to the circulation of false information about flood zones on WhatsApp and Facebook. Similarly, during the 2011 Fukushima Daiichi nuclear disaster in Japan, false claims about radioactive contamination triggered mass panic buying, resulting in shortages that impaired supply chains [18]. In contrast, *long-term effects* are often subtler but more corrosive. Misinformation that undermines trust in climate science or technologies (e.g., renewables) can result in sustained public scepticism, resistance to adaptation policies, and hesitancy in adopting resilience-enhancing infrastructure[6]. The 2024 Global Risks Report ranks misinformation and disinformation as the most significant short-term risk to human society, with extreme weather events as the top long-term risk, highlighting the critical impact of obscuring facts about climate risks [19].

Misinformation can also cause cascading failures across the cyber-physical-social nexus and exacerbate vulnerabilities in critical infrastructure, such as power grids, water systems, and healthcare facilities, particularly when these infrastructures are already under stress due to climate impacts. For instance, false information about an impending natural disaster can lead to unnecessary panic, resulting in premature and uncoordinated evacuations that strain resources and may hinder genuine emergency response efforts. Misinformation about resource availability or emergency procedures during climate hazards can also lead to mismanagement of resources, further destabilizing already vulnerable systems and potentially leading to catastrophic failures. Similarly, campaigns that target climate risks data can lead to scepticism about the severity of climate threats, reducing public support for essential policy measures and investments in resilience-enhancing infrastructure.

13.2.2 Misinformation and Policy Resistance

Misinformation fuels policy resistance when stakeholders (regardless of whether they are individuals, organizations, or governments) oppose regulatory changes aimed at enhancing resilience [20, 21]. The large-scale cross-country survey by International Monetary Fund (IMF) in 2023 shows that public support for climate policies increases by up to 7% when co-benefits are communicated, while cost-emphasizing messages reduce support by 9% [22]. This resistance stems from economic concerns, political ideologies, distrust in government or technology, and reluctance to shift from established practices.

Misinformation about the implications of new policies or technologies exacerbates this resistance, leading to delays or the complete blocking of necessary upgrades critical to ensuring the resilience of cyber and physical infrastructure [20]. For example, in 2022–24, fossil-fuel interests used denial and greenwashing campaigns to delay EU climate regulations and block renewable energy policies. These disinformation efforts directly slowed policy rollouts[7]. Similarly, fake narratives that misattribute disaster causes (e.g., blaming High-frequency Active Auroral Research Program (HAARP) for hurricanes) complicate attempts to implement evidence-based disaster policies, hindering resilience planning and systematic reforms[8]. Such policy

[6] https://www.theguardian.com/environment/2025/jun/19/climate-misinformation-turning-crisis-into-catastrophe-ipie-report.

[7] https://www.theguardian.com/environment/2025/jun/19/climate-misinformation-turning-crisis-into-catastrophe-ipie-report.

[8] https://disa.org/bloomberg-discusses-disaster-misinformation-with-uc-professor/.

resistance delays the implementation of necessary upgrades, thereby increasing vulnerability to future disasters. Over time, misinformation embedded within public discourse may institutionalise inaction or erode political will.

The spread of misinformation contributes to gaps in community awareness and trust, which are crucial for addressing vulnerabilities within the nexus [23]. Many individuals and organizations lack a comprehensive understanding of the risks associated with weak cybersecurity hygiene or the importance of proactive climate adaptation, leading to poor cybersecurity practices and inadequate preparation for extreme weather events. A pertinent example is the opposition to wind farm developments in New South Wales, Australia, where false claims about adverse health effects, environmental degradation, and economic inefficiency delayed several major projects. Misinformation disseminated via online forums contributed to local councils withdrawing support, illustrating how disinformation can stall renewable energy transitions [24, 25].

Enhanced community awareness and strengthened policy frameworks can counter the detrimental effects of misinformation and disinformation [25, 26]. Awareness and open discussions for community involvement are also pivotal; increasing digital literacy and raising awareness about climate adaptation prepare stakeholders to recognize and reject misinformation. Building a culture of trust and transparency in government and technology adoption can help mitigate resistance to necessary changes. Engaging communities in the policymaking process and providing clear, factual information about the benefits and challenges of new policies and technologies fosters a supportive environment for systemic resilience [27]. By addressing the misinformation that often underlies policy resistance and public unawareness, stakeholders across the cyber-physical-social-climate nexus can enhance their collective ability to withstand and adapt to emerging challenges, ensuring a more resilient infrastructure and a better-prepared society.

13.3 Impacts of Misinformation on Critical Infrastructure

Misinformation has become a significant threat to critical infrastructure, particularly during climate hazards and natural disasters. Specifically due to the broad scope of interconnected infrastructure sectors and the wide reach of associated essential services, any inaccurate information shared could have far-reaching impacts, particularly if the users do not take the time to verify and validate the information before sharing [28]. Not only does it undermine public trust and disrupt emergency responses, but it also amplifies vulnerabilities in essential systems such as power grids, water supplies, and healthcare, and unduly influences commodity consumption. When these essential systems are involved, people can very easily become riled up, causing false and misleading information to spread even further and faster. This can lead to increasing misperceptions and knowledge resistance [29], which pose significant threats due to increased commodity consumption and resultant cascading failures.

13.3.1 Misinformation During Natural Disasters

Rapid information dissemination during natural disasters is vital for managing public response and ensuring safety. While social media can spread lifesaving information, it is also a source of misinformation [30]. Misinformation during natural disasters has repeatedly demonstrated its potential to cause significant disruptions and harm, despite users believing that they are helping by sharing the information. Misinformation during a time in which people are particularly vulnerable and susceptible to external influences could cause public panic, hinder effective communication, and exacerbate the challenges faced by emergency responders [30]. Key global incidents provide stark examples of the chaos and confusion that can ensue when false information is disseminated. These examples underscore the critical need for accurate information during natural disasters and highlight the detrimental effects misinformation can have on disaster management and public safety.

A notable instance took place during the 2025 wildfires in Los Angeles County, US when a software glitch resulted in an evacuation alert being sent to nearly 10 million residents, far beyond the actual danger zone. Compounding the error, outdated alerts were rebroadcast when cell towers resumed service after being offline, leading to widespread panic, traffic congestion and public confusion, delaying real evacuation orders to those in need [31]. The chaos included injuries from rushed evacuations and clogged routes that significantly delayed emergency services from reaching those in actual need.[9] Although system communication failures, such as software glitches or erroneous alerts, are not intended to mislead and thus fall outside the strict definition of misinformation, they can still erode public trust in emergency systems. This erosion may increase susceptibility to actual misinformation, as individuals become desensitized and in the long-run may choose to ignore future legitimate alerts.

Another example occurred in 2024 when misinformation complicated emergency responses during Hurricanes Helene and Milton. Conspiracy theories circulated, falsely claiming the hurricanes were government-engineered or that aid was being selectively withheld from certain political groups. These allegations threatened the safety of federal aid workers, delayed relief efforts and exacerbated tensions in communities already suffering from the hurricanes' impacts[10]. Another instance of harmful misinformation was during the 2019/20 bushfires in Australia (refer to Case Study 1), where the *#ArsonEmergency* hashtag inaccurately attributed the widespread devastation to arson, diverting attention from the emergency response and public safety. In this instance, the authority's ability to manage the crisis was undermined because resources were misallocated based on incorrect assumptions. This disinformation severely disrupted emergency coordination efforts and contributed to a significant erosion of public trust in climate policies [32].

---

[9] https://smdp.com/news/after-false-alerts-panic-millions-officials-move-to-state-system-for-evacuation-notification/.

[10] https://www.bbc.com/news/articles/cx2lyzw7xwxo.

13.3.2 Vulnerabilities Amplified by Misinformation

Misinformation can significantly worsen vulnerabilities within critical infrastructure sectors, leading to delayed or inadequate responses to emerging challenges and crises. False information can cause confusion and disrupt critical operations due to unnecessary shutdowns or system overloads, or misinformation can be used specifically to exploit vulnerabilities, for example, a denial of service can occur after false information indicates a system vulnerability on a critical service.

Several examples demonstrate how misinformation can impact various critical infrastructures, underscoring the importance of accurate information for effective management and resilience. Studies such as the 2023 Edelman Trust Barometer [33] show a significant decline in institutional trust following major misinformation events. Likewise, the Notre Dame Global Adaptation Initiative (ND-GAIN) country Index[11] provides a benchmark for assessing climate adaptation readiness, which can be correlated with resilience to misinformation-induced delays. In Africa, misinformation about climate-related healthcare risks, such as those linked to malaria and malnutrition, undermines efforts to implement effective adaptive measures during extreme weather events. Public confusion about the causes and preventive strategies for these health crises significantly delays effective responses, compromizing public health outcomes during critical periods [34]. Misinformation also disrupts water management, as false narratives about water scarcity can lead to ineffective or misguided strategies during droughts and floods. In vulnerable regions, conflicting information about water usage policies causes delays in the implementation of sustainable water management solutions, exacerbating the effects of water scarcity on communities and ecosystems [34].

In Taiwan, the energy sector faced challenges from misinformation campaigns that fuel distrust in renewable energy projects and the reliability of power grids. Misleading claims exaggerating the limitations of solar power or promoting nuclear energy as a more viable alternative have undermined public confidence in energy transitions necessary for achieving climate resilience [35]. During the 2019 European heatwave, disinformation campaigns exaggerated the inefficacy of renewable energy sources, claiming they were failing to provide the necessary power, stirring public and political unrest (refer to Case Study 2) [36, 37]. Furthermore, cybersecurity risks are increasingly exacerbated by misinformation campaigns that target critical infrastructure systems, exploiting vulnerabilities in interconnected networks such as power grids and transportation systems. These misinformation campaigns can significantly delay responses to real threats or amplify damage during cybersecurity crises, affecting the security and efficiency of critical operational processes [17]. Consider the 2017 WannaCry ransomware attack, where misinformation fuelled fears, amplifying the impact of the attack and negatively impacting productivity because users were hesitant to switch on their computers for fear of being ransomed[12].

---

[11] https://gain.nd.edu/our-work/country-index/rankings/.

[12] https://www.cbsnews.com/news/wannacry-ransomware-attacks-wannacry-virus-losses/.

Case Study 1: 2019/2020 Australian wildfires

The 2019/2020 Australian wildfires season [9], or *Black Summer*, was one of the most devastating bushfire events in Australian history, burning over 18 million hectares, destroying thousands of homes, and causing significant loss of life and biodiversity. As the crisis unfolded, misinformation spread rapidly across digital platforms, with one of the most damaging falsehoods claiming that arson, rather than climate risks and hazards, was the primary cause of the fires. This misleading narrative was amplified by social media, some political groups, and climate sceptics, undermining scientific consensus that linked the severity of the fires to extreme climate conditions, including prolonged droughts and rising temperatures.

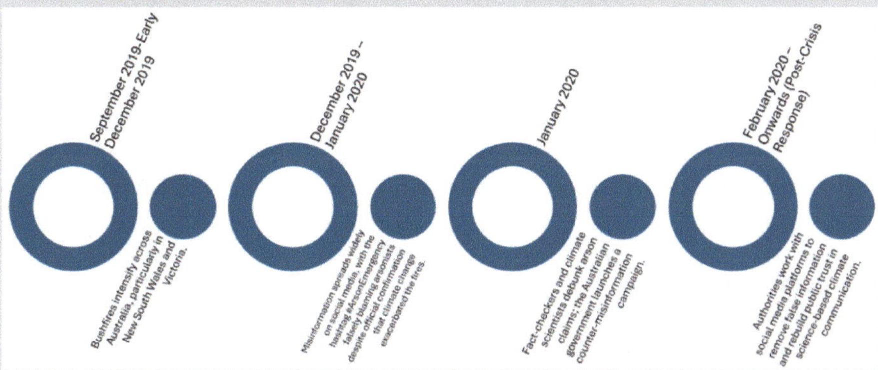

Misinformation had widespread effects across the cyber-physical-social-climate nexus, influencing public perception, emergency response efforts, and climate policy, e.g., adaptation measures such as prescribed burning practices, and early warning systems, rather than emissions mitigation strategies. The effects included:

Cyber dimension: Digital networks and the spread of misinformation

- **Social media disinformation** (*#ArsonEmergency*) falsely shifted blame from climate risk to arson, amplified by bot networks and foreign actors, fueling polarization and distrust in climate science.
- **Emergency response hindered** as authorities diverted resources to counter misinformation, while fake donation scams misled donors, diverting aid from real relief efforts.

Physical dimension: Infrastructure and environmental impacts

- **Misallocation of emergency resources**, delaying real responses. Distrust in evacuation orders caused confusion—some ignored warnings, others evacuated unnecessarily.
- **Weakened climate adaptation**, as misinformation reduced support for controlled burns and land reforms, contributing to long-term vulnerabilities.

Social dimension: Public perception, policy resistance, and trust

- **Erosion of trust in authorities** complicated crisis management, as misinformation fuelled climate scepticism and delayed adaptation policies.
- **Misinformation weaponized by interest groups**, particularly fossil fuel lobbyists, to downplay climate risks and stall policy reforms.

During the Australian wildfires, misinformation disrupted crisis response and delayed climate action by eroding trust in science and weakening support for preventive measures like controlled burns. This underscores the need for stronger public trust, clear communication, and proactive misinformation management.

Key takeaways

- **Misinformation strains crisis response** by diverting resources and increasing confusion.
- **Digital platforms amplify misinformation**, spreading both facts and falsehoods.
- **Public trust is vital** as misinformation undermines science, policy, and emergency response.
- **Delays in climate adaptation and policy reforms** result from persistent misinformation.
- **Proactive misinformation management is essential**, requiring coordinated efforts across governments, media, and scientific institutions.

Case Study 2: 2019 European heatwave

In 2019, Europe experienced one of its most intense heatwaves, with temperatures reaching record highs in countries such as France, Germany, and Spain [36, 37]. This extreme weather event put immense pressure on energy grids, increased health risks, and disrupted daily life. Disinformation campaigns emerged, falsely claiming that renewable energy sources were failing and incapable of meeting energy demands. As the heatwave intensified, disinformation campaigns spread through social media and certain media outlets, arguing that solar and wind energy were unreliable and responsible for power shortages. This was despite evidence that renewable energy sources continued to perform effectively, and the actual strain on the grid was caused by increased electricity demand for air conditioning and cooling systems.

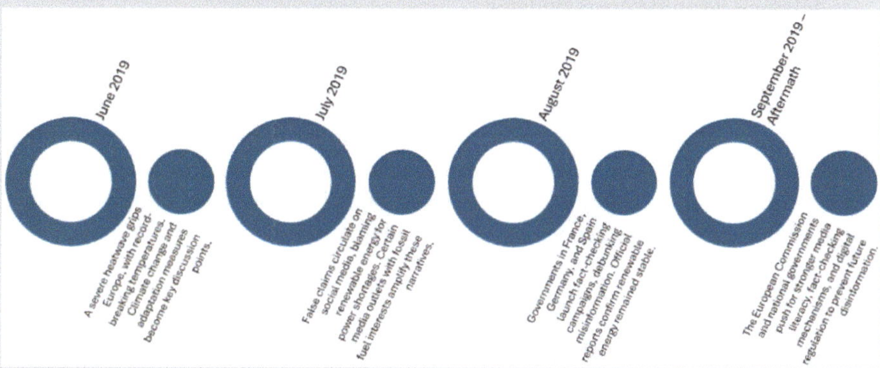

Misinformation had widespread effects across the cyber-physical-social-climate nexus, contributing to public distrust in renewable energy, exacerbated political divisions, and influenced policy discussions on energy security. The effects included:

Cyber dimension: Digital networks and the spread of misinformation

- **Social media disinformation:** False claims on Facebook, Twitter, and YouTube blamed renewable energy for power shortages. Fossil fuel interest groups and climate sceptics spread doubts about solar and wind reliability.
- **Energy communication disruptions:** False reports about electricity prioritization caused panic and erratic energy use.

Physical dimension: Infrastructure and energy system impacts

- **Grid strain:** Misleading claims led to increased energy use and overloading power networks.
- **Weakened climate adaptation:** Disinformation shifted blame from climate conditions to renewables, reducing support for sustainability, fueling policy debates and stalling green energy infrastructure.

Social dimension: Public perception, policy resistance, and trust

- **Trust erosion:** False narratives framed renewable energy policies as failures, lowering confidence in climate strategies.
- **Media distraction:** Fact-checkers focused on debunking false claims, delaying accurate crisis updates.

The case highlights cyber-enabled disinformation's role in undermining public confidence in climate-related infrastructure. The social dimension was particularly affected, as public mistrust fuelled political opposition to renewable energy projects. These disinformation narratives threatened to delay investment in critical energy transition policies, showing how false narratives can have long-term consequences on climate adaptation.

Key takeaways

- **Disinformation can shape political and economic decisions**, particularly in climate-related sectors.
- Social media plays a critical role in amplifying climate falsehoods, **requiring proactive counter-misinformation strategies**.
- **Government-led fact-checking initiatives** are essential in maintaining public trust in renewable energy transitions.
- **Improving media literacy** can help individuals critically assess false claims related to climate adaptation efforts.

13.3.3 Increased Commodity Consumption and Cascading Failure

Misinformation can fuel behavior that has unintended and often unforeseen consequences. Electric power utilities are a prime target of these commodity-driven misinformation campaigns due to their demand-response programs to reduce or shift electricity usage during peak periods in response to time-based rates or other financial incentives. Consider a scenario where a deceptive social media post is shared by a realistic looking fraudulent account. The message might incentivize increased electricity usage during peak times, citing a customer loyalty bonus. As the message spreads and more people attempt to specific electricity consumption, the likelihood of blackouts increases due to system overloads. This can lead to widespread impacts on public health and safety. If enough users in a specific region act on this misinformation, it can disrupt the power network [28]. Due to the recognized interdependency of most other critical infrastructure sectors on the energy sector [38], cascading infrastructure failures would follow with these interconnected systems particularly vulnerable to misinformation-induced disruptions. For example, false narratives about power grid reliability can lead to public panic and overloading of energy systems during extreme weather events.

A real-world example of misinformation leading to increased commodity consumption and resulting in cascading failures across critical infrastructure is the 2011 Fukushima, Japan disaster. Following the earthquake and tsunami, misinformation and rumours about radiation levels and food and water safety spread rapidly. This led to panic buying and hoarding of essential commodities like bottled water, food, and fuel. The increased demand led to widespread supply shortages and strained supply chains, leading to cascading failures in the distribution of these critical resources. The misinformation amplified the impact of the disaster, making it more challenging for authorities to manage the situation and ensure the availability of essential supplies to those in need [18].

13.4 Government Strategies to Counter Misinformation

Governments' responses are critical in combating misinformation that undermines public safety and security, particularly about critical infrastructure. Effective policy-making and governance are essential in addressing the challenges of misinformation, especially as they pertain to public awareness and resistance to necessary changes. The gaps in public awareness and widespread policy resistance pose significant obstacles to the resilience of the cyber-physical-social-climate nexus. These challenges are often exacerbated by misinformation that leads to public mistrust and hinders the adoption of necessary cybersecurity practices and climate adaptation strategies. In response, governments and policy-makers are tasked with devizing strategies that not only counter misinformation but also foster a more informed public and a more responsive infrastructural system.

Efforts to combat climate misinformation are gaining significant momentum globally as governments and intergovernmental organizations recognize accurate information's profound impact on societal resilience and critical infrastructure planning. To effectively combat the pervasive influence of misinformation, governments are deploying a range of strategic responses that aim to enhance public safety and bolster infrastructure resilience. These strategies include strengthening public awareness, increasing transparency, fostering collaborative governance, and leveraging technology and data [39].

13.4.1 Strategic Awareness and Governance Initiatives

Enhancing public awareness and fostering collaborative governance is pivotal. By investing in comprehensive awareness initiatives, governments can significantly improve public understanding of cybersecurity risks and the urgency of climate adaptation. Globally, several initiatives exemplify this approach.

For example, initiatives like the UN's *Verified* campaign, launched in May 2020, aimed to combat COVID-19 misinformation by disseminating accurate, science-based information [7]. Its primary mission was to disseminate credible, science-based content on science, solidarity, and solutions. It further supported the "digital first responders" to amplify UN-verified content across social media and messaging platforms. The campaign achieved substantial reach and engagement in its first three years, partnering with 250 organizations, collaborating with 200 influencers, and generating 1.4 billion engagements across social media platforms. It reportedly reached over 1 billion people globally by 2023. Panel testing of the *Verified Champions*' content in 2022 showed significant increases in knowledge levels and intentions to act on climate misinformation[13]. While explicit sentiment metrics reported by the UN are limited, insights can be drawn from the WHO's broader "infodemic management" frameworks, which highlight sentiment change as a key evaluation factor. The campaign also successfully recruited millions of volunteer "digital first responders", encompassing grassroots to national-level networks, including initiatives led by country-specific hubs.

The *Global Initiative for Information Integrity on Climate Change*, launched by UNESCO, the UN and Brazil, aims to strengthen global information integrity and combat climate-related mis- and disinformation [40]. This initiative includes a Global Fund dedicated to research on disinformation and supports campaigns designed to safeguard accurate climate information. Specific goals include exposing disinformation, strengthening global cooperation, and enhancing public communication campaigns. Similarly, the European Commission's *Communication on Managing Climate Risks* commits to combating climate disinformation by addressing its impacts on public trust, policy implementation, and international cooperation [41]. Australia's *National Climate Resilience and Adaptation Strategy* (2021–2025) emphasizes collaboration to improve public understanding of climate risks and counter misinformation as part of broader adaptation efforts [42]. These initiatives focus on developing accessible and engaging programs that provide clear, accurate information about the threats posed by misinformation. Such efforts are crucial for empowering citizens to

[13] https://www.purpose.com/case-study/verified/.

critically evaluate information and make informed decisions. UNESCO's *Global Initiative for Information Integrity on Climate Change* seeks to address climate-related disinformation by supporting research and communication campaigns. The initiative aims to produce evidence on the impact of disinformation and supports policies that protect information integrity [43].

Collaborative governance plays a crucial role in strengthening these efforts. This involves a multi-sectoral approach where governments collaborate with international bodies, private sectors, and non-governmental organizations [39]. Building public trust through transparent governance and regular, clear communication helps to demystify policy decisions and infrastructure developments. Governments providing regular updates and engaging directly with citizens through open forums and discussions help reduce the opportunities for misinformation to take root. The *Global Covenant of Mayors for Climate and Energy* exemplifies how cities can combat disinformation through transparency, community engagement, and partnerships with media outlets to debunk false narratives. Examples include Cape Town's real-time updates on water levels and Nairobi's transparent communication during development projects [16] and the G20 Summit Initiative, announced in 2024, involving countries like the UK, France, Chile, Morocco, Denmark, and Sweden working to address climate disinformation through shared strategies and resources [44]. The World Meteorological Organization also contributes by disseminating scientifically rigorous information and developing tools to enhance public understanding of climate science [45]. Intergovernmental security-focused programs such as the NATO Science for Peace and Security (SPS) Programme[14] also present strategic opportunities to strengthen information integrity and societal resilience against climate-related misinformation.

13.4.2 Technology and Data Integration

Leveraging technology and data is crucial in the fight against misinformation. Advanced technologies, such as data analytics and machine learning, enable governments to monitor trends in misinformation, identify sources of false information, and develop precise responses that neutralize misleading narratives before they spread widely. These technological tools are indispensable for swiftly countering misinformation in an increasingly connected world.

Multiple nations and organizations have called for holding social media platforms accountable for spreading climate misinformation, urging the adoption of universal definitions of climate disinformation and promoting information integrity [46]. Consider the monitoring systems at COP Conferences that actively track and counteract misinformation during global climate summits to ensure accurate narratives dominate discussions [47]. Complementing these efforts, projects like *defalsif-AI* are

[14] https://www.nato.int/cps/en/natohq/78209.htm.

developing advanced media-forensic tools to evaluate the credibility of online content. Focused on detecting politically motivated disinformation, *defalsif-AI* aims to safeguard democratic processes and public trust [13]. These tools utilize audio-visual media forensics, text analysis, and the fusion of these technologies with AI to identify and assess the authenticity of digital content.

To reinforce technological resilience against misinformation, a robust suite of cybersecurity tools is emerging to address the full spectrum of mis- and disinformation threats. These tools span legal, technical, operational, and collaborative domains, enabling not only the detection of harmful content but also the protection of information integrity in communication infrastructures. Integrating these mechanisms into national and organizational resilience strategies is essential to safeguard public discourse and critical infrastructure. These tools exemplify the intersection of cybersecurity and misinformation response, providing both reactive and proactive mechanisms to detect, disrupt, and deter false information in the digital landscape. Key cybersecurity tools to counter misinformation include:

- **Digital Services Act (DSA)** – An EU legislative framework that mandates large online platforms to identify, assess, and mitigate systemic misinformation risks, with enforceable accountability measures[15].
- **Cyabra** – An AI-powered platform that collaboratively detects fake accounts, bot networks, and coordinated disinformation campaigns in real time[16] to enhance situational awareness and coordinated responses.
- **Fighting Disinformation Online** – A database of open-source web tools like factcheck.org[17], that uses detection algorithms to identify hashtags, links, accounts, and media being amplified by likely bots in a coordinated fashion, helping researchers and journalists track emerging disinformation campaigns[18].
- **SimSearchNet++ (Meta/Facebook)** – An advanced AI model that enhances image matching to detect variations and manipulations in images, such as cropping or blurring, which are common in misinformation campaigns. It also uses optical character recognition to group and analyze text-based images[19].

Such multifaceted strategies of leveraging technology and data underscore a proactive and integrated approach by governments to mitigate the impacts of misinformation. Through education, policy enforcement, transparency, collaboration, and technological innovation, governments are better equipped to safeguard public safety and enhance the resilience of critical infrastructure against the challenges posed by misinformation.

---

[15] https://commission.europa.eu/strategy-and-policy/priorities-2019-2024/europe-fit-digital-age/digital-services-act.

[16] https://cyabra.com/.

[17] https://www.factcheck.org/.

[18] https://www.rand.org/research/projects/truth-decay/fighting-disinformation/search.html.

[19] https://ai.meta.com/blog/heres-how-were-using-ai-to-help-detect-misinformation/.

13.5 Leveraging Social Infrastructure for Community Resilience

To effectively counter misinformation within critical infrastructure sectors, social infrastructure – including the extended network of professionals involved in maintaining the cyber, physical, and social facets of these systems – can be leveraged. This broad community encompasses various stakeholders like utility workers, IT professionals, emergency responders, and local government officials, all pivotal in driving resilience efforts across different sectors. It is imperative that mechanisms are put in place to combat misinformation effectively to ensure that during times of crisis, truthful and helpful information prevails, aiding in efficient and safe disaster responses. There are many social infrastructure and community engagements that can be employed in this regard.

13.5.1 Integrating Human Factors Into Resilience Planning

Resilience within the cyber-physical-social-climate nexus is heavily influenced by human factors, particularly those related to misinformation, communication failures, and lack of preparedness. Poor information-sharing, policy misalignment, and inadequate risk awareness can significantly undermine the ability of critical infrastructure systems to withstand and recover from disruptions.

Failures in governance, policy misalignment, and institutional gaps have significantly impacted the resilience of critical infrastructures in past crises. One example is the 2021 ransomware attack on the Colonial Pipeline in the US, which disrupted fuel supply across the East Coast. The attack exposed weaknesses in cybersecurity policies for privately operated infrastructure, where existing guidelines were merely advisory rather than mandatory, resulting in inconsistent security measures across sectors[20]. This incident highlights the necessity for enforceable cybersecurity policies that standardize best practices across all critical infrastructure systems.

The integration of cyber, physical, and social dimensions is essential to addressing these vulnerabilities. Decision-makers must adopt strategies that enhance preparedness, improve coordination, and boost awareness to manage the interconnected risks posed by cyber threats, physical disruptions, and climate risks [48, 49]. For instance, research on mapping social networks for climate adaptation in Shoalhaven, Australia, has revealed the importance of key nodes in disseminating information effectively [50]. Such findings emphasize the necessity of robust coordination among local actors to mitigate misinformation risks, infrastructure failures, and climate-related threats [51].

---

[20] https://www.cisa.gov/news-events/news/attack-colonial-pipeline-what-weve-learned-what-weve-done-over-past-two-years.

Climate-induced risks also demonstrate the need to integrate human factors into resilience planning. For example, extreme flooding events have repeatedly overwhelmed urban drainage systems due to outdated flood management policies. Many cities lack the necessary infrastructure upgrades suggested by evolving climate models, leaving them vulnerable to extensive damage and economic loss. The absence of proactive climate resilience planning underscores the risks of inadequate adaptation measures. Similarly, the 2019/2020 Australian wildfire crisis showcased how poor coordination and communication between various government agencies exacerbated disaster response efforts. Delays in information-sharing and unclear jurisdictional responsibilities led to inefficient resource allocation and hindered evacuation efforts [9]. These cases illustrate how human vulnerabilities – whether due to policy gaps, communication failures, or lack of preparedness – can result in cascading failures that jeopardize public safety and infrastructure integrity. Transparent communication about the benefits of cybersecurity measures and climate adaptation strategies is essential to building public trust and overcoming resistance. Organizations can foster confidence by openly sharing how these measures mitigate risks and protect critical infrastructure, particularly in the face of climate-induced disruptions [52, 53].

Addressing human vulnerabilities in resilience planning requires a *multi-faceted approach* that strengthens policies, enhances communication, and fosters systemic preparedness. Implementing *enforceable regulations* is essential to standardizing cybersecurity measures and climate adaptation strategies, ensuring that misinformation does not hinder critical decision-making. Additionally, *improving coordination and communication* between government agencies, infrastructure operators, and local stakeholders can help prevent disaster mismanagement and mitigate misinformation-driven delays. Strengthening interagency collaboration and establishing clear communication channels ensures that accurate information reaches the public and decision-makers in a timely manner.

Beyond policy and coordination, fostering systemic resilience is crucial to equipping governments, infrastructure operators, and communities with the necessary *risk mitigation frameworks, training programs, and crisis response mechanisms.* Proactive adaptation to emerging threats, combined with improved public education and institutional preparedness, enhances infrastructure resilience and community decision-making, along with organizational-level integrated security planning, where IT teams and senior management collaborate, can significantly reduce financial and operational impacts[21]. Furthermore, businesses that incorporate climate risks into their cybersecurity frameworks demonstrate a commitment to resilience, thereby strengthening stakeholder confidence and setting a precedent for sustainable operations[22]. By integrating these strategies, stakeholders can *minimize vulnerabilities, counter misinformation, and build a more resilient system* capable of withstanding and recovering from evolving risks.

[21] https://www.rutherfordsearch.com/blog/2021/09.

[22] https://www.earth-scan.com/blog/climate-risk.

13.5.2 Strengthening Social Behavior and Collective Action for Resilience

Governments and institutions are instrumental in building trust and encouraging participation through inclusive decision-making processes. By promoting public awareness campaigns and facilitating partnerships between sectors, these bodies help to reinforce societal norms that promote cooperation and shared responsibility. This collective action at all levels—individual, community, corporate, and governmental—ensures that resilience strategies are well-designed and effectively implemented. For example, community workshops on renewable energy or waste reduction have been shown to enhance collective action and strengthen societal norms around sustainability [54].

Community-driven solutions and local leadership are crucial in disseminating accurate information and educating the public. Local leaders serve as trusted sources of information, making them key players in countering misinformation and enhancing community resilience. For example, in Namibia, agriculture sector networks have effectively utilized stakeholder collaboration to disseminate accurate climate information, showcasing the power of local initiatives in enhancing resilience against misinformation [55]. Similarly, community workshops focusing on sector-specific issues like renewable energy or waste management can amplify societal norms around sustainability and resilience, further enhancing community engagement. The integration of community engagement and social infrastructure into resilience strategies is indispensable. By harnessing the collective power of communities and leveraging robust educational frameworks, societies can enhance their ability to combat misinformation effectively. This approach strengthens community resilience and supports the broader efforts to safeguard critical infrastructure from the myriad threats posed by misinformation and other emerging challenges.

Public education initiatives, such as cybersecurity training programs and climate preparedness workshops, are pivotal in equipping individuals and organizations to respond to evolving threats. These programs help professionals understand how interconnected risks, such as compromized digital infrastructure or manipulated climate data, can undermine mitigation efforts. Workplaces and public institutions are particularly critical in embedding these practices into daily routines. For example, regular incident response drills and cybersecurity awareness campaigns have been implemented in the ISO 27001 framework, ensuring that organizations remain prepared for cyber threats while maintaining operational continuity during climate-related disruptions [56]. By fostering a culture of awareness and resilience through these measures, organizations can effectively safeguard both their digital assets and critical infrastructure from emerging challenges. In addition, by integrating cybersecurity considerations into climate policies and encouraging collaboration between government agencies, businesses, and communities, organizations can foster a more resilient ecosystem that is resistant to misinformation.

Finally, individuals should be encouraged to build their own social identity (i.e., perceive oneself as belonging to social groups), collective efficacy (i.e., people's shared beliefs about their group's capability to accomplish collective tasks), and positive deviance (i.e., a behavior-change approach that deepens the successful actions of existing community members) while working towards these bigger social connections [57]. Collectively, these government, community and individual initiatives can strengthen social behaviors that will make individuals within those communities more resistant against misinformation.

13.5.3 Addressing Policy and Institutional Gaps for Enhanced Resilience

To enhance the resilience of the cyber-physical-social-climate nexus effectively, a comprehensive and coordinated approach is essential to address existing policy and institutional gaps. This involves modernizing regulations to reflect the evolving threat landscape, incorporating clear enforcement mechanisms for operators of critical infrastructure, and integrating long-term climate resilience measures into infrastructure planning. Policies need to be comprehensive, enforceable, and adaptable, ensuring new projects are designed with climate risk assessments and appropriate adaptation strategies [58, 59]. Furthermore, enhancing cross-sector coordination is crucial to break down silos and foster collaboration among various sectors to align policies across cyber, physical, and social domains enhances system-wide resilience [60].

Establishing robust accountability and compliance mechanisms is also vital to ensure that resilience measures are more than just recommendations – they must be mandatory and effectively enforced [52, 58]. By implementing stricter monitoring and compliance protocols, regulatory agencies can guarantee that resilience practices are adhered to, significantly reducing the risks of system failures. Collectively, these strategies ensure that cyber, physical, and social systems are not only prepared to handle current threats but are also equipped to deal with future challenges. Addressing these policy and institutional gaps allows governments and stakeholders to build a more resilient infrastructure framework, ready to prevent and withstand cascading failures that could arise from both anticipated and unforeseen challenges [39, 48].

Improving decision-making within the cyber-physical-social-climate nexus necessitates a sophisticated and nuanced approach to managing the growing interdependencies between these systems, particularly as the frequency of climate-related disruptions increasingly challenges them. Decision-makers must shift away from traditional static risk assessments towards dynamic, data-driven strategies that utilize advanced decision-support tools and predictive analytics. Such tools enable real-time risk assessments that account for cyber vulnerabilities, climate risks, and infrastructural interdependencies. For example, AI-driven tools allow scenario-based modeling that helps visualize potential cascading failures, offering decision-makers the ability to test and refine response strategies proactively [48, 59].

Furthermore, breaking down organizational silos through enhanced cross-sector collaboration is essential for effective decision-making. Integrating efforts across cybersecurity, physical infrastructure management, and climate adaptation through public-private partnerships and interdisciplinary task forces ensures a cohesive approach to risk assessment and management [60]. Training programs and scenario-based exercises are crucial in mitigating cognitive biases and enhancing decision-makers' ability to handle complex, multi-sector disruptions. Additionally, adopting flexible governance models and integrating decision-making with real-time data significantly enhance the responsiveness to emerging threats, ensuring that infrastructure systems are robust and resilient against both current and future challenges [48, 59].

By closing institutional and policy gaps, while embedding resilience in both governance frameworks and public consciousness, stakeholders can ensure long-term adaptability. Strengthening public awareness and integrating cybersecurity and climate preparedness into routine practices are essential components of a proactive resilience strategy.

13.6 Technological Solutions and Innovations to Fight Against Misinformation

In the ongoing battle against misinformation, technological solutions and innovations play a crucial role. By leveraging advancements in AI, machine learning (ML), and cybersecurity, organizations and governments can significantly enhance their ability to identify, analyze, and counteract misinformation. These technologies are not only essential for maintaining the integrity of information but also for protecting digital platforms from the pernicious effects of misinformation.

13.6.1 AI and ML in Countering Misinformation

AI and ML are leading the charge in technological interventions against misinformation, utilizing their capability to process vast amounts of data with unparalleled speed and accuracy. These AI and ML tools aid in detecting patterns of misinformation, tracking narratives, and supporting informed decision-making, although they require human oversight and continuous retraining [61]. These tools enable the identification of fake news articles, and manipulated images and videos.

A notable application within this domain is natural language processing (NLP), which empowers AI systems to scrutinize news articles, social media posts, and other digital content for misleading cues or factually incorrect statements by analyzing human language [62]. An example is the use of AI by researchers in Australia and the UK to combat climate misinformation. They are developing models that detect and debunk false climate claims on social media, using data from sources known for spreading misinformation, such as denial blogs, biased opinion sites, and politicized media outlets. These models are trained for real-time fact-checking, providing a proactive tool in the fight against misinformation [63, 64]. Another key innovation is the Augmented Computer Assisted Recognition of Denial and Scepticism (CARDS) model, a sophisticated hierarchical machine learning tool. It specifically targets contrarian climate claims on Twitter, analyzing large datasets to identify misinformation patterns and categorize claims for automated responses. These AI technologies are adept at tracking the sources and spread of misinformation, which equips analysts with detailed insights necessary for targeted interventions [64].

In a related advancement, a techno-cognitive approach combines topic modeling with network analysis to detect specific contrarian claims about climate hazards. This method illuminates the associations between these claims and corporate funding, shedding light on the strategic dissemination of misinformation [65]. The adaptability of machine learning models is also crucial as they continuously learn from new data and evolve to counteract the changing tactics of those spreading misinformation. This ongoing adaptation is vital for maintaining an effective defence against the dynamic nature of misinformation threats.

13.6.2 Cybersecurity Measures for Information Integrity

Cybersecurity measures play a pivotal role in safeguarding information integrity across digital platforms, particularly vital in an era where misinformation frequently exploits vulnerabilities in information systems. Establishing robust cybersecurity protocols is essential for protecting data integrity, preventing unauthorized access, and ensuring that information remains untampered with and reliable across dissemination channels. Key to these efforts are encryption technologies, which secure data both at rest and in transit, preventing unauthorized interception and alteration. Additionally, robust authentication processes are crucial, as they prevent unauthorized access to systems where sensitive information is stored, thereby reducing the risk of internal leaks or the spread of misinformation.

Internationally, the UN is leading initiatives such as the proposed *Code of Conduct*, which aims to protect information integrity on digital platforms by addressing false narratives, coordinated inauthentic behavior, and harmful content while safeguarding freedom of expression [66]. Similarly, Research ICT Africa's *Information Integrity Programme* delves into the impacts of digital misinformation and explores regulatory solutions to ensure transparency, accountability, and trustworthiness in digital environments. This programme highlights the essential role of cybersecurity in maintaining credible information ecosystems[23]. The UN Secretary-General issued a policy brief on information integrity, which underscores the need for coordinated international action to address threats to information integrity, including the implementation of cybersecurity measures to guard against online hate speech,

[23] https://researchictafrica.net/programme/information-integrity/.

misinformation, and harmful content. These measures are critical in enhancing the reliability of digital communications [52].

Moreover, cybersecurity is fundamental in preventing data manipulation and securing digital platforms from coordinated attacks, ensuring that automated tools such as AI models used for fact-checking operate effectively without interference [61]. To support these defences, cybersecurity teams must employ active monitoring systems capable of detecting and responding to threats in real time. These systems help identify unusual activities that could indicate an information breach or attempts to inject misinformation. Once a potential threat is detected, immediate countermeasures are deployed to isolate and neutralize the threat, thereby minimizing potential damage and maintaining the integrity of information systems.

While emerging cybersecurity frameworks (e.g., DSA, SimSearchNet++)[24,25] offer promising avenues for combating misinformation, practical challenges persist in their global deployment. These include enforcement limitations, varying platform compliance, and technological gaps in detecting sophisticated disinformation such as deepfakes. Their efficacy varies based on institutional readiness, platform cooperation, and technological sophistication. Policy enforcement, interoperability, and public adoption remain significant barriers.

13.7 Lessons Learned and Best Practice Recommendations

Managing human error, particularly misinformation susceptibility, in the cyber-physical-social-climate nexus requires a multifaceted approach. The potential for widespread disruption is significant, and therefore addressing the challenges posed by misinformation in critical areas requires a concerted effort to enhance public awareness, promote accurate information dissemination, and implement robust countermeasures against misinformation. By ensuring that truthful and helpful information prevails, especially in scenarios involving critical infrastructure, stakeholders can better safeguard these essential systems against the disruptive impacts of misinformation, thereby enhancing overall resilience and security.

13.7.1 Lessons Learned

The following lessons highlight the interplay between human error, systemic interdependencies, and cascading failures, offering insights into the challenges faced in managing misinformation effectively.

[24] https://commission.europa.eu/strategy-and-policy/priorities-2019-2024/europe-fit-digital-age/digital-services-act.

[25] https://ai.meta.com/blog/heres-how-were-using-ai-to-help-detect-misinformation/.

Human error and susceptibility as a major factor

Misinformation thrives on cognitive biases, limited digital literacy, and emotional responses, making individuals particularly susceptible during crises. The rapid dissemination of false narratives often leads to misguided decision-making, delayed response efforts, and reduced trust in official sources. The two presented case studies illustrate how false claims about arson and renewable energy failures contributed to widespread confusion, demonstrating that misinformation can exploit human cognitive vulnerabilities, even when factual information is readily available.

Interdependency between human and systemic vulnerabilities

Misinformation does not operate in isolation; it interacts with infrastructure, governance, and digital platforms, leading to widespread consequences. As seen with the Australian wildfires and European heatwaves, misleading claims did not just alter public opinion but disrupted emergency communications, strained resources, and fueled political polarization. This underscores the need for holistic countermeasures that integrate human behavioral understanding with systemic resilience strategies.

System interdependencies and cascading failures

False information can trigger a chain reaction of failures across interconnected systems, intensifying the scale of disruption. Misinformation in the cyber domain affects social perceptions, which in turn influence policy decisions and physical infrastructure resilience. The failure to curb disinformation about energy grid stability during the heatwave contributed to panic consumption, further destabilizing an already strained system. Similarly, misinformation about fire origins in Australia redirected resources away from critical climate adaptation efforts, delaying long-term resilience-building strategies. The highlights that misinformation exacerbates crises by distorting decision-making, undermining policy, and eroding trust. Addressing it requires understanding human biases, systemic interdependencies, and cascading failures. These insights are crucial for developing targeted interventions to enhance resilience and prevent future disruptions.

13.7.2 Recommendations

Building on the lessons learned, the following recommendations aim to address misinformation challenges within the cyber-physical-social-climate nexus. Each recommendation is aligned with a key lesson and includes actionable best practices to mitigate misinformation risks and enhance systemic resilience.

Addressing human susceptibility to misinformation by enhancing digital literacy and public awareness

Human error and cognitive biases play a significant role in the spread of misinformation, particularly in high-stress situations where individuals rely on heuristics rather than critical analysis [8]. To counteract this, targeted interventions should equip individuals with the skills and tools necessary to critically assess information and reduce susceptibility to false narratives.

- **Digital and media literacy upskilling**: Introduce educational programs that teach individuals how to critically evaluate online information and recognize disinformation tactics [7].
- **Behavioral interventions**: Implement nudges, warnings, and credibility indicators in digital platforms to alert users about potentially misleading content [67].
- **Community-based misinformation monitoring**: Empower local communities and trusted organizations to detect and counter misinformation in real time [16].
- **Crisis communication training**: Equip emergency responders and policymakers with tools to recognize, debunk, and respond effectively to misinformation in real-time scenarios [8].

Strengthening systemic resilience against misinformation by integrating misinformation resilience into crisis management and policy frameworks

Misinformation interacts with infrastructure, governance, and digital platforms, leading to widespread consequences, including emergency response failures and political polarization [27]. To mitigate these risks, resilience strategies must be embedded within crisis management and policy frameworks, ensuring a coordinated and proactive approach to counter false narratives.

- **Early warning systems for misinformation**: Develop AI-driven misinformation monitoring tools to identify and flag disinformation in real time [13].
- **Multi-stakeholder coordination**: Foster collaboration between government agencies, technology companies, scientific institutions, and media to ensure rapid response to misinformation [40, 68].
- **Transparent communication strategies**: Ensure that official sources provide timely, transparent, and consistent updates to preempt false narratives [41].
- **Fact-checking and rapid response units**: Establish dedicated teams within emergency agencies to detect, verify, and counter misinformation during disasters [17, 47].

Building on the systemic vulnerabilities discussed throughout this chapter, the NATO SPS Programme plays a critical role in fostering coordinated, transnational responses to climate-related misinformation. As NATO continues to refine its response to information threats[26], some relevant areas for strategic development and collaboration are identified:

- **Cross-national misinformation monitoring**: Support interoperable tools for real-time tracking, sentiment analysis, and early warning of misinformation across NATO member states.
- **Crisis communication protocols**: Promote standardized protocols for verifying and disseminating official information during cyber-physical-climate crises.
- **Cyber hygiene integration**: Fund training and exercises to embed cyber hygiene practices, such as multi-factor authentication, access controls, and security audits, across civil and military systems.
- **Misinformation intelligence consortium**: Establish a NATO-led consortium to coordinate disinformation threat analysis and share countermeasures across sectors.
- **Content provenance pilots**: Invest in blockchain and watermarking pilots to verify the authenticity of official digital content during high-risk events.

By investing in these strategic lines of effort, NATO SPS would not only help counter the destabilizing effects of misinformation but also reinforce transnational trust, operational cohesion, and public awareness – key ingredients for effective climate adaptation and security resilience across the Alliance.

Mitigating cascading failures from system interdependencies by developing proactive misinformation management strategies

Misinformation can cause cascading failures across interconnected systems, amplifying disruptions in cyber, physical, social, and climate infrastructures [28]. To prevent misinformation from escalating into systemic crises, proactive strategies must be embedded into misinformation management efforts, ensuring early intervention and infrastructure resilience.

- **Pre-bunking over debunking**: Implement pre-bunking strategies that educate the public on common misinformation tactics before crises occur [67].
- **Resilient infrastructure planning**: Integrate misinformation resilience into the design of smart grids, emergency communication systems, and disaster management plans [39].
- **Accountability measures for disinformation actors**: Implement regulatory frameworks to hold individuals and organizations accountable for deliberate misinformation dissemination[15].
- **Empirical testing of counter-misinformation measures**: Conduct real-world simulations and behavioral studies to refine misinformation counter-strategies based on empirical evidence [64].

[26] https://www.nato.int/cps/en/natohq/topics_219728.htm.

To summarize, countering misinformation demands proactive strategies across individual, institutional, and systemic levels. Strengthening digital literacy, embedding resilience in policies, and reinforcing infrastructure will help mitigate risks. Implementing countermeasures ensures robust crisis management, infrastructure security, and climate adaptation efforts.

13.8 Challenges and Future Directions

Misinformation poses a severe threat to critical infrastructure by undermining public trust, delaying emergency responses, and exacerbating vulnerabilities during climate-related hazards and natural disasters. Addressing this challenge requires proactive measures such as improving communication accuracy, debunking false narratives swiftly, and building public resilience through education on disaster preparedness and climate science. By effectively countering misinformation, governments and organizations can enhance infrastructure resilience and ensure more robust responses to future crises, safeguarding lives and essential services from preventable failures [19]. The following key challenges highlight the primary obstacles in managing misinformation effectively.

- **Political and institutional barriers**: Climate misinformation is often driven by political and economic interests, leading to hesitancy in implementing countermeasures due to censorship concerns and geopolitical tensions.
- **Technological limitations**: AI tools struggle to keep pace with evolving disinformation tactics like deepfakes and bot networks, while social media's decentralized nature allows false narratives to spread before fact-checking can intervene.
- **Social and behavioral challenges**: Cognitive biases, ideological beliefs, and low digital literacy make misinformation difficult to counter, as false narratives often reinforce pre-existing opinions despite factual corrections.
- **Interdependencies and cascading failures**: Misinformation interacts with cyber, physical, and social systems, triggering widespread disruptions such as panic-driven overconsumption, policy resistance, and delayed crisis responses, necessitating a coordinated approach.

Considering the current landscape and future challenges, effective misinformation management must go beyond high-level goals and be grounded in domain-specific interventions. Table 13.1 summarizes these actionable strategies across the cyber, physical, social, and climate dimensions of the cyber-physical-social-climate nexus. For each domain, it outlines the primary impacts of misinformation, targeted countermeasures, and relevant data sources for monitoring progress. These strategies respond directly to the challenges identified above: political and institutional barriers are addressed through transparency and governance, technological limitations through AI-driven detection and platform accountability, social and behavioral challenges via literacy, trusted messengers, and behavioral nudges, and cascading failures

Table 13.1 Misinformation impact and possible way forward

Dimension	Impact of misinformation	Possible way forward	Example metrics or data sources
Cyber	Spread of false narratives through social media, disruption of emergency alerts, fake donation scams.	Strengthen misinformation monitoring with AI-driven detection, enforce platform accountability, and improve digital literacy initiatives.	EU DSA [15]; number of verified fact-checks flagged [17].
Physical	Misallocation of emergency resources, public non-compliance with evacuation orders, infrastructure strain.	Implement early warning misinformation response systems, integrate misinformation awareness in emergency training, and enhance crisis communication strategies.	Response lag time [16, 31]; ND-GAIN Infrastructure Indicators [9]; public adherence to evacuation orders [16].
Social	Erosion of trust in authorities, increased climate scepticism, political polarization, resistance to climate adaptation.	Promote transparency in government communication, increase public engagement in policy decisions, and strengthen media literacy education.	Edelman Trust Barometer [33]; social media sentiment analysis [7, 10, 11]; participation rates in climate policy forums [22].
Climate	Delay in climate adaptation policies, undermining of climate science, increased vulnerability to future wildfires.	Integrate misinformation resilience into climate policies, enhance fact-checking collaborations with scientific institutions, and conduct pre-bunking campaigns to counter climate misinformation.	ND-GAIN Readiness Score[11]; IPCC policy adoption lags [2].

through systemic integration of misinformation resilience. By operationalizing these responses, stakeholders can move from reactive misinformation control to proactive resilience-building across critical systems.

13.9 Conclusion

The cyber-physical-social-climate nexus intricately connects diverse domains that underpin the functionality and resilience of modern society. Each domain – cyber, physical, social, and climate – plays a distinct role, yet their interdependencies mean that disruptions in one can cascade across the others. For instance, digital components of physical infrastructure may rely on data connectivity or environmental monitoring, both of which are vulnerable to climate-driven disruptions. These systems are further compromized by misinformation, which complicates public understanding, weakens institutional trust, and impairs critical decision-making.

This chapter systematically addressed these challenges through the four primary objectives introduced in Sect. 13.1. First, the impact of climate misinformation on interconnected systems was examined in detail in Sects. 13.2 and 13.3. Second, Sect. 13.3.1 analyzed how communication failures, whether due to misinformation or technological faults, can escalate crises. Third, countermeasures including governance strategies, technological interventions, and cybersecurity frameworks were explored in Sects. 13.4 and 13.6. Finally, Sects. 13.5 and 13.7 outlined actionable resilience strategies that leverage social infrastructure and institutional reforms to improve systemic adaptability and public trust.

Table 13.1 provides a comprehensive overview of the challenges presented by misinformation within the cyber-physical-social-climate nexus, as discussed in this chapter. This offers insights into how these challenges can be effectively managed and mitigated through coordinated efforts across multiple domains. It examined the impact of climate misinformation on the cyber-physical-social nexus, highlighting how communication failures and misinformation increase vulnerabilities in critical infrastructure. It discussed combating climate misinformation through governance, community initiatives, and technology. Finally, it presented actionable strategies to enhance critical infrastructure resilience and societal functions against climate misinformation.

This chapter emphasized the need for a holistic, cross-domain approach to protect societal functions such as energy supply, healthcare delivery, and public safety, against the growing threats posed by climate risks and information disruption. Resilience in this context means not only securing individual systems but also safeguarding the interconnections between them. Achieving this requires a comprehensive strategy that integrates proactive engagements, cross-sector collaboration, and informed decision-making. To support this, the chapter explores vulnerabilities rooted in human behavior within the cyber-physical-social-climate nexus, examining how actions, decisions and behaviors can undermine the resilience and adaptability of these interconnected systems.

Acknowledgements The research supporting this chapter was funded by CSIRO's Critical Infrastructure Protection and Resilience (CIPR) mission, which aims to strengthen resilience and foster informed dialogue around the critical infrastructure that underpins economic, environmental, and societal systems.

References

1. Haque MA, Shetty S, Gold K, Krishnappa B (2021) Realizing cyber-physical systems resilience frameworks and security practices. In: Security in cyber-physical systems: foundations and applications, pp 1–37
2. Intergovernmental Panel on Climate Change (IPCC) (2021–2023) Sixth Assessment Report (AR6). https://www.ipcc.ch/assessment-report/ar6/
3. LSE Grantham Institute (2024) What are climate misinformation and disinformation and what is their impact? https://www.lse.ac.uk
4. Spampatti T, Hahnel UJJ, Trutnevyte E, Brosch T (2023) Psychological inoculation strategies to fight climate disinformation. Nature Human Behav 7:1136–1148
5. Grobler M, Aamir T (2024) Building cognitive resilience for enhanced cyber governance. In: Psybersecurity, pp 52–72. CRC Press
6. Cook J (2019) Understanding and countering misinformation about climate change. In: Chiluwa I, Samoilenko S (eds) Handbook of research on deception, fake news, and misinformation online. IGI Global, pp 281–306
7. United Nations Department for Global Communications (2020) The UN globally launches 'Verified' campaign to combat misinformation about COVID-19. https://www.undp.org/kyrgyzstan/press-releases/un-globally-launches-verified-campaign-combat-misinformation-about-covid-19
8. Zhuo S, Biddle R, Recomendable JD, Russello G, Lottridge DM (2024) Eyes on the phish(er): towards understanding users' email processing pattern and mental models in phishing detection. In: Karegar F, Farooq A (eds) Proceedings of the 2024 European symposium on usable security, EuroUSEC 2024, Karlstad, Sweden, 30 Sept 2024–1 Oct 2024. ACM, pp 15–29
9. NSW Bushfire Inquiry (2020) Final report of the NSW Bushfire inquiry. State inquiry report, NSW State Government
10. Islam MS, Kamal A-HM, Kabir A, Southern DL, Khan SH, Hasan SM, Sarkar T, Sharmin S, Das S, Roy T et al (2021) Covid-19 vaccine rumors and conspiracy theories: the need for cognitive inoculation against misinformation to improve vaccine adherence. PloS One 16(5):e0251605
11. Pertwee E, Simas C, Larson HJ (2022) An epidemic of uncertainty: rumors, conspiracy theories and vaccine hesitancy. Nature Med 28(3):456–459
12. Edelman (2023) 2023 Edelman trust barometer special report: trust and climate. https://www.edelman.com/sites/g/files/aatuss191/files/2023-11/2023%20Edelman%20Trust%20Barometer%20Special%20Report%20Trust%20and%20Climate.pdf
13. Schreiber D, Picus C, Fischinger D, Boyer M (2021) The defalsif-ai project: protecting critical infrastructures against disinformation and fake news. e&i Elektrotechnik und Informationstechnik 138(7):480–484
14. Hameleers M (2023) Disinformation as a context-bound phenomenon: toward a conceptual clarification integrating actors, intentions and techniques of creation and dissemination. Commun Theory 33(1):1–10
15. Maertens R, Anseel F, van der Linden S (2020) Combatting climate change misinformation: Evidence for longevity of inoculation and consensus messaging effects. J Environ Psychol 70:101455
16. Global Covenant of Mayors for Climate & Energy (2024) Addressing rampant climate disinformation, report, Global Covenant of Mayors for Climate & Energy, October 2024

17. Cyber and Infrastructure Security Centre (2024) Critical infrastructure annual risk review: Second edition, report, Department of Home Affairs, Australian Government, November 2024. Available https://www.cisc.gov.au/resources-subsite/Documents/critical-infrastructure-annual-risk-review-2024.pdf. Accessed on 07 Feb 2025
18. Owen LH (2019) Nuclear disasters, information vacuums: how a lack of data in Fukushima led to the spread of fake health news. https://www.niemanlab.org/2019/07/nuclear-disasters-information-vacuums-how-a-lack-of-data-in-fukushima-led-to-the-spread-of-fake-health-news/
19. World Economic Forum (2024) The global risks report 2024, report. World Economic Forum, Geneva, Switzerland, January 2024
20. Ecker UK, Lewandowsky S, Cook J, Schmid P, Fazio LK, Brashier N, Kendeou P, Vraga EK, Amazeen MA (2022) The psychological drivers of misinformation belief and its resistance to correction. Nat Rev Psychol 1(1):13–29
21. Kozyreva A, Smillie L, Lewandowsky S (2023) Incorporating psychological science into policy making. Eur Psychol
22. Dabla-Norris E, Khalid S, Magistretti G, Sollaci A (2023) Public support for climate change mitigation policies: a cross country survey. IMF Working Pap 223:1–48
23. OECD (2022) Building trust and reinforcing democracy: preparing the ground for government action. Tech Rep
24. OilPrice.com (2024) The impact of misinformation on wind energy development. https://oilprice.com/Alternative-Energy/Wind-Power/The-Impact-Of-Misinformation-On-Wind-Energy-Development.html. Accessed 2025-05-08
25. The Long View Group (2014) Summary of the literature: wind farms and health. https://eisdocs.dsdip.qld.gov.au/Coopers%20Gap%20Wind%20Farm/Final%20EIS/Final%20EIS%20-%20Appendices/appendix-h-summary-of-the-literature-wind-farms-and-health.pdf. Accessed 2025-05-08
26. Roozenbeek J, Culloty E, Suiter J (2023) Countering misinformation. In: European psychologist
27. Trijsburg I, Sullivan H, Park E, Bonotti M, Costello P, Nwokora Z, Pejic D, Peucker M, Ridge W (2024) Disinformation in the city: response playbook, Tech rep. The University of Melbourne
28. Jamalzadeh S, Barker K, González AD, Radhakrishnan S (2022) Protecting infrastructure performance from disinformation attacks. Sci Rep 12(1):16832
29. Broda E, Strömbäck J (2024) Misinformation, disinformation, and fake news: lessons from an interdisciplinary, systematic literature review. Ann Int Commun Assoc 48(2):139–166
30. Hilberts S, Evers S, Govers M, Petelos E (2024) The impact of misinformation on social media in the context of natural disasters. Eur J Public Health 34
31. Down To Earth (2025) False alerts like the one sent during the Greater Los Angeles wildfires can undermine trust and provoke anxiety. https://www.downtoearth.org.in/climate-change/false-alerts-like-the-one-sent-during-the-greater-los-angeles-wildfires-can-undermine-trust-and-provoke-anxiety
32. Graham T (2020) Bushfires, bots and arson claims: Australia flung in the global disinformation spotlight. In: The conversation. https://theconversation.com/bushfires-bots-and-arson-claims-australia-flung-in-the-global-disinformation-spotlight-129556
33. Edelman (2023) 2023 Edelman trust barometer. https://www.edelman.com/trust/2023/trust-barometer
34. Heffernan A (2024) Disinformation in Africa, digital policy hub—working paper. Centre for International Governance Innovation (CIGI). https://www.cigionline.org/static/documents/DPH-paper-Heffernan_fxkEQFh.pdf
35. Liu JC-E, Lee C-F (2025) Climate and energy misinformation in Taiwan. Frontiers Commun 9:1531126
36. Strauss N, Painter J, Ettinger J, Doutreix M-N, Wonneberger A, Walton P (2024) Reporting on the 2019 European heatwaves and climate change: journalists' attitudes, motivations and role perceptions. In: Journalism and reporting synergistic effects of climate change. Routledge, pp 226–249

37. Casero-Ripollés A, Tuñón J, Bouza-García L (2023) The European approach to online disinformation: geopolitical and regulatory dissonance. Humanities Soc Sci Commun 10(1):1–10
38. Kaur ST, Aamir T, Trewin S, Grobler M (2024) REAG (Resilience Expert Advisory Group) pre-workshop survey quantitative analysis findings
39. Critical 5 (2024) Adapting to evolving threats: a summary of critical 5 approaches to critical infrastructure security and resilience. Government Report PS9-35/2024E-PDF, Public Safety Canada, Ottawa, Canada, June 2024. In partnership with the governments of Australia, Canada, New Zealand, United Kingdom, and United States
40. UNESCO. Global initiative for information integrity on climate change. https://www.unesco.org/en/information-integrity-climate-change
41. European Commission. Climate disinformation. https://climate.ec.europa.eu/eu-action/climate-disinformation_en
42. Department of Agriculture, Water and the Environment (2021) National climate resilience and adaptation strategy 2021–2025: positioning Australia to better anticipate, manage and adapt to our changing climate. https://www.agriculture.gov.au/sites/default/files/documents/national-climate-resilience-and-adaptation-strategy.pdf
43. UNESCO (2024) Global initiative for information integrity on climate change. https://www.unesco.org/en/information-integrity-climate-change
44. United Nations (2024) G20 Leaders' Summit Brazil: UN and UNESCO launch global initiative on information integrity. https://www.un.org/en/climatechange/page/g20-leaders%E2%80%99-summit-brazil-un-and-unesco-launch-global-initiative-information
45. World Meteorological Organization (WMO) (2024) WMO joins global initiative for information integrity to fight climate disinformation. WMO. https://wmo.int/media/news/wmo-joins-global-initiative-information-integrity-fight-climate-disinformation
46. Climate Action Against Disinformation (CAAD) coalition and climate experts. Open letter: governments should act now to curb climate disinformation. https://caad.info/wp-content/uploads/2024/11/Open-Letter_-Governments-Should-Act-Now-to-Curb-Climate-Disinformation-1.pdf
47. Climate Action Against Disinformation (CAAD) (2025) Climate action against disinformation's response to the senate inquiry on greenwashing. https://www.aph.gov.au/DocumentStore.ashx?id=0ed5ee92-5eee-44f5-a288-724aad574f55&subId=744106
48. Papamichael M, Dimopoulos C, Boustras G (2024) Performing risk assessment for critical infrastructure protection: an investigation of transnational challenges and human decision-making considerations. In: Sustainable and resilient infrastructure, pp 1–19
49. Zhuo S, Biddle R, Koh YS, Lottridge DM, Russello G (2023) SoK: human-centered phishing susceptibility. ACM Trans Priv Secur 26(3):24:1–24:27
50. Cunningham R, Cvitanovic C, Measham T, Jacobs B, Dowd A-M, Harman B (2014) A preliminary assessment into the utility of social networks for engaging local communities in climate adaptation policy. Tech rep. Institute for Sustainable Futures and CSIRO, Nov 2014. Working Paper Prepared for: NSW Office of Environment and Heritage
51. Satizábal P, Cornes I, Zurita MLM, Cook BR (2022) The power of connection: navigating the constraints of community engagement for disaster risk reduction. Int J Disaster Risk Reduction 68:102699
52. United Nations (2023) Code of conduct for information integrity on digital platforms. Tech rep
53. UNICEF. Baku guiding principles on human development for climate resilience. https://www.unicef.org/media/165126/file
54. Djinlev V, Pearce BJ (2024) Collective action lessons for the energy transition: learning from social movements of the past. Sustain Sci 19:847–863
55. Ofoegbu C, New M (2021) Collaboration relations in climate information production and dissemination to subsistence farmers in Namibia. Environ Manage 67(1):133–145
56. ISO 9001 Auditing Practices Group (2024) Guidance on: auditing climate change issues in ISO 9001,1st edition. International Organization for Standardization (ISO) and International Accreditation Forum (IAF)

57. Cosentino M, Gal-Oz R, Safer DL (2024) Community-based resilience: the influence of collective efficacy and positive deviance on climate change-related mental health. In: Storytelling to accelerate climate solutions. Springer International Publishing, Cham, pp 319–338
58. Department of Home Affairs (2023) 2023–2030 Australian cyber security strategy. Government report. Australian Government, Canberra, Australia November 2023
59. Australian Government Department of Home Affairs (2023) Critical infrastructure annual risk review: first edition. Available https://www.cisc.gov.au/resources-subsite/Documents/critical-infrastructure-annual-risk-review-first-edition-2023.pdf
60. Zhu T, Haugen S, Liu Y (2021) Risk information in decision-making: definitions, requirements and various functions. J Loss Prevention Process Ind 72:104572
61. Brundage M, Avin S, Clark J, Toner H, Eckersley P, Garfinkel B, Dafoe A, Scharre P, Zeitzoff T, Filar B, Anderson H, Roff H, Allen G, Steinhardt J, Flynn C, Kagan R, Leung G, Maharaj T, Trager M et al (2020) Toward trustworthy ai development: mechanisms for supporting verifiable claims, arXiv preprint arXiv:2004.07213
62. Rojas C, Algra-Maschio F, Andrejevic M, Coan T, Cook J, Li Y-F (2024) Hierarchical machine learning models can identify stimuli of climate change misinformation on social media. Commun Earth Environ 5(1):436
63. Coan TG, Boussalis C, Cook J, Nanko MO (2021) Computer-assisted classification of contrarian claims about climate change. Sci Rep 11(1):22320
64. Rojas C, Algra-Maschio F, Andrejevic M, Coan T, Cook J, Li Y-F (2024) Augmented cards: a machine learning approach to identifying triggers of climate change misinformation on twitter. arXiv preprint arXiv:2404.15673
65. Zanartu F, Cook J, Wagner M, García J (2024) A technocognitive approach to detecting fallacies in climate misinformation. Sci Rep 14(1):27647
66. United Nations (2023) Information integrity on digital platforms: our common agenda policy brief 8. Tech rep
67. Beiser-McGrath LF, Bernauer T, Gampfer R, Sisco MR, van der Linden S (2023) Climate action with, by and for the people. Nat Human Behav 7:1655–1667
68. NATO CPSICC Workshop (2024) Cpsicc nexus workshop 2024: addressing climate change impacts on cyber-physical-social systems. https://cpsiccnexusworkshop2024.org

Part VI
Risk Assessment and Decision Making

Chapter 14
Cyber Jackal: Opportunistic and Malicious Behavior in Times of Extreme Climate Conditions—A Position Paper

Hicham Hammouchi and Gabriele Lenzini

Abstract Extreme climate conditions, amplified by the climate-change, may create opportunities and grounds for cyber exploitation. Drawing from episodes of looting or plundering during crisis, this chapter examines how opportunistic attackers could exploit societal and infrastructural vulnerabilities during disasters, which we refer to as Cyber Jackal. We explore the interplay between stressed socio-technical systems and cyber risks during disasters, and how it may increase the possibility of cascading infrastructure failure. We propose a view on a conceptual integrated resilience strategy combining early warning systems, adaptive cybersecurity protocols, and cross-agency coordination to mitigate these emerging risks. We introduce a resilience score to quantify the interplay between environmental readiness and cyber posture, providing a measurable framework for assessing and mitigating compound risks.

Keywords Cybersecurity · Natural disasters · Critical infrastructure · Socio-technical systems · Cyber resilience

14.1 Introduction

Recent analyses from the Intergovernmental Panel on Climate Change (IPCC) indicate that extreme weather events, from catastrophic floods to intense heatwaves, are becoming more frequent and severe [1, 2]. Meanwhile, the World Economic Forum's Global Risks Report has consistently highlighted cyber threats as a top global concern [3]. This raises an important question: what happens when these two areas of risk converge during large-scale natural disasters?

H. Hammouchi (✉) · G. Lenzini
SnT, University of Luxembourg, Esch-sur-Alzette, Luxembourg
e-mail: hicham.hammouchi@uni.lu

G. Lenzini
e-mail: Gabriele.Lenzini@uni.lu

M. Huber et al. (eds.), *Critical Infrastructure Systems and Extreme Threat Nexus: Preparing for Resilience and Security*, NATO Science for Peace and Security Series C: Environmental Security, https://doi.org/10.1007/978-3-032-13170-6_14

When a crisis strikes, the dynamics of societal and infrastructural systems shift dramatically. Power grids fail, communication networks collapse, and emergency services become overwhelmed, creating a fertile ground for digital vulnerabilities to multiply [4]. Attackers capitalize on this chaos, targeting moments when organizational defenses are weakest or under extreme stress to launch ransomware, data theft, or sabotage attempts [5]. This intersection of climate change and cybersecurity has emerged as a critical area of concern, as opportunistic malicious behavior thrives when disaster response systems are pushed to their limits. In fact, early case studies of hurricanes and wildfires demonstrate that malicious actors exploit these situations much like jackals in the wild, circling and looting upon weakened targets [5, 6].

The changing dynamics during disasters disrupt normal behavior and create systemic disturbances that amplify vulnerabilities. For example, extreme weather events such as hurricanes or floods can cause surges and overload in critical infrastructure such as power grids, communication networks and emergency response systems. Malicious actors may target these already overloaded systems with attacks such as Distributed Denial-of-Service (DDoS) amplifying chaos and compromising response efforts. Additionally, compound events such as cyberattacks coinciding with power grid failures, can lead to sudden surges in unauthorized access attempts or irregular traffic patterns, further complicating recovery of systems [4, 5]. These disruptions highlight the importance of integrating environmental and cybersecurity data to detect threats and mitigate cascading failures effectively.

To address this challenge, a coordinated interdisciplinary approach is necessary, one that integrates extreme climate events, disaster risk management, human vulnerabilities, and cybersecurity [7]. This approach requires a deep understanding of the interdependencies between climate-induced disasters, human behavior under stress, and how adversaries exploit systemic vulnerabilities heightened by these conditions. The role of information technologies, particularly AI-aided systems [8], is critical in providing real-time insights and predictive analytics to prioritize services and alerts that present high risk and support decision-making [9].

This position paper invokes the metaphor of the Cyber Jackal to describe how opportunistic malicious behavior thrives when disaster response systems are pushed to the edge. The Cyber Jackal metaphor, as used in this chapter, refers to opportunistic cyber adversaries who, much like jackals in nature that prey on weakened animals, systematically identify and exploit the heightened vulnerabilities of critical infrastructure and societal systems when they are already stressed or compromised, particularly during and in the aftermath of extreme weather events. This behavior capitalizes on diminished defenses, resource diversion, and the general chaos of a crisis. During a climate crisis, network traffic can change and adversaries could attempt to target particular services which may be critical. By observing such changes, one could reveal potential Cyber Jackal situations. The aim is twofold: first, to understand whether and how extreme weather amplifies existing cyber threats by observations on recorded incidents; and second, to propose an interdisciplinary resilience blueprint that integrates climate science, socio-technical safeguards, and cyber intelligence to anticipate and mitigate the risks of cyberattacks during climate-related disasters.

In this context, the development of a cyber-attack resiliency framework is essential, one that integrates cybersecurity with disaster management and resilience [10]. This framework includes guidelines and best practices for improving cyber-attack resiliency, as well as tools and resources for assessing and mitigating cyber risks [5]. The rationale for such a framework lies in the interconnected nature of climate-related threats: while extreme weather does not directly cause cyberattacks, it amplifies systemic vulnerabilities as adversaries exploit these conditions. The framework spans preparedness, response, and recovery, interconnecting disaster management, cybersecurity, and human factors.

This chapter is structured as follows: Sect. 14.2 examines cyber opportunism through ecological and socio-technical lenses, highlighting how crises amplify vulnerabilities. Section 14.3 presents an empirical analysis using blackhole network data to assess cyber threats during extreme weather events. Section 14.4 explores the impact of disasters on infrastructure and societal stability, emphasizing malicious exploitation. Section 14.5 discusses countermeasures for enhancing resilience with an integrated framework linking climate intelligence and cybersecurity to mitigate emerging risks. Finally, a conclusion is presented in Sect. 14.6.

14.2 Ecological and Socio-Technical Perspectives

This section explores how extreme weather events disrupt societal and technical systems, creating fertile ground for cyber threats. Drawing inspiration from ecological principles, we examine socio-technical systems as dynamic entities that shift from periods of stability to vulnerabilities under stress. Just as jackals exploit weakened prey in nature, cyber adversaries capitalize on disrupted infrastructures and cognitive overload during disasters.

14.2.1 Ecological and Behavioral Parallels

Ecologists have long observed that opportunistic predators, such as jackals, flourish when environmental stress weakens their targets' natural defenses [11, 12]. Similarly, human behavior during disasters often includes looting. For instance, following Hurricane Katrina, there were numerous reports of individuals taking advantage of the chaos to steal from abandoned stores and unsecured residences [6]. This dynamic mirrors how cyber attackers target organizations overwhelmed by large-scale emergencies. When floods or wildfires disrupt power grids, emergency services, or communication networks, security teams are stretched thin, much like weakened targets become vulnerable in nature.

From a human-behavior standpoint, crisis conditions cause cognitive overload, reduce situational awareness, and erode trust in standard protocols [13]. Just as jackals loot more aggressively during droughts, malicious actors could launch phishing attacks on individuals or organizations, ransomware through disguised emails or pretending humanitarian aid, or data breaches on organizations when staff attention is diverted to immediate disaster response [14]. Research in organizational psychology indicates that sustained stress impairs decision-making and team performance, leading to overlooked anomalies and bypassed security checks [15]. Aligning these ecological and behavioral insights with the Cyber-Physical-Social-Climate (CPSIC) lens shows how physical and social vulnerabilities, magnified by climate disruptions, create ideal conditions for cyber predators [7]. For example, during Hurricane Ida in 2021, Jefferson Parish in Louisiana experienced extensive power and internet outages. Cybercriminals reportedly capitalized on the chaos by targeting the JeffNet online database used by local courts, leading to prolonged closures and significant disruptions [16]. This real-world scenario illustrates the "jackal effect," which we define as the observed pattern where malicious actors opportunistically exploit operational chaos and reduced vigilance during crises.

These parallels highlight a central dynamic and challenge: ecological opportunism aligns closely with human-system vulnerabilities under crisis. To examine how this vulnerability increases across interconnected networks and infrastructures, it is necessary to look at other factors such as the socio-technical factors that intensify risk in extreme weather contexts.

14.2.2 *Socio-Technical Complexity*

Socio-technical systems are intricate networks of human decisions, technological infrastructures, and organizational structures. During extreme weather events, the interdependence of these components becomes both a strength and a liability. This section examines how stress amplifies vulnerabilities in socio-technical systems, explores the interplay of human and technological factors.

14.2.2.1 The Socio-Technical Ecosystem Under Stress

Socio-technical systems are interconnected networks that operate and evolve through a dynamic interplay of human processes, technical artifacts, and institutional frameworks [17, 18]. When extreme climate events strike, interdependencies in power grids, communication networks, and supply chains can trigger cascading failures, as a single disruption often destabilizes the larger ecosystem [4, 9]. Research indicates that when critical infrastructure is compromised by weather-induced disruptions, organizations reallocate manpower to immediate response, leaving cyber defenses understaffed.

Human factors, particularly decision-making under stress, play a critical role in such conditions of crisis. Research shows that under excessive information load, time pressure, and complexity, cybersecurity analysts struggle to detect and respond to threats effectively [19, 20]. Also, organizational structures and processes, may be compromized by reduced communication or non-adherence to internal policies or cybersecurity standards, which can further slow or delay emergency responses, amplify vulnerabilities across interconnected subsystems and increasing the risk of systemic failures as well as external actors exploiting these flaws [21]. However, the protocols and cybersecurity standards should be clear and non-ambigous [22].

These observations highlight the strength and, at the same time, the fragility of socio-technical ecosystems under crisis. As operational stress intensifies, vulnerabilities in human, technical, and organizational domains intensify and reinforce one another [23]. Inspired by [24, 25], a simplified stress-induced vulnerability function might be expressed as follows:

$$V_s = \alpha H_s + \beta T_s + \gamma O_s \tag{14.1}$$

where H_s, T_s, and O_s capture human, technological, and organizational vulnerabilities, respectively, and α, β, γ are scaling factors. V_s represents the overall stress-induced vulnerability of the socio-technical system. This formulation quantifies the synergistic compounded nature of risk when extreme weather interacts with pre-existing cyber weaknesses including Human, Technical and Organizational. This could be useful in cases of simulations to assess the compounded risk generated by these three factors.

14.2.2.2 Stress-Induced Vulnerabilities

Crises amplify vulnerabilities through a combination of factors, often a mix for technical and human such as reduced vigilance, cognitive bias, and overload in critical systems [13, 26]. In the face of hurricanes or wildfires, individuals and organizations prioritize immediate survival, allowing attackers to exploit emotional responses and lapse in security measures [4]. During Hurricane Harvey, for instance, malicious actors launched a surge of fraudulent donation websites, netting an estimated $2 million in illicit funds [27–29].

High-stress environments also intensify decision-making biases like confirmation bias and overconfidence [30, 31]. In fact, prolonged crises can sharply reduce decision accuracy, a phenomenon rooted in both psychological heuristics and neurobiological stress responses [32]. Under these conditions, security teams may disregard early threat indicators, assuming they are false alarms or attributing anomalies to system overload [15, 33]. Simultaneously, systemic overload in which communication networks and cloud services operate at or above capacity, creates fertile ground for DDoS attacks. A 2022 threat intelligence report documented a substantial spike in DDoS attempts targeting critical infrastructure during major storm events [34]. This cyclical interaction where crisis-induced stress fuels attacker success, demonstrates how socio-technical vulnerabilities can morph from isolated incidents into full-scale systemic risks.

14.2.2.3 Strategies for Enhancing Socio-Technical Resilience

Building resilience requires a comprehensive approach spanning human factors, technological safeguards, and organizational processes [35]. Regular cybersecurity drills, such as NATO's Locked Shields exercise [36], enable experts to enhance their skills in defending national IT systems and critical infrastructure under cyberattacks. These exercises simulate realistic multi-hazard scenarios, allowing teams to practice incident response and improve their readiness. To assess the effectiveness of these preparedness efforts, the relative improvement from such exercises, we introduce a conceptual Exercise Drill Effectiveness (E_{drill}) metric, which quantifies the percentage improvement in a team's incident response time after a cybersecurity drill. This provides a tangible assessment of training impact and skill enhancement, enabling organizations to objectively track readiness improvements and refine preparedness strategies based on measurable performance gains. The formula is expressed as:

$$E_{\text{drill}} = \frac{T_{\text{baseline}} - T_{\text{drill}}}{T_{\text{baseline}}} \times 100\,(\%) \tag{14.2}$$

where T_{baseline} is the average response time before training, and T_{drill} is the time post-training. This simple time efficiency formula should allow organizations conducting regular preparedness trainings quantify gains and efficiency and adapt scenarios. Additionally, other metrics could be considered to evaluate the team resilience such as time to recovery and containment.

On the other hand, redundancy in technological systems further mitigates single points of failure [7]. By adopting multi-cloud architectures and redundant distributed backup solutions across different geographical regions, organizations ensure services availability and that essential functions remain operational [37] in case of one fails. Also, having AI-based and automated adaptive technologies, including anomaly detection powered by machine learing, tools like Security Information and Event Management (SIEM) augmented with correlation with up-to-date threat intelligence can augment human capabilities, enhance decision making and aid in reducing workload by autonomously flagging and prioritizing suspicious events [38]. However, in such systems, explainability is crucial, as black-box algorithms can erode operator trust, leading to misinterpretation of alerts [39].

Dynamic organizational protocols also reinforce resilience. Establishing pre-approved incident playbooks for critical threat levels expedites decision-making, avoiding delays due to hierarchical bottlenecks. By assigning escalation paths for different emergencies (e.g., power grid failures versus DDoS attacks) ensures that teams act swiftly rather than awaiting upper-level approvals.

In this context, agent-based simulations, where virtual entities replicate both human and machine behaviors, offer a powerful lens for analyzing these cascading effects. In fact, with the advances in Large Language Models (LLMs), the agent-based simulations powered by LLMs further enhance the analysis of complex interactions between cyberattacks, critical infrastructure and extreme climate events by modeling and understanding human actions in various scenarios [40, 41]. For instance, the

European Union Agency for Cybersecurity's (ENISA) recurring "Cyber Europe" exercises provide concrete examples of such simulations. The 2022 edition [42] targeted the healthcare sector, testing its resilience against a sophisticated ransomware campaign. This was followed by the 2024 exercise, which focused on the cascading effects of a large-scale cyberattack on the European energy sector [43]. This emphasis on proactive measures is supported by resilience modeling research, which concludes that investing in robust defense is more vital for improving resilience than investing in post-breach recovery capabilities [44].

14.3 Empirical Technical Case Study

Blackhole (also known as telescope) networks provide a distinctive perspective on network traffic by capturing packets destined for "null" addresses, with no services running behind. Such blackhole sinks can help reveal malicious behaviors, such as opportunistic scanning and denial-of-service attempts, that might otherwise go unnoticed in conventional netflow logs [45].

To illustrate, we examine here data from a Luxembourg-based blackhole network consisting of 29 IP addresses for the period of July 2021, during which a flood was recorded on July 14. The data from the blackhole is aggregated hourly to capture traffic volumes, used protocols, packet-size distributions, and potential scanning indicators. Our aim is to assess whether the local flood event on July 14, 2021 correlates with any measurable shifts in blackhole traffic patterns (e.g., scanning or DDoS-like signatures) that could indicate heightened probing.

To process the raw logs data from the blackhole, we extract and normalize fields that reflect traffic rates, TCP handshake anomalies, and IP/port distributions. Key features include the approximate counts of RST (reset) packets and half-open connections (SYN, SYN-ACK) that help reveal scanning behavior. We also compute entropy to evaluate how narrowly or broadly traffic was concentrated among particular IPs or ports, adapting Shannon's formula:

$$H = -\sum_{i=1}^{n} p_i \log_2(p_i) \tag{14.3}$$

where each p_i denotes the fraction of traffic associated with category i. The use of $\log_2$ means that the resulting entropy H is measured in bits. Variations in entropy can suggest either widespread probing or localized, repetitive connections.

Time-series visualization of overall volumes revealed a continuous up-and-down pattern throughout July. No large spike is visible on July 14 itself, slight increases in overall traffic volume and in incomplete TCP handshakes occur during the surrounding days. Similar findings have been reported in other blackhole-based studies, where environmental or human-driven disruptions can coincide with opportunistic scans [4]. Figure 14.1 (an example of one such visualization) illustrates the measured entropy of source IPs and ports over time. Higher entropy indicates a wider distribution of IPs or ports, possibly reflective of broad scanning or varied legitimate usage. Lower entropy suggests concentrated traffic patterns (e.g., repeated queries from a single source or to a single port). Although these entropies fluctuate, no single day stands out as an overwhelming outlier.

Given the flood on July 14, the dataset was partitioned into flood-day observations (encompassing the 24-hour window of July 14) and non-flood days. As shown in Table 14.1 Mean differences between these two subsets showed modest shifts in overall volume, along with a slight upturn in reset flags and scanning activity. Such changes are consistent, though not conclusive, with the idea that crisis conditions can coincide with reduced oversight or altered routing. Yet several metrics linked to incomplete handshakes (e.g., SYN-ACK counts) decline during the flood window, suggesting that at least some of the variation may stem from benign traffic disruptions rather than deliberate probing. This ambiguity underscores the challenge of attributing short-term changes in blackhole traffic to malicious exploitation alone and suggests in-depth analysis of other security data, including a larger blackhole network.

Because blackhole sinks capture malicious or misrouted packets by design, they are valuable tools for detecting emergent scans or denial-of-service attempts. Nevertheless, they do not inherently distinguish opportunistic attacks from legitimate surges triggered by misconfiguration or heightened user queries. Environmental emergencies, such as floods, can also prompt unexpectedly high benign traffic (e.g., local inquiry spikes) that might appear suspicious. An important caveat, therefore, is that blackhole logs alone lack deeper contextual clues, whether in the form of meteorological data alignment or cross-sensor correlation. Observing slight changes

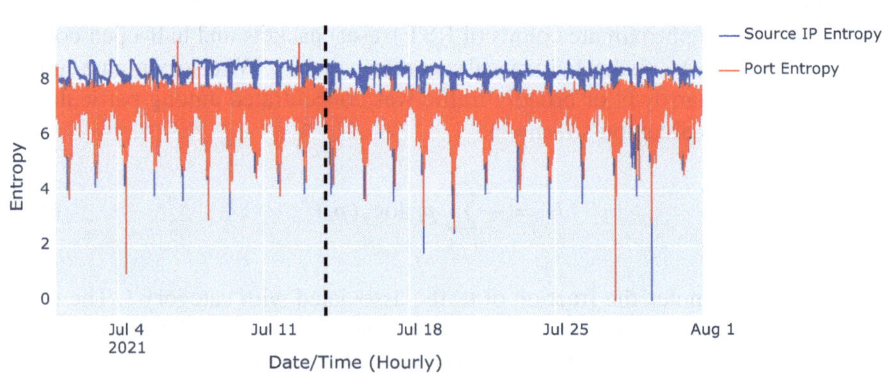

Fig. 14.1 Line chart showing Shannon entropy over time (July 1–August 1, 2021) for source IPs (blue) and destination ports (red). Higher entropy values indicate more evenly distributed traffic across many IPs or ports, while lower values indicate concentration on fewer sources or services. Both measures exhibit recurrent fluctuations, with source IP entropy generally higher. The vertical dashed black line marks July 14 (flood date)

Table 14.1 Averaged indicators for flood day versus non-flood days (July 2021)

Indicator	Flood day (07/14)	Non-flood days	% change (%)
Approx. volume (packets/hour)	35.873	33.453	+ 7.22
Avg. traffic rate (packets/sec)	128.70	120.73	+ 6.60
Partial scanning (count/hour)	2.46	1.62	+ 51.85
TCP SYN flags (count/hour)	1.152	992	+ 16.13
TCP RST flags (count/hour)	508	349	+ 45.56
TCP SYN-ACK (count/hour)	7.20	35.43	−79.68
IP distribution entropy	8.40	8.33	+ 0.84
Port distribution entropy	7.21	7.37	−2.17
SYN ratio	0.035	0.037	−5.41
RST ratio	0.016	0.012	+ 33.33
SYN-ACK ratio	0.0002	0.0013	−84.62

Note that each indicator is aggregated hourly across the specified subset of dates. The % Change shows the relative difference from Non-Flood to Flood Day

during a flood is not conclusive of malicious exploitation unless combined with external intelligence, such as recognized threat IP lists or netflow records from adjacent networks.

The analyzed data, although limited in size and coverage, blackhole observations capture variations in scanning indicators and reset flags that may become more pronounced during periods of environmental stress. However, given the modest magnitude of the changes and the ordinary variability of the baseline, the evidence is insufficient to claim any malicious intent in the flood period that. Future investigations could reinforce these preliminary findings by integrating a broader set of meteorological feeds, netflow logs from multiple vantage points, and threat intelligence repositories to more conclusively link climate-related disasters and hostile cyber behaviors.

14.4 Cyber Opportunism in Disaster Scenarios

Disasters create a complex interplay of physical disruption, operational strain, and heightened societal vulnerability. In such environments, cyber adversaries thrive by capitalizing on infrastructure weaknesses and exploiting social chaos. This section explores how core infrastructure systems and societal pressures converge to form a dual threat, thus emphasizing the need for integrated technical, organizational, and policy-based countermeasures.

14.4.1 Infrastructure Weaknesses

Critical infrastructure elements such as power grids, emergency services, and communication networks, provide foundational services, especially during crises. Yet these systems may become prime targets for cyber opportunists when natural disasters worsen existing vulnerabilities. Recent analyses suggest that the operational challenges inherent in emergencies offer adversaries additional windows of exploitation [9].

Extreme weather events often disrupt telecommunication networks, electrical grids, forcing utilities to rely on backup systems or isolated segments, known to as the Islanded operation in which the microgrid is disconnected from the main grid. This operational strain may translate to misconfigurations and delayed patching, thereby expanding the potential attack surface. Also, incidents of power outages may coincide with increased scam attempts targeting utility customers, as the example of Oncor, a Texas-based utility company, that reported a rise in fraudulent activities, including imposters demanding payment and threatening disconnection [46, 47]. These incidents highlight how opportunistic actors exploit moments of peak operational stress to conduct social engineering attacks, which creates multi-facets issues by hampering logistical support and at the same time eroding public trust in official channels.

In order to model the interconnectivity of infrastructure components, a conceptual model of cascading failures would provide a clearer view on the issues to address. Such models have been previously studied in the context of interdependent networks [48] and critical infrastructure systems [49]. These studies highlight the importance of accounting for both the individual functionality of each component and the dependencies between them, as well as the potential vulnerabilities introduced by cyberattacks [50]. To formalize this, let $F_i \in [0, 1]$ denote the functionality and service-availability of component i. We define $C_{ij} \in [0, 1]$ drawing on the dependency taxonomy of Rinaldi et al. [49] and the quantitative interdependency matrix of Haimes et al. [51], as the coupling strengths between components i and j, which quantifies the fraction of i's service if j is unavailable, and $V_{ij} \in [0, 1]$ as the probability that the $i - j$ dependency is disabled under a cyber attack scenario, following the SCADA vulnerability index of Ten et al. [50]. The overall system functionality F_s can be approximated in Eq. 14.4 as follows:

$$F_s = \prod_{i=1}^{n} F_i \times \prod_{i,j} \left(1 - C_{ij} \cdot V_{ij}\right) \tag{14.4}$$

This multiplicative formulation is conservative and represents the worst-case scenario which entails that the overall system functionality (F_s) is dependent on the individual functionality of each component (F_i) and the integrity of their interconnections. A failure in a single component (an F_i approaching zero) or a successfully exploited vulnerability in a critical coupling (a V_{ij} leading to $C_{ij} \cdot V_{ij}$ approaching (1) can significantly degrade or entirely compromise the overall system functionality. This mirrors the classical series-system reliability law by Fussell [52], where system success equals the product of individual reliabilities. The formulation therefore provides an upper bound on disruption—appropriate for safety-critical applications in which even one hard failure is intolerable. When a softer aggregation is more realistic and appropriate, two established substitutions can replace the hard-AND rule without altering the remainder of the framework: (1) the k-out-of-n model—formalised in Rausand and Hoyland, Chap. 7 [53] and used for infrastructure survivability by Zio et al. [54] shows that a critical-infrastructure system can be modelled as functional as long as at least k of its $n + m$ individual elements and dependency links remain operative, allowing explicit tolerance to multiple simultaneous losses. (2) Exponential inoperability penalty by Haimes et al. [51] embed interdependency effects in an Inoperability Input/Output Model, which has a tuning parameter that lets functionality decline gradually instead of collapsing when any single $C_{ij} V_{ij}$ reaches unity.

Even under these softer rules, small weaknesses can propagate widely. For example, a compromised backup-power controller (low F_i) increases the load on neighbouring grids (large C_{ij}), boosting the risk of further outages if the associated cyber vulnerability V_{ij} is left unmitigated. The framework therefore captures the cascading-failure dynamics described in power-grid studies such as Dobson et al., [55] while explicitly integrating cyber-attack likelihoods [50].

While layered defense, including network segmentation and real-time anomaly detection, helps constrain infiltration, these defenses require continuous updates and robust drills. For instance, consider a scenario where a backup power controller (component k) is compromised, leading to its F_k becoming very low. This not only affects its direct function but also increases the load and stress on interlinked grid components (represented by high C_{kj} values for linked components j). If this stress also makes the couplings more vulnerable to cyber exploitation (increasing V_{kj}), the term $(1 C_{kj} \cdot V_{kj})$ could decrease significantly, further reducing F_s and potentially triggering cascading failures even if other components initially had high F_i values. Implementing zero-trust architectures can further constrain unauthorized lateral movement, particularly in telecommunication hubs. However, purely technical measures are insufficient if not supported by strong organizational preparedness and coordinated policy initiatives [56]. The 2024 CrowdStrike incident, for example, demonstrated how even well-designed systems can falter due to human error and operational unpreparedness [57]. Ensuring resilience, therefore, requires an integrated approach—combining best-practice cybersecurity, redundant infrastructure planning, and pre-approved disaster protocols designed to function under stress.

14.4.2 Societal Chaos and Exploitation

Beyond infrastructure weaknesses, disasters induce societal turmoil, characterized by panic, misinformation, and resource scarcity. Cyber adversaries often exploit these stressors to escalate attacks such as phishing, fraud, and ransomware, compounding humanitarian crises and undermining public confidence [14].

When individuals are seeking aid or critical information, opportunistic phishing campaigns pose as government agencies or relief NGOs, capturing credentials or financial data. During the COVID-19 outbreak, malicious emails impersonating health authorities or vaccine providers effectively exploited on heightened public anxiety [58]. Similarly, after Hurricane Harvey, a proliferation of fraudulent charity websites diverted donor funds, illustrating how minimal oversight and heightened urgency fuel deception. The same strategy of tagetting of people in areas affected by natural disasters and continues to occur during the California Wildfires 2025, where many phishing campaigns appear to be trying to mimic official governments agencies, fake donation pages and other means of deception [59].

These examples illustrate how societal chaos, driven by panic, misinformation, and resource scarcity, creates fertile ground for cyber exploitation. To better understand the dynamics of this relationship, we conceptualize the interplay between resource scarcity (R_s) and institutional capacity (I_c) in jointly modulating the prevalence of cyber attacks (A_p). Let A_p be attack prevalence, and β scenario-dependent opportunistic constant:

$$A_p = \beta \cdot \frac{R_s}{I_c} \tag{14.5}$$

As resource scarcity (R_s) rises implying intensified competition for limited aid or services, and institutional capacity (I_c) falls signalling lower maturity and overstretched response teams due to systemic stress or shortage of skilled responders, the expected prevalence of opportunistic cyber-attacks (A_p) grows in direct proportion to the ratio (R_s/I_c). The structure of Eq. 14.5 conceptually inspired by national-level frameworks that link maturity to resilience. We adapt the principles of the Cybersecurity Capacity Maturity Model for Nations (CMM) [60, Factor D1.2, pp. 14–15] to define the institutional Capacity (I_c). While the CMM is a national-level blueprint for resilience, its core dimensions can be scalable to organizational level. These dimensions include the capacity to develop a clear strategy, maintain robust incident response capabilities, ensure skilled personnel through education and training, and adhere to technical standards. A decline in these capacities represents an increase in vulnerability, aligning with the ($1/I_c$) term in the Eq. 14.5. As for the resource scarcity (R_s), we model it as a driver of threat likelihood, which is inspired by the cyber risk as the product of threat likelihood and vulnerability [61], and ($1/I_c$) as a proxy for vulnerability. This formulation illustrates how even small disruptions in institutional capacity can lead to significant increases in the probability of the prevalence of attacks,

Defensive strategies should aim to support both institutional security (e.g., multifactor authentication, real-time fraud monitoring) and community awareness (e.g., targeted anti-phishing campaigns). In such scenarios timely takedown of fraudulent donation sites reduces victim losses and preserves donor confidence, which needs an active monitoring and classification of such websites. Thus, collaboration across government, NGOs, and private sectors is essential for rapid intelligence sharing: a single malicious domain identified in one relief effort can be proactively blocked elsewhere.

14.5 Integrated Resilience Blueprint

A practical blueprint for multi-hazard resilience in cyber-physical-social systems requires an interdisciplinary approach that makes it possible to anticipate, withstand, and recover from interconnected threats during extreme weather events. This needs to consider all combined the cybersecurity protocols, environmental intelligence as well as organizational readiness. Recent multi-hazard research identifies data silos (i.e., isolated repositories of information that do not interoperate or share data seamlessly) as a key bottleneck in aligning climate forecasting with cyber threat alerts [62]. Traditional weather agencies (e.g., National Oceanic and Atmospheric Administration (NOAA), European Centre for Medium-Range Weather Forecasts (ECMWF)) generate storm trajectories or flood probabilities at temporal and spatial resolutions insufficient for security operation centers (SOCs). Conversely, logs reports from SIEM platforms or Information Sharing and Analysis Centers (ISACs) typically emphasize IP-based threat indicators without referencing environmental vulnerabilities [63]. To overcome these mismatches, there need to be a data fusion layer that ingests near-real-time forecast updates and correlates them with threat feeds. This aligns with emerging CPS resilience frameworks that emphasize multilayered threat correlation [63]. This system should map meteorological footprints (e.g., projected flood zones) onto subnets that may host critical services and are most susceptible to compound hazards, and ranks them by priority for real-time anomaly detection.

In parallel, we introduce stress-aware response loops to address the dynamic resource strain conditions that arise during crises. Unlike static security baselines, these loops adapt to real-time stressors, such as spikes in emergency calls or system overloads, by triggering defensive escalations (e.g., restricting high-risk ports, enforcing stricter authentication). This concept is inspired by deep reinforcement learning (DRL) approaches for cascading failures, which optimize response strategies dynamically under resource constraints. For example, Liu et al. [64] demonstrated that adaptive protocols can mitigate the impact of concurrent disruptions by dynamically reallocating resources based on measured strain. These loops are formalized through a Resource Strain Index (Γ_{strain}), which quantifies the fraction of personnel and physical resources consumed by environmental response efforts. By logging

Γ_{strain} during simulated or real crises, organizations can identify how environmental stresses inhibit routine cyber defenses, such as patching or alert review.

In order to have a quantifiable view on the efficiency of resilience strategy, we can include these metrics as a way to gauge the different aspects. Time to Detection (T_{det}), Time to Recovery(T_{rec}), and the Resource Strain Index (Γ_{strain}), reflecting the fraction of available personnel and physical resources already consumed by environmental response efforts, which all can contribute to a resilience score $\mathcal{R}$ that encompasses environmental readiness $\mathcal{R}_{\text{env}}$ and cyber posture $\mathcal{R}_{\text{cyb}}$:

$$\mathcal{R} = w_{\text{env}} \mathcal{R}_{\text{env}} + w_{\text{cyb}} \mathcal{R}_{\text{cyb}} - w_{\text{int}} \mathcal{V}_{\text{env-cyb}}, \tag{14.6}$$

where $\mathcal{R}_{\text{env}}$ (environmental readiness) quantifies the ability of critical computational systems to absorb and adapt to climate-induced disruptions while maintaining critical operational functions [65]. $\mathcal{V}_{\text{env-cyb}}$ captures vulnerability overlap the extent to which climate disruptions drain off resources needed to respond to cyber intrusions. One of the metrics that can be used to evaluate $\mathcal{V}_{\text{env-cyb}}$, is Γ_{strain}, which measures how are staff and equipment are allocated to disaster response. If the Γ_{strain} is high, the security team has less capacity for normal cyber defenses (e.g., fewer staff available to monitor logs). Based on this as the Γ_{strain} increases, and makes $\mathcal{V}_{\text{env-cyb}}$ increase as well, and overall $\mathcal{R}$ go down if the resources are not managed efficiently in times of crisis. The two T_{det} and T_{rec} both contribute to the $\mathcal{R}_{\text{cyb}}$ for the cybersecurity responce. The factors w_{env}, w_{cyb}, and w_{int} can be determined through multi-criteria decision methods, experimental studies from simulations or real exercises, or stakeholder consensus [66]. The subtraction term highlights that environmental hazards add risk and also multiply systemic vulnerabilities in a socio-technical ecosystem.

The resilience score $\mathcal{R}$ is deeply intertwined with three previously discussed metrics: the System Function (F_s), the Stress-Induced Vulnerability Function (V_s), and the Adaptive Performance Metric (A_p). The System Function F_s quantifies the operational resilience of critical infrastructure under stress (e.g., maintaining critical services during disruptions). However, F_s alone cannot account for vulnerabilities amplified by stress, captured by V_s, where human (H_s), technical (T_s), and organizational (O_s) factors interact. A high Γ_{strain} amplifies H_s (e.g., cognitive overload in overburdened staff) and T_s (e.g., single-point IoT failures), directly increasing $\mathcal{V}_{\text{env-cyb}}$ and reducing $\mathcal{R}$. The Adaptive Performance Metric A_p bridges this gap by enhancing adaptive capacity, for example, pre-approved incident playbooks reduce T_{rec}, while AI-assisted log analysis lowers T_{det}. These improvements in A_p can therefore increase F_s lower V_s, as streamlined workflows reduce cognitive load (H_s) and technical errors (T_s). When A_p is optimized $\mathcal{R}$ increases, reflecting a system's capacity to adapt to compound threats.

Achieving a consistent level of resilience also demands policy alignment to join forces towards a unified resilience strategy. Our proposed blueprint thus advocates for regional pilot projects in high-risk areas (e.g., flood-prone coastal cities) where local utilities, emergency services, and SOCs collaborate in an integrated scenario [67]. By observing how quickly critical infrastructure can revert to stable operation under dual hazards such as storms plus targeted intrusions, these pilots would refine estimates for $\mathcal{V}_{\text{env-cyb}}$ and validate or adjust the weighting factors in $\mathcal{R}$. The results would then help shape broader policies and improve the robustness of socio-technical systems under stress.

In essence, this resilience blueprint consolidates quantifiable metrics across multiple dimensions of climate-related cyber threats, from multi-hazard data fusion, stress-adaptive protocols, and socio-technical indicators into a unified conceptual architecture. Rather than considering climate hazards and cyberattacks as separate problems, it underscores their coupled nature: environmental disruptions can exacerbate cyber vulnerabilities, and adversarial actions may exploit degraded operational states. The capacity to respond effectively in a compound crisis requires integrated monitoring, adaptive control mechanisms, and governance models that coordinate across technical, organizational, and institutional boundaries. Formal methods such as attackâŁ"defense trees (ADTs) [68], can provide a structured basis for modeling how environmental stressors might amplify cyber intrusion pathways. This quantitative modeling could be further enhanced by employing Bayesian Networks (BNs) [69], which are particularly well-suited for fusing disparate data from environmental and cyber domains to dynamically update risk probabilities. Additionally, System Dynamics (SD) [70] can be considered to simulate the complex feedback loops between adaptive responses, resource strain, and system recovery over the course of a crisis, exploring "what-if" scenarios for different adaptive responses.

14.6 Conclusion

This chapter examines how extreme climate events and cyber threats converge, using the metaphor of the Cyber Jackal to illustrate how opportunistic attackers exploit vulnerabilities in weakened socio-technical systems during natural disasters. Rather than direct causation, we highlight how climate-related crises such as floods, hurricanes and wildfires, amplify systemic vulnerabilities, creating opportunities for adversarial threats to launch disruptive attacks, exploiting resource scarcity, institutional strain, and systemic fragility. We discuss an integrated resilience blueprint that consists of a data fusion layer that integrates real-time meteoroligical forecasts with cyber intelligence and organizational preparadness. In addition, we discuss the stress-aware response loops that dynamically adapt to resource strain and trigger defensive measures. Additionally, we introduce a resilience risk score $\mathcal{R}$ designed to quantify the interplay between environmental readiness and cyber posture through measurable indicators. While the formulation of $\mathcal{R}$ is conceptual at this stage, it constitutes a structured research proposition and a basis for future empirical validation and computational refinement.

In addition, human factors remain critical: security teams must be prepared for multi-hazard scenarios through targeted training, as cognitive overload and operational bottlenecks can impair effective response. Tools such as AI-assisted triage and decision support systems are increasingly being explored and used in security operations, though their effectiveness in high-stress, multi-hazard environments remains an open question and should be examined with empirical rigor. Addressing these challenges will require structured escalation pathways and multi-hazard training, as well as a clear research agenda that tests and iterates on these proposed mechanisms in realistic scenarios.

Acknowledgements We would like to thank Restena and the Computer Incident Response Center Luxembourg (CIRCL) from the Luxembourg House of Cybersecurity for providing access to the data used in the empirical technical case study. Data access is provided in the scope of NGSOTI project which is funded by the European Union's Digital Europe programme (DIGITAL) - Project: 101127921 under the agreement: DIGITAL-ECCC-2022-CYBER-03, and which is supported by the European Cybersecurity Competence Centre.

References

1. IPCC: (2021) Climate change 2021: the physical science basis. Contribution of working group I to the sixth assessment report of the intergovernmental panel on climate change
2. IPCC (2022) Climate change 2022: impacts, adaptation, and vulnerability. Contribution of working group II to the sixth assessment report of the intergovernmental panel on climate change
3. The Global Risks Report 2022. https://www.weforum.org/reports/global-risks-report-2022
4. Avraam C, Ceferino L, Dvorkin Y (2023) Operational and economy-wide impacts of compound cyber-attacks and extreme weather events on electric power networks. Appl Energy 349:121577 https://doi.org/10.1016/j.apenergy.2023.121577
5. Chakraborty S, Mombeshora EM, Clark KP, Mbavarira TS (2024) Understanding of cyber-attack vulnerabilities during natural disasters and discussing a cyber-attack resiliency framework. In: SoutheastCon 2024, pp 466–471. https://doi.org/10.1109/SoutheastCon52093.2024.10500233
6. Sheikhi RA, Javanbakhtian R, Heidari M (2025) Looting and antisocial behavior after disasters: a systematic review. BMC Public Health 25(1):309. https://doi.org/10.1186/s12889-025-21285-7
7. Cassotta S, Pettersson M (2019) Climate change, environmental threats and cyber threats to critical infrastructures in multi-regulatory sustainable global approach with sweden as an example. Beijing Law Rev 10(3):616–636. https://doi.org/10.4236/blr.2019.103035
8. Nasirov J, Cheng L (2024) Enhancing cybersecurity in climate technology transfer: The interaction of environmental law with digital security. Int J Digit Law Gov 1(1):187–206. https://doi.org/10.1016/j.crm.2021.100387
9. Hossin MA, Chen L, Asante IO, Boadi EA, Adu-Yeboah SS (2023) Climate change and cop26: role of information technologies in disaster management and resilience. Environ Dev Sustain 1–27 https://doi.org/10.1007/s10668-023-04134-8
10. Panda A, Bower A (2020) Cyber security and the disaster resilience framework. Int J Disaster Resil Built Environ 11(4):507–518. https://doi.org/10.1108/IJDRBE-07-2019-0046
11. Allee WC (1931) Animal aggregations: a study in general sociology. University of Chicago Press, Chicago, IL
12. Sih A, Crowley PH, McPeek MA, Petranka JW, Strohmeier K (1985) Predation, competition, and prey communities: a review of field experiments. Annu Rev Ecol Syst 16: 269–311 https://doi.org/10.1146/annurev.es.16.110185.001413
13. Reason J (1990) Human error. Cambridge University Press, Cambridge
14. National Institute of Standards and Technology (NIST) (2021) Cybersecurity Framework: Manufacturing Profile. U.S. Department of Commerce. Available at https://nvlpubs.nist.gov/nistpubs/CSWP/NIST.CSWP.112021.pdf

15. Driskell JE, Salas E, Johnston J (1999) Does stress lead to a loss of team perspective? Group Dyn: Theory Res Pract 3(4):291. https://doi.org/10.1037/1089-2699.3.4.291
16. Burlee T (2021) Cybersecurity Challenges During Hurricane Ida. SecureData . https://www.securedata.com/blog/malware-attack-follows-hurricane-ida-landfall
17. Mumford E (2006) The story of socio-technical design: reflections on its successes, failures and potential. Inf Syst J 16(4):317–342. https://doi.org/10.1111/j.1365-2575.2006.00221.x
18. Baxter G, Sommerville I (2011) Socio-technical systems: from design methods to systems engineering. Interact Comput 23(1):4–17. https://doi.org/10.1016/j.intcom.2010.07.003
19. Phillips-Wren G, Adya M (2020) Decision making under stress: the role of information overload, time pressure, complexity, and uncertainty. J Decis Syst 29(sup1):213–225 https://doi.org/10.1080/12460125.2020.1768680
20. Chowdhury NH, Adam MTP, Teubner T (2020) Time pressure in human cybersecurity behavior: theoretical framework and countermeasures. Comput Secur 97:101963 https://doi.org/10.1016/j.cose.2020.101963
21. Mott G, Nurse JR, Baker-Beall C (2023) Preparing for future cyber crises: lessons from governance of the coronavirus pandemic. Policy Des Pract 6(2):160–181 https://doi.org/10.1080/25741292.2023.2205764
22. Pollini A, Callari TC, Tedeschi A, Ruscio D, Save L, Chiarugi F, Guerri D (2022) Leveraging human factors in cybersecurity: an integrated methodological approach. Cogn Technol Work 24(2):371–390. https://doi.org/10.1007/s10111-021-00683-y
23. Moran J, Pijpers FP, Weitzel U, Bouchaud J-P, Panja D (2023) Critical fragility in socio-technical systems. https://doi.org/10.48550/arXiv.2307.03546
24. Alexander D (2012) Models of social vulnerability to disasters. RCCS Annu Rev 4 https://doi.org/10.4000/rccsar.412
25. Bergstrand K, Mayer B, Brumback B, Zhang Y (2015) Assessing the relationship between social vulnerability and community resilience to hazards. Soc Indic Res 122: 391–409 https://doi.org/10.1016/j.ijdrr.2024.104774
26. Weick KE, Sutcliffe KM (2007) Managing the Unexpected: resilient performance in an age of uncertainty, 2nd edn. Jossey-Bass, San Francisco, CA. https://doi.org/10.1017/S1833367200003072
27. Federal Bureau of Investigation (FBI) (2017) Hurricane harvey: avoid fraudulent charitable contribution schemes. https://www.fbi.gov/news/stories/avoid-hurricane-harvey-charity-schemes
28. Taodang D, Gundur R (2023) How frauds in times of crisis target people. Vict Offenders 18(5):889–914 https://doi.org/10.1080/15564886.2022.2043968
29. Acharya B, Lazzaro D, Cinà AE, Holz T (2024) Pirates of charity: exploring donation-based abuses in social media platforms. arXiv preprint arXiv:2412.15621
30. Kahneman D (2011) Thinking. Fast and slow. Farrar, Straus and Giroux, New York
31. Driskell JE, Salas E (1991) Group decision making under stress. J Appl Psychol 76(3):473–478. https://doi.org/10.1037/0021-9010.76.3.473
32. Sarmiento LF, Cunha PL, Tabares S, Tafet G, Gouveia A (2024) Decision-making under stress: a psychological and neurobiological integrative model. Brain, Behav, Immun-Health 38:100766 https://doi.org/10.1016/j.bbih.2024.100766
33. Endsley MR (1995) Measurement of situation awareness in dynamic systems. Human Factors 37(1):65–84. https://doi.org/10.1518/001872095779049499
34. NETSCOUT (2022) Threat Intelligence Report, 2H 2022. Online resource. https://www.netscout.com/threatreport
35. National Institute of Standards and Technology (NIST) (2018) Framework for Improving Critical Infrastructure Cybersecurity, Version 1.1. Technical report, U.S. Department of Commerce . https://doi.org/10.6028/NIST.CSWP.04162018 . https://doi.org/10.6028/NIST.CSWP.04162018
36. NATO (2023) Cooperative Cyber Defence Centre of Excellence: World's largest cyber defense exercise Locked Shields brings together over 3000 participants . https://ccdcoe.org/news/2023/6016/

37. Patel HB, Kansara N (2024) Dynamic orchestration of multi-cloud resources for scalable and resilient ai/ml workloads: Strat Fram J. https://doi.org/10.2139/ssrn.5070162
38. Molnar C (2020) Interpretable machine learning: a guide for making black box models explainable. Leanpub, Canada
39. Doshi-Velez F, Kim B (2017) Towards a rigorous science of interpretable machine learning. https://arxiv.org/abs/1702.08608
40. Gao C, Lan X, Li N, Yuan Y, Ding J, Zhou Z, Xu F, Li Y (2024) Large language models empowered agent-based modeling and simulation: a survey and perspectives. Hum Soc Sci Commun 11(1):1–24. https://doi.org/10.1057/s41599-024-03611-3
41. Park JS, Zou CQ, Shaw A, Hill BM, Cai C, Morris MR, Willer R, Liang P, Bernstein MS (2024) Generative agent simulations of 1000 people. arXiv preprint arXiv:2411.10109
42. European Union Agency for Cybersecurity (ENISA) (2022) Cyber Europe 2022: Official After-Action Report. Technical Exercise Report . https://www.enisa.europa.eu/publications/cyber-europe-2022-after-action-report
43. European Union Agency for Cybersecurity (ENISA) (2024) Cyber Europe 2024: Official After-Action Report. Technical Exercise Report . https://www.enisa.europa.eu/publications/cyber-europe-2024-after-action-report
44. Dong K, Zhen J, Xie Z, Chen L (2024) Building cybersecurity resilience: integrating defense and recovery investment strategies in an expected resilience framework. J Enterp Inf Manag (ahead-of-print) https://doi.org/10.1108/JEIM-04-2023-0189
45. Kori M, Kumar VA, Pachouri R, Jain A, Dutt, H.: Quantitative and qualitative evaluation of TCP target ports through active network telescope. Int J Inf Technol 1–15 (2024) https://doi.org/10.1007/s41870-024-01816-y
46. Oncor Electric Delivery Company LLC: Oncor Raises Awareness of Utility Scams During National Consumer Protection Week. https://www.oncor.com/content/oncorwww/wire/en/home/safety/oncor_raises_awareness_of_utility_scams_during_national_consumer_protection_week.html Accessed 2025-05-08
47. KXAN News: A Year Later: Looking Back at Texas' February 2021 Winter Storm. https://www.kxan.com/news/texas/a-year-later-looking-back-at-texas-february-2021-winter-storm/. Accessed 08 May 2025
48. Buldyrev SV, Parshani R, Paul G, Stanley HE, Havlin S (2010) Catastrophic cascade of failures in interdependent networks. Nature 464(7291):1025–1028. https://doi.org/10.1038/nature08932
49. Rinaldi SM, Peerenboom JP, Kelly TK (2001) Identifying, understanding, and analyzing critical infrastructure interdependencies. IEEE Control Syst Mag 21(6):11–25. https://doi.org/10.1109/37.969131
50. Ten C-W, Liu C-C, Manimaran G (2008) Vulnerability assessment of cybersecurity for scada systems. IEEE Trans Power Syst 23(4):1836–1846. https://doi.org/10.1109/TPWRS.2008.2002298
51. Haimes YY, Horowitz BM, Lambert JH, Santos JR, Lian C, Crowther KG (2005) Inoperability input-output model for interdependent infrastructure sectors. i: Theory and methodology. J Infrastruct Syst 11(2):67–79
52. Fussell J (2009) How to hand-calculate system reliability and safety characteristics. IEEE Trans Reliab 24(3):169–174
53. Rausand M, Hoyland A (2003) System reliability theory: models, statistical methods, and applications, vol 396. Wiley
54. Zio E (2016) Challenges in the vulnerability and risk analysis of critical infrastructures. Reliab Eng Syst Saf 152:137–150 https://doi.org/10.1016/j.ress.2016.02.009
55. Dobson I, Carreras BA, Lynch VE, Newman DE (2007) Complex systems analysis of series of blackouts: cascading failure, critical points, and self-organization. Chaos: Interdiscip J Nonlinear Sci 17(2) https://doi.org/10.1063/1.2737822
56. Neri M, Niccolini F Rosario P et al (2022) Assessing smes' cybersecurity organizational readiness: findings from an italian survey. Online J Appl Knowl Manag 10(2): 1–22 https://doi.org/10.36965/OJAKM.2022.10(2)1-22

57. Cyber I 2024 in Retrospect: lessons learned and cyber strategies shaping future of critical infrastructure. https://industrialcyber.co/features/2024-in-retrospect-lessons-learned-and-cyber-strategies-shaping-future-of-critical-infrastructure/. Accessed 08 May 2025
58. World Health Organization (WHO) (2020) Beware of criminals pretending to be WHO. Online resource . https://www.who.int/about/cybersecurity
59. Charles McFarland: scammers exploit California wildfires: how to stay safe (2025). https://www.mcafee.com/blogs/internet-security/scammers-exploit-california-wildfires-how-to-stay-safe/. Accessed 08 May 2025
60. Centre GCSC (2021) Cybersecurity capacity maturity model for nations. Technical report, University of Oxford. https://gcscc.ox.ac.uk/cmm-2021-edition
61. Strupczewski G (2021) Defining cyber risk. Saf Sci 135:105143
62. Dagnas R, Barbeau M, Boutin M, Garcia-Alfaro J, Yaich R (2023) Exploring the quantitative resilience analysis of cyber-physical systems. In: 2023 IFIP networking conference (IFIP Networking), pp 1–6. https://doi.org/10.23919/IFIPNetworking57963.2023.10186355
63. Segovia-Ferreira M, Rubio-Hernan J, Cavalli A, Garcia-Alfaro J (2024) A survey on cyber-resilience approaches for cyber-physical systems. ACM Comput Surv 56(8). https://doi.org/10.1145/3652953
64. Zhong K, Yang Z, Yu S, Li K (2024) Deep reinforcement learning-based multi-layer cascaded resilient recovery for cyber-physical systems. IEEE Trans Serv Comput 17(6):3330–3344. https://doi.org/10.1109/TSC.2024.3478729
65. Linkov I, Eisenberg DA, Plourde K, Seager TP, Allen J, Kott A (2013) Resilience metrics for cyber systems. Environ Syst Decis 33(4):471–476
66. Padgett J, Rincon R, Panakkal P (2024) Future cities demand smart and equitable infrastructure resilience modeling perspectives. npj Nat Hazards 1(1):28. https://doi.org/10.1038/s44304-024-00028-5
67. Amin F, Verma K, Acharya P (2024) Multi-hazard risk and integrated approach to resilience. In: Disaster risk and management under climate change, pp 581–592. Springer, Singapore. https://doi.org/10.1007/978-981-99-4105-6_28
68. Kordy B, Mauw S, Radomirović S, Schweitzer P (2014) Attack-defense trees. J Log Comput 24(1):55–87
69. Huang K, Zhou C, Tian Y-C, Tu W, Peng Y (2017) Application of Bayesian network to data-driven cyber-security risk assessment in scada networks. In: 2017 27th International telecommunication networks and applications conference (ITNAC), pp 1–6. IEEE
70. Naugle A, Langarudi S, Clancy T (2024) What is (quantitative) system dynamics modeling? defining characteristics and the opportunities they create. Syst Dyn Rev 40(2):1762

MIX
Papier aus verantwortungsvollen Quellen
Paper from responsible sources
FSC® C105338

If you have any concerns about our products,
you can contact us on
ProductSafety@springernature.com

In case Publisher is established outside the EU,
the EU authorized representative is:
Springer Nature Customer Service Center GmbH
Europaplatz 3, 69115 Heidelberg, Germany

Printed by Libri Plureos GmbH
in Hamburg, Germany